Control Systems

The State Variable Approach
(Conventional and MATLAB)

Control Systems

The State Variable Approach
(Conventional and MATLAB)

Anoop K. Jairath

Ane Books India

CRC Press
Taylor & Francis Group
Boca Raton London New York

CRC Press is an imprint of the
Taylor & Francis Group, an **Informa** business
A TAYLOR & FRANCIS BOOK

Control Systems : The State Variable Approach (Conventional and MATLAB)

First published 2008 by Ane Books India for CRC Press

CRC Press
Taylor & Francis Group
6000 Broken Sound Parkway NW, Suite 300
Boca Raton, FL 33487-2742

© 2008 by Taylor & Francis Group, LLC
CRC Press is an imprint of Taylor & Francis Group, an Informa business

No claim to original U.S. Government works

ISBN 13: 978-1-4200-7757-5 (hbk)

Visit the Taylor & Francis Web site at
http://www.taylorandfrancis.com

and the CRC Press Web site at
http://www.crcpress.com

For distribution in rest of the world other than the Indian sub-continent

British Library Cataloguing in Publication Data
A catalogue record for this book is available from the British Library

Dedicated

To my wife

Geeta

"whose arrival in my life changed my fortunes"

and

my son

Varun

"who is also my best friend"

PREFACE

Control System Engineering as an electrical subject has always held a fascination for me ever since I was a student of this discipline in the College of Military Engineering, Pune. Later, when I was an Instructor at the same Institute; the guidance, encouragement and advise rendered by our Head of Department, Prof. Dr C K Murty infused further interest in this subject.

Students generally find Control Systems a difficult subject to understand and learn as its comprehension requires sound knowledge of mathematics. Keeping this in view, I have endeavoured to present the subject in a simple way so that the students find this book an easy learning experience. The analysis of the Control Systems presented in this book is restricted to modern approach; popularly known as State Variable Analysis. This book will be useful to the undergraduate and post graduate students pursuing engineering degrees in electrical, electronics, mechanical, aerospace etc.

The book is organised into eight chapters in such a way that, the students can proceed to read from start to the end, in a clear way.

Chapter 1: Besides introducing the subject, comparison of the modern approach with the classical approach has been presented in brief.

Chapter 2: Since the modern approach completely depends upon matrices, computation and manipulation with the help of matrices has been covered in detail so that the students do not have to consult any other book for the same.

Chapter 3: It is devoted to Laplace Transform and includes associated theorems for solving the mathematical equations.

Chapter 4: This chapter is assigned to MATLAB which is an extremely powerful technical computing environment. Being an array based software language, it is most suited for analysis of complicated and complex control systems based on the modern approach. Students can understand the basics of MATLAB and apply the same while studying subsequent chapters.

Chapter 5: It will help the students in formulation of state space models and diagrams as it revises the concepts of block diagrams and signal flow graphs. Also introduces the concept of open-loop and closed-loop control systems

Chapter 6: Concepts related to state space analysis has been included in this chapter.

Chapter 7: It is devoted to formulation of different types of state space models. Also focuses on the transformation techniques which help in transforming the state space representation and the associated state space model, from one form to the another.

Chapter 8: Concepts of controllability and observability has been included which finally leads to design of controllers and observers.

This book has several explicatory solved examples to strengthen and stress the concepts presented and assimilate the basic theory related to the modern approach. It also contains a number of solved problems to further strengthen the theory.

Analysis by way of the modern approach by using the control system toolbox of the MATLAB has been included in the book so that students intending to undertake higher studies in this discipline and finally deciding to do advanced research, are well versed in analysing control systems later on.

Every chapter commences with the learning objectives and ends with the summary and the key points learnt. This will help in better understanding. Review exercises appended at the end of each chapter has been designed in such a way to ensure students practice what they have learnt.

I wish to express my sincere thanks to Ane Books India who by offering this book for writing, has helped in realising my dream. In addition, I wish to express my heartfelt thanks to my parents who always wanted their children to attain pinnacles of success in the field of education. After my father's death, my mother Prem and brothers, Ashok and Subhash are responsible for what I am today.

Anoop K. Jairath

CONTENTS

1

INTRODUCTION

1.1 INTRODUCTION

Control systems perform a very important role in our day-to-day life and find applications in various fields such as industry, medicine, computer, electronics, weaponry, robotics etc. Control system engineering has become a very important science dealing with automatic mechanisms replacing manual control. It lowers the cost of production and incorporates quality in the finished product thereby making the entire process more economical and globally competitive. Recent advancements and technological developments present a challenge to the control system engineer to develop a control system that is aligned to human imagination. The aim of this book is not to cite various examples of automatic control systems, as students interested in this book would have already grasped the preliminaries and the associated basic theory. What we will study in this book is the *state variable analysis* of control systems, a modern approach, that has many inherent advantages.

LEARNING OBJECTIVES

- ◆ To understand basic steps involved in design and analysis of a control system.
- ◆ Advantages of state variable analysis over classical approach.

1.2 BASIC STEPS TO DESIGN A CONTROL SYSTEM

The study and design of a control system consists of many well defined steps including the requirements, specifications, modelling, simulation, analysis and optimisation. Only after this, a control system engineer can physically evolve a stable and desirable control system. Basic steps in design of a control system are illustrated in Fig. 1.1. The user presents its *requirements* to the control system engineer. The need is accompanied by the specifications. *Specifications* are followed by *physical modelling* (i.e. what should constitute a control system so that the needs and requirements are met).

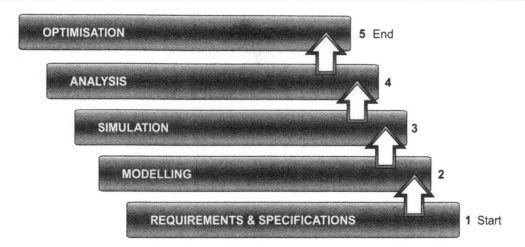

Fig. 1.1 *Basic steps in design of a control system*

After developing a model, mathematical equations are evolved by applying fundamental physical laws and principles namely Kirchoff's current and voltage laws, Newton's laws, laws of thermodynamics etc. The equations derived may be linear, non-linear, differential, integral, or partial-differential. *Analysis* of the mathematical model is the next step towards physical realisation of the control system. The response of the control system is obtained by subjecting the model to standard input signals and the stability, controllability and observability studied. Computer *simulation* techniques are helpful in faster and accurate analysis. If the response of the model is found to be unsatisfactory, the system is further improved and *optimised*. This may require adjustment of system parameters. Compensation techniques are used to design and incorporate a suitable compensation device.

1.3 MATHEMATICAL REPRESENTATION

Mathematical modelling is the first step towards analysis of a control system. Individual system components are described by their physical characteristics and converted into equations. The interconnection of the physical components constituting a control system are also linked with mathematical equations. The equations thus formed describe the behaviour of individual components and the input-output characteristics of a control system.

Two approaches are available to the control system engineer to analyse and physically realise the desired control system. The first approach is called the *classical or frequency domain approach*. In this approach, all variables are eliminated from mathematical equations. As a result, a set of differential or difference equations are obtained. These equations are converted into transfer function by relating the input and output using Laplace or Z-transforms. The classical approach has applicability only to linear time-invariant systems. Systems that can be approximated can also be analysed using this method. Various classical graphical tools' namely *bode plots, nyquist plots, root locus plots* aided with computer simulation techniques help in faster analysis and design of control systems. However, multi-input-multi-output (MIMO) systems cannot be designed accurately, using the classical approach. In a nutshell classical approach has the following disadvantages:

- Applicability only to single-input single-output (SISO) control systems that are linear and time-invariant.
- Analysis of MIMO control systems is time-consuming and complicated.
- Internal behaviour of components is not reflected in the output response. Analysis reveals the output behaviour of the known input.
- Stability study is restricted to output analysis. The behaviour of the components and exceeding of their specified ratings and parameters, cannot be visualised.

Development of state-variable analysis forms the basis of *modern control theory*. State-space analysis overcomes the disadvantages inherited in the classical approach analysis and is thus referred to as *modern approach* or *time-domain approach* which can be used to analyse wide-ranging control systems. State variable analysis has the following advantages:

- Helps analyse all systems that can be modelled using classical approach.
- State-space approach involves characterisation of higher-order systems using a set of first-order differential equations. The solution of first-order differential equations is easily obtained in comparison to the solution of equivalent higher-order differential equations.
- State-space approach is an effective mathematical tool for analysing systems described by linear, non-linear, time-invariant and time-varying equations. The system may have zero or nonzero initial conditions.
- MIMO control systems can be analysed easily.
- State-space approach utilises the vector matrix notation which simplifies its representation.
- Initial conditions get accounted in the state-space analysis and the intermediate variables remain focussed.
- State-space analysis can be easily programmed and hence suitable for analysis using modern computer methods and techniques.

State-space approach of analysing the control systems requires the basic knowledge of Laplace transform and matrices. The next two chapters have been devoted to understanding basic knowledge on these topics.

SUMMARY

- Role of control systems.
- Steps involved in design of a control system.
- Importance of development of the mathematical model.
- Control system analysis approaches i.e. classical and modern.
- Disadvantages of the classical approach.
- Advantages of the state variable approach.

REVIEW EXERCISE

1. What are the basic steps in designing a control system. Illustrate with the help of a diagram.

2. What different approaches can be applied for analysing the control systems. What are the disadvantages of the classical approach.

3. List the advantages of using the state-space approach.

2

MATRICES

2.1 INTRODUCTION

State-variable approach employs matrices as an important tool in the analysis of the control systems. Though the students must have learnt matrices while studying mathematics, it is assumed that the basic and elementary operations, definitions, characteristics, properties and manipulations of the matrices must have been forgotten. This chapter has been included in the book so that students need not refer to any other book to revise and refresh.

LEARNING OBJECTIVES

- ◆ To understand basic forms of matrices commonly employed in state variable analysis.
- ◆ To revise elementary matrix operations and manipulations.

2.2 DEFINITION OF MATRIX

Matrix is an ordered rectangular array of elements. It is represented by

$$A = \begin{bmatrix} a_{11} & a_{12} & \cdots & a_{1n} \\ a_{21} & a_{22} & \cdots & a_{2n} \\ \cdots & \cdots & \cdots & \cdots \\ a_{m1} & a_{m2} & \cdots & a_{mn} \end{bmatrix} \tag{2.1}$$

The elements a_{11}, $a_{12} \ldots a_{2n} \ldots a_{mn}$ arranged in m rows and n columns are enclosed in a set of brackets and equated with a bold faced capital alphabet signifying matrix A equivalent to $m \times n$ matrix. A matrix with m rows and n columns is said to be of order ($m \times n$). It is also referred as (m-by-n) matrix. Order of a matrix refers to the total number of rows and columns of the matrix number a_{ij} also called *elements* of the matrix. The subscript i denotes the row and the subscript j denotes the column. In this fashion, matrix is defined by their elements and locations.

2.3 SQUARE MATRIX

In a matrix when number of rows equals the number of columns i.e. $m = n$, the matrix is called a *square matrix* of the order n.

2.4 COLUMN MATRIX OR VECTOR

A matrix consisting of only one column and more than one row i.e. $n = 1$ is termed as *column matrix* or *column vector* and is represented and denoted by lower case letter as shown below

$$a = \begin{bmatrix} a_1 \\ a_2 \\ \dots \\ a_m \end{bmatrix} \tag{2.2}$$

The elements of a column vector or matrix requires only one subscript.

2.5 ROW MATRIX OR VECTOR

A matrix consisting of only one row and more than one column i.e. $m = 1$ is called *row matrix* or *row vector*. It is represented by lower case letter in the following manner

$$a = [a_1 \ a_2 \ \dots \ a_n], \text{ with } n \text{ elements} \tag{2.3}$$

The elements of a row matrix or vector requires only one subscript.

2.6 DIAGONAL MATRIX

A diagonal matrix is a square matrix having all elements as zero except for those occupying positions of the *principal diagonal*. Therefore, matrix A of eqn (2.1) would be square if $m = n$ and would be diagonal if represented as

$$A = \begin{bmatrix} a_{11} & 0 & 0 \\ 0 & a_{22} & 0 \\ 0 & 0 & a_{33} \end{bmatrix} \tag{2.4}$$

2.7 IDENTITY MATRIX

It is also referred as *unity matrix*. It is a diagonal matrix having all the diagonal elements $a_{12} = a_{22} = a_{nn} = 1$. An *identity matrix* is represented as

$$I = \begin{bmatrix} 1 & 0 & \dots & 0 \\ 0 & 1 & \dots & 0 \\ \dots & \dots & \dots & \dots \\ 0 & 0 & \dots & 1 \end{bmatrix} \tag{2.5}$$

An identity matrix is often denoted as I or U.

2.8 NULL MATRIX

It is also called *zero matrix*. It has all its elements equal to zero.

2.9 SYMMETRIC MATRIX

It is a square matrix having equal number of rows and columns and exhibit a relationship in which $a_{ij} = a_{ji}$. Therefore, if the elements of the rows are interchanged with that of the columns of a given matrix, the same matrix is obtained. Matrices A and B are the examples of *symmetric matrix*.

$$A = \begin{bmatrix} -3 & 4 & 1 \\ 4 & 2 & 5 \\ 1 & 5 & -1 \end{bmatrix}$$

$$B = \begin{bmatrix} 2 & -7 \\ -7 & 2 \end{bmatrix}$$

2.10 DETERMINANT OF MATRIX

Determinant is associated with a square matrix and has the same elements and order. If matrix

$$A = \begin{bmatrix} a_{11} & a_{12} & a_{13} \\ a_{21} & a_{22} & a_{23} \\ a_{31} & a_{32} & a_{33} \end{bmatrix} \tag{2.6}$$

then determinant of square matrix A is obtained by enclosing the elements of matrix A within vertical lines as shown below

$$\Delta_A = \det A = \begin{vmatrix} a_{11} & a_{12} & a_{13} \\ a_{21} & a_{22} & a_{23} \\ a_{31} & a_{32} & a_{33} \end{vmatrix} \tag{2.7}$$

If the determinant of the square matrix A is zero, then it is *singular matrix*, otherwise, it is referred to as *nonsingular matrix*. The value of the determinant is obtained by determining the *cofactors* and the *minors*.

The *minor* of an element a_{ij} of a determinant is designated as M_{ij} or m_{ij} and is related with the *cofactor* by

$$\text{Cofactor of } a_{ij} = (-1)^{i+j} M_{ij} \tag{2.8}$$

The *minor* is obtained by removing the row i and column j of the determinant. For example, M_{12} of determinant A represented in eqn (2.7) is obtained by removing the elements contained in row 1 and column 2 and is

$$M_{12} = \begin{vmatrix} a_{21} & a_{23} \\ a_{31} & a_{33} \end{vmatrix} \tag{2.9}$$

Then the *cofactor* of element a_{12} of determinant A is

$$\alpha_{12} = (-1)^{1+2} \, M_{12} = (-1)^3 \begin{vmatrix} a_{21} & a_{23} \\ a_{31} & a_{33} \end{vmatrix} \tag{2.10}$$

or
$$\alpha_{12} = - \begin{vmatrix} a_{21} & a_{23} \\ a_{31} & a_{33} \end{vmatrix} \tag{2.11}$$

Let us consider a (2×2) matrix A

$$A = \begin{bmatrix} a_{11} & a_{12} \\ a_{21} & a_{22} \end{bmatrix} \tag{2.12}$$

The determinant is $\quad \det A = \begin{vmatrix} a_{11} & a_{12} \\ a_{21} & a_{22} \end{vmatrix} \tag{2.13}$

The value of $\det A$ is

$$\begin{vmatrix} a_{11} & a_{12} \\ a_{21} & a_{22} \end{vmatrix} = (a_{11}a_{12} - a_{21}a_{12}) \tag{2.14}$$

In general, if A is $n \times n$ matrix; then the $\det A$ can be expressed in terms of the cofactor of any row *i.e.*

$$\det A = \sum_{j=1}^{n} a_{ij} A_{ij} \qquad \text{where } i = 1, \text{ or } 2, \ldots\ldots, \text{ or } n.$$

or the cofactor of any column *i.e.*,

$$\det A = \sum_{i=1}^{n} a_{ij} A_{ij} \qquad \text{where } j = 1, \text{ or } 2, \ldots\ldots, \text{ or } n.$$

Example 2.1 Determine the value of the determinant of a (2×2) order square matrix A consisting of elements

$$A = \begin{vmatrix} 2 & 5 \\ 3 & 1 \end{vmatrix}$$

Solution Determinant of matrix A is given as
$$\det A = \begin{vmatrix} 2 & 5 \\ 3 & 1 \end{vmatrix}$$

The value of $\det A$ is $\quad = (2 \times 1) - (5 \times 3) = 2 - 15 = -13$

Example 2.2 Determine the value of determinant of matrix A of the order (3×3) having elements

$$A = \begin{bmatrix} 2 & 5 & 3 \\ 1 & 1 & 0 \\ 2 & 3 & 2 \end{bmatrix}$$

Solution The determinant of the given (3×3) square matrix A

$$\det A = \begin{vmatrix} 2 & 5 & 3 \\ 1 & 1 & 0 \\ 2 & 3 & 2 \end{vmatrix}$$

The value of determinant is

$$= 2\begin{vmatrix} 1 & 0 \\ 3 & 2 \end{vmatrix} - 1\begin{vmatrix} 5 & 3 \\ 3 & 2 \end{vmatrix} + 2\begin{vmatrix} 5 & 3 \\ 1 & 0 \end{vmatrix}$$
$$= 2(1 \times 2 - 3 \times 0) - 1(5 \times 2 - 3 \times 3) + 2(5 \times 0 - 3 \times 1)$$
$$= 4 - 1 - 6 = -3$$

Example 2.3 Prove that the given matrix A is nonsingular

$$A = \begin{bmatrix} 2 & 5 & 3 \\ -1 & -1 & 2 \\ 2 & 3 & 2 \end{bmatrix}$$

Solution

$$\det A = \begin{vmatrix} 2 & 5 & 3 \\ -1 & -1 & 2 \\ 2 & 3 & 2 \end{vmatrix}$$

The value of det A is

$$= 2\begin{vmatrix} -1 & 2 \\ 3 & 2 \end{vmatrix} - (-1)\begin{vmatrix} 5 & 3 \\ 3 & 2 \end{vmatrix} + 2\begin{vmatrix} 5 & 3 \\ -1 & 2 \end{vmatrix}$$
$$= 2(-1 \times 2 - 2 \times 3) + 1(5 \times 2 - 3 \times 3) + 2(5 \times 2 - 3 \times -1)$$
$$= 2(-2 - 6) + 1(10 - 9) + 2(10 + 3)$$
$$= -16 + 1 + 26 = 11$$

Since the value of the determinant is not zero, the matrix A is nonsingular.

2.11 MATRIX ALGEBRA

Matrix algebra is defined in terms of simple mathematical operations like addition, subtraction, multiplication and division.

2.11.1 Addition

Addition of the given matrix is possible provided the matrices are of the same order. The sum of the matrices is obtained by adding the corresponding elements. If

$$A = \begin{bmatrix} a_{11} & a_{12} & a_{13} \\ a_{21} & a_{22} & a_{23} \\ a_{31} & a_{32} & a_{33} \end{bmatrix} \text{ and } B = \begin{bmatrix} b_{11} & b_{12} & b_{13} \\ b_{21} & b_{22} & b_{23} \\ b_{31} & b_{32} & b_{33} \end{bmatrix}$$

then
$$C = A + B = \begin{bmatrix} a_{11} + b_{11} & a_{12} + b_{12} & a_{13} + b_{13} \\ a_{21} + b_{21} & a_{22} + b_{22} & a_{23} + b_{23} \\ a_{31} + b_{31} & a_{32} + b_{32} & a_{33} + b_{33} \end{bmatrix}$$

or
$$C_{ij} = A_{ij} + B_{ij} \tag{2.15}$$

The addition operation of matrices is associative i.e.
$$A + B = B + A \tag{2.16}$$

and also follows the commutative rule i.e.
$$(A + B) + C = A + (B + C) \tag{2.17}$$

2.11.2 Subtraction

Subtraction operation is carried out by multiplying the matrix to be subtracted by -1 and then carrying out the addition operation. If $A = \begin{bmatrix} 4 & 1 \\ 3 & 2 \end{bmatrix}$ and $B = \begin{bmatrix} 1 & -1 \\ 2 & 1 \end{bmatrix}$

$$C = A + B = \begin{bmatrix} 4 & 1 \\ 3 & 2 \end{bmatrix} + (-1) \begin{bmatrix} 1 & -1 \\ 2 & 1 \end{bmatrix}$$

$$= \begin{bmatrix} 4 & 1 \\ 3 & 2 \end{bmatrix} + \begin{bmatrix} -1 & 1 \\ -2 & -1 \end{bmatrix}$$

$$= \begin{bmatrix} 4-1 & 1+1 \\ 3-2 & 2-1 \end{bmatrix} = \begin{bmatrix} 3 & 2 \\ 1 & 1 \end{bmatrix}$$

2.11.3 Multiplication

Multiplication of two matrices A and B is possible, if the number of columns of matrix A are equal to number of rows of matrix B. Thus, if matrix A is of that order $(m \times p)$ and matrix B is of the order $(p \times n)$, then the resultant product $C = AB$ is of order $(m \times n)$. The elements contained in the resultant product designated as matrix C are formed by multiplying elements of i th row of matrix A with j th column of matrix 'B' and then adding up the products to give the elements of C_{ij}

$$C = AB \tag{2.18}$$

$$C_{ij} = \sum_{k=1}^{p} a_{ik} b_{kj} \text{ where } i = 1, 2, \dots, n \text{ and } j = 1, 2, \dots, m. \tag{2.19}$$

$$= a_{i1} b_{1j} + a_{i2} b_{2j} + \dots + a_{ip} b_{pj} \tag{2.20}$$

Example 2.4 Perform multiplication operation on matrices A and B to obtain $C = AB$. Verify that the matrix multiplication is not commutative.

$$A = \begin{bmatrix} 1 & 3 & 2 \\ 2 & 5 & 3 \\ 0 & 2 & 3 \end{bmatrix}; B = \begin{bmatrix} 1 & 1 & 1 \\ 2 & 2 & 2 \\ 3 & 0 & 3 \end{bmatrix}$$

Solution

$$AB = \begin{bmatrix} 1 & 3 & 2 \\ 2 & 5 & 3 \\ 0 & 2 & 3 \end{bmatrix} \begin{bmatrix} 1 & 1 & 1 \\ 2 & 2 & 2 \\ 3 & 0 & 3 \end{bmatrix}$$

Multiplication operation is performed by multiplying the elements of row 1 of matrix A with corresponding elements in the column 1 of matrix B and summed up. Same operation is proceeded with the remaining columns of matrix B. Thereafter, rows 2 and 3 of matrix A are operated with columns in matrix B one by one.

$$AB = \begin{bmatrix} 1\times1+3\times2+2\times3 & 1\times1+3\times2+2\times0 & 1\times1+3\times2+2\times3 \\ 2\times1+5\times2+3\times3 & 2\times1+5\times2+3\times0 & 2\times1+5\times2+3\times3 \\ 0\times1+2\times2+3\times3 & 0\times1+2\times2+3\times0 & 0\times1+2\times2+3\times3 \end{bmatrix}$$

$$= \begin{bmatrix} 13 & 7 & 13 \\ 21 & 12 & 21 \\ 13 & 4 & 13 \end{bmatrix}$$

Check

$$BA = \begin{bmatrix} 1 & 1 & 1 \\ 2 & 2 & 2 \\ 3 & 0 & 3 \end{bmatrix} \begin{bmatrix} 1 & 3 & 2 \\ 2 & 5 & 3 \\ 0 & 2 & 3 \end{bmatrix} = \begin{bmatrix} 3 & 10 & 8 \\ 6 & 20 & 16 \\ 3 & 15 & 15 \end{bmatrix}$$

Inference

Since $AB \neq BA$, matrix multiplication is not commutative.

Example 2.5   Verify the associative and distributive properties for the following matrices for multiplication operation.

$$A = \begin{bmatrix} 1 & 2 \\ 2 & 1 \end{bmatrix}; B = \begin{bmatrix} 2 & 3 \\ 3 & 2 \end{bmatrix}; C = \begin{bmatrix} 1 & 3 \\ 3 & 1 \end{bmatrix}$$

Solution

Associative Law   The associative law states that

$$A(BC) = (AB)C$$

$$(AB)C = \left(\begin{bmatrix} 1 & 2 \\ 2 & 1 \end{bmatrix} \begin{bmatrix} 2 & 3 \\ 3 & 2 \end{bmatrix} \right) \begin{bmatrix} 1 & 3 \\ 3 & 1 \end{bmatrix} = \begin{bmatrix} 8 & 7 \\ 7 & 8 \end{bmatrix} \begin{bmatrix} 1 & 3 \\ 3 & 1 \end{bmatrix} = \begin{bmatrix} 29 & 31 \\ 31 & 29 \end{bmatrix}$$

and

$$A(BC) = \begin{bmatrix} 1 & 2 \\ 2 & 1 \end{bmatrix} \left(\begin{bmatrix} 2 & 3 \\ 3 & 2 \end{bmatrix} \begin{bmatrix} 1 & 3 \\ 3 & 1 \end{bmatrix} \right) = \begin{bmatrix} 1 & 2 \\ 2 & 1 \end{bmatrix} \begin{bmatrix} 11 & 9 \\ 9 & 11 \end{bmatrix} = \begin{bmatrix} 29 & 31 \\ 31 & 29 \end{bmatrix}$$

Since $(AB)C = A(BC)$; the associative law for multiplication operation is verified.

Distributive Law The distributive law states that $A(B + C) = AB + AC$.

$$A(B + C) = \begin{bmatrix} 1 & 2 \\ 2 & 1 \end{bmatrix} \left(\begin{bmatrix} 2 & 3 \\ 3 & 2 \end{bmatrix} + \begin{bmatrix} 1 & 3 \\ 3 & 1 \end{bmatrix} \right)$$

$$= \begin{bmatrix} 1 & 2 \\ 2 & 1 \end{bmatrix} \begin{bmatrix} 3 & 6 \\ 6 & 3 \end{bmatrix} = \begin{bmatrix} 15 & 12 \\ 12 & 15 \end{bmatrix}$$

and

$$AB + AC = \begin{bmatrix} 1 & 2 \\ 2 & 1 \end{bmatrix} \begin{bmatrix} 2 & 3 \\ 3 & 2 \end{bmatrix} + \begin{bmatrix} 1 & 2 \\ 2 & 1 \end{bmatrix} \begin{bmatrix} 1 & 3 \\ 3 & 1 \end{bmatrix}$$

$$= \begin{bmatrix} 8 & 7 \\ 7 & 8 \end{bmatrix} \begin{bmatrix} 7 & 5 \\ 5 & 7 \end{bmatrix} = \begin{bmatrix} 15 & 12 \\ 12 & 15 \end{bmatrix}$$

Since $A(B + C) = AB + AC$; distributive law for multiplication operation is verified.

2.12 TRANSPOSE OF A MATRIX

The *transpose* of a given matrix is obtained by interchanging the elements of the rows and columns. If the given matrix A is

$$A = \begin{bmatrix} 2 & 5 & 3 \\ 1 & 4 & 2 \\ 6 & -5 & 3 \end{bmatrix}; \text{ then the transpose of matrix } A \text{ which is designated as } A^T \text{ or } A' \text{ is}$$

$$A^T \text{ or } A' = \begin{bmatrix} 2 & 1 & 6 \\ 5 & 4 & -5 \\ 3 & 2 & 3 \end{bmatrix}$$

Example 2.6 For the matrix given in Example 2.5, verify the following

(a) $(A + B)' = A' + B'$, (b) $(AB)' = B'A'$

Solution

(a) $$(A + B) = \begin{bmatrix} 1 & 2 \\ 2 & 1 \end{bmatrix} + \begin{bmatrix} 2 & 3 \\ 3 & 2 \end{bmatrix} = \begin{bmatrix} 3 & 5 \\ 5 & 2 \end{bmatrix}$$

Therefore, $$(A + B)' = \begin{bmatrix} 3 & 5 \\ 5 & 2 \end{bmatrix}$$

Given $$A = \begin{bmatrix} 1 & 2 \\ 2 & 1 \end{bmatrix} \text{ and } B = \begin{bmatrix} 2 & 3 \\ 3 & 2 \end{bmatrix}$$

Therefore, $A' = \begin{bmatrix} 1 & 2 \\ 2 & 1 \end{bmatrix}$ and $B' = \begin{bmatrix} 2 & 3 \\ 3 & 2 \end{bmatrix}$

and $A' + B' = \begin{bmatrix} 1 & 2 \\ 2 & 1 \end{bmatrix} + \begin{bmatrix} 2 & 3 \\ 3 & 2 \end{bmatrix} = \begin{bmatrix} 3 & 5 \\ 5 & 2 \end{bmatrix}$

Since, $(A+B)' = A' + B'$, the requirement given in the problem is verified.

(b) $AB = \begin{bmatrix} 1 & 2 \\ 2 & 1 \end{bmatrix} \begin{bmatrix} 2 & 3 \\ 3 & 2 \end{bmatrix} = \begin{bmatrix} 8 & 7 \\ 7 & 8 \end{bmatrix}$

and $(AB)' = \begin{bmatrix} 8 & 7 \\ 7 & 8 \end{bmatrix}$

Also $B'A' = \begin{bmatrix} 2 & 3 \\ 3 & 2 \end{bmatrix} \begin{bmatrix} 1 & 2 \\ 2 & 1 \end{bmatrix} = \begin{bmatrix} 8 & 7 \\ 7 & 8 \end{bmatrix}$

Since, $(AB)' = B'A'$, the requirement given in the problem is verified.

KEY POINTS LEARNT

- Matrices addition is associate and commutative, provided the matrices are of same order
  - ✓ $A + B = B + A$; Associative
  - ✓ $(A + B) + C = A + (B + C)$; Commutative
- Matrix multiplication may not be commutative. The associative and distributive laws are:
  - ✓ $A(BC) = (AB)C$; Associative
  - ✓ $A(B + C) = AB + BC$; Distributive
- Inverse of a singular matrix does not exist. Only nonsingular square matrices have inverse.
- The following is true for the transpose of matrix.
  - ✓ $(AB)' = B'A'$
  - ✓ $(A + B) = A' + B'$

2.13 ADJOINT OF A MATRIX

Adjoint of a square matrix A is a matrix formed by replacing elements a_{ij} by their respective cofactors α_{ij} or A_{ij} and by transposing the matrix so formed. The adjoint is designated as *adjoint A* or *adj A*. If

$$A = \begin{bmatrix} a_{11} & a_{12} & \cdots & a_{1n} \\ a_{21} & a_{22} & \cdots & a_{2n} \\ \vdots & \vdots & \cdots & \vdots \\ a_{n1} & a_{n2} & \cdots & a_{nn} \end{bmatrix}$$

then $\quad \alpha_{ij} = \begin{bmatrix} \alpha_{11} & \alpha_{12} & \cdots & \alpha_{1n} \\ \alpha_{21} & \alpha_{22} & \cdots & \alpha_{2n} \\ \vdots & \vdots & \cdots & \vdots \\ \alpha_{n1} & \alpha_{n2} & \cdots & \alpha_{nn} \end{bmatrix}$; and $\quad adj\ A = \begin{bmatrix} \alpha_{11} & \alpha_{21} & \cdots & \alpha_{n1} \\ \alpha_{12} & \alpha_{22} & \cdots & \alpha_{n2} \\ \vdots & \vdots & \cdots & \vdots \\ \alpha_{1n} & \alpha_{2n} & \cdots & \alpha_{nn} \end{bmatrix}$

2.14 INVERSE OF A MATRIX

Inverse of a nonsingular square matrix A designated by A^{-1} is defined by the following relationship

$$A^{-1} = \frac{adj\ A}{\det A}$$

The inverse of a matrix A also satisfies the following relationships

(a) $AA^{-1} = A^{-1}A = I$

(b) $(A^{-1})^{-1} = A$

(c) $(AB)^{-1} = B^{-1}A^{-1}$ $\qquad$ where A and B are square and nonsingular matrices.

2.15 RANK OF A MATRIX

The *rank* of a given matrix is the order of the largest nonsingular matrix contained in the given matrix. Consider a matrix A of order $m \times n$, then the useful properties associated in determining the rank of the matrix A are

(a) Rank of A' = Rank of A

(b) Rank of AA' = Rank of A

(c) Rank of $A'A$ = Rank of A

2.16 TRACE OF A MATRIX

Trace of a square matrix A designated as *trace* (A) or *tr* A is defined as the sum of the elements of the diagonal of matrix A. The trace of a square matrix A is equal to the trace of transpose of matrix A. Also, trace of summation of two square matrices of the order $n \times n$ equals the sum of the trace of matrix A and matrix B. In other words

$$tr(A + B) = tr(A) + tr(B)$$

2.17 DERIVATIVE OF A MATRIX

The *derivative* of a matrix A is defined as the derivative of each element $a_{ij}(t)$ of the matrix.

$$\frac{d}{dt}[A(t)] = \begin{bmatrix} \dfrac{da_{11}(t)}{dt} & \dfrac{da_{12}(t)}{dt} & ..... & \dfrac{da_{1n}(t)}{dt} \\ \dfrac{da_{21}(t)}{dt} & ..... & ..... & \dfrac{da_{2n}(t)}{dt} \\ \vdots & \vdots & ..... & \vdots \\ \dfrac{da_{n1}(t)}{dt} & ..... & ..... & \dfrac{da_{nn}(t)}{dt} \end{bmatrix}$$

2.18 EXPONENTIAL OF A MATRIX

Exponential of a matrix A is defined as

$$\exp[A] = e^A = I + \frac{A}{1!} + \frac{A^2}{2!} + \cdots + \frac{A^K}{K!} + \ldots = \sum_{K=0}^{\infty} \frac{A^K}{K!}$$ where, A^K means A multiplied by K times A

matrix exponential of an $n \times n$ matrix which is a function of time is defined as

$$e^{At} = \frac{d}{dt} e^{At}$$

$$= I + At + \frac{1}{2!} A^2 t^2 + \ldots\ldots + \frac{1}{K!} A^K t^K + \ldots\ldots$$

Properties: Matrix Exponential

- $e^{A0} = I$
- $eA(t + \tau) = e^{At} e^{A\tau} = e^{A\tau} e^{At}$
- $(e^{At})^{-1} = e^{-At}$
- $\dfrac{d}{dt} e^{At} = A e^{At} = e^{At} A$

PROBLEMS AND SOLUTIONS

Problem 2.1

Express the following sets of algebraic eqns in matrix form, $Ax = B$.

(a)
$$x_1 + x_2 - x_3 = 1$$
$$-x_1 + 3x_2 - x_3 = 1$$
$$3x_1 - 5x_2 - 3x_3 = 0$$

(b)
$$x_1 + x_2 - x_3 = 1$$
$$-x_1 + 3x_2 - x_3 = 1$$
$$2x_1 - 3x_2 = 0$$

Solution

The symbols A, x and B are defined as matrices. These matrices contain the coefficients and variables of the given algebraic eqns as their elements. The given algebraic eqns in the problem are required to be represented as the product of the matrices A and x equal to matrix B

(a) The three variables are x_1, x_2 and x_3. Hence

$$x = \begin{bmatrix} x_1 \\ x_2 \\ x_3 \end{bmatrix}$$

The co-efficient of the variables x_1, x_2 and x_3 in the given algebraic eqns are 1, 1, –1; –1, 3, –1 and 3, –5, –3. In matrix form

$$A = \begin{bmatrix} 1 & 1 & -1 \\ -1 & 3 & -1 \\ 3 & -5 & -3 \end{bmatrix}$$

Matrix B is represented by a column vector containing elements on the right hand side of the algebraic eqns.

$$A = \begin{bmatrix} 1 \\ 1 \\ 0 \end{bmatrix}$$

Expressing in the required form i.e. $Ax = B$

$$\begin{bmatrix} 1 & 1 & -1 \\ -1 & 3 & -1 \\ 3 & -5 & -3 \end{bmatrix} \begin{bmatrix} x_1 \\ x_2 \\ x_3 \end{bmatrix} = \begin{bmatrix} 1 \\ 1 \\ 0 \end{bmatrix}$$

(b) Based on the logic explained in part (a), the required expression is

$$\begin{bmatrix} 1 & 1 & -1 \\ -1 & 3 & 1 \\ 2 & -3 & 0 \end{bmatrix} \begin{bmatrix} x_1 \\ x_2 \\ x_3 \end{bmatrix} = \begin{bmatrix} 1 \\ 1 \\ 0 \end{bmatrix}$$

Problem 2.2

Determine if the following matrices are conformable for the products AB and BA. Find the valid products

(a) $A = \begin{bmatrix} 3 \\ 2 \\ 2 \end{bmatrix}$, $B = \begin{bmatrix} 6 & 2 & 0 \end{bmatrix}$

(b) $A = \begin{bmatrix} -1 & 3 \\ 0 & 2 \end{bmatrix}$, $B = \begin{bmatrix} 0 & 5 & -5 \\ -1 & 1 & 1 \end{bmatrix}$

Solution

(a)
$$AB = \begin{bmatrix} 3 \\ 2 \\ 2 \end{bmatrix} [6 \ \ 2 \ \ 0] = \begin{bmatrix} 18 & 6 & 0 \\ 12 & 4 & 0 \\ 12 & 4 & 0 \end{bmatrix}$$

$$BA = [6 \ \ 2 \ \ 0] \begin{bmatrix} 3 \\ 2 \\ 2 \end{bmatrix} = 18 + 4 + 0 = 22$$

Matrices A and B have been multiplied to form the product AB since they are *conformable* which implies that number of columns of matrix A is equal to the number of rows in matrix B. Matrix A has one column and matrix B has one row. Matrices A & B are also conformable for the product BA, since matrix B has three columns and equals the number of rows in matrix A which are also ®three.

(b)
$$AB = \begin{bmatrix} -1 & 3 \\ 0 & 2 \end{bmatrix} \begin{bmatrix} 0 & 5 & -5 \\ -1 & 1 & 1 \end{bmatrix} = \begin{bmatrix} -3 & -2 & 8 \\ -2 & 2 & 2 \end{bmatrix}$$

Matrices A and B are conformable to product AB since number of columns in matrix A equal number of rows in matrix B.

The matrices B and A are not conformable to product BA as number of columns in matrix B which are three in numbers is not equal to number of rows in matrix A which are two in numbers.

Problem 2.3

Find the inverse of the following matrices, if they exist.

(a) $A = \begin{bmatrix} 2 & 5 \\ 10 & -3 \end{bmatrix}$ 　　 (b) $A = \begin{bmatrix} 1 & 3 & 4 \\ -1 & 1 & 0 \\ -1 & 0 & -1 \end{bmatrix}$ 　　 (c) $A = \begin{bmatrix} 0 & 1 & 0 \\ 2 & -2 & 3 \\ 0 & 1 & 5 \end{bmatrix}$

Solution

$$A^{-1} = \frac{adj \ A}{\det A}$$

Since the value of determinant is nonzero, the matrix A is nonsingular.

The adjoint of a 2×2 matrix is obtained by interchanging the elements of the main diagonal and changing the signs of the other diagonal. Therefore,

(a)
$$A = \begin{bmatrix} 2 & 5 \\ 10 & -3 \end{bmatrix} = \begin{bmatrix} a_{11} & a_{12} \\ a_{21} & a_{22} \end{bmatrix}$$

$$\det A = |A| = a_{11}a_{22} - a_{12}a_{21}$$

$$= (2 \times -3) - (5 \times 10) = -6 - 50 = -56$$

$$\text{adj } A = \begin{bmatrix} -3 & -5 \\ -10 & 2 \end{bmatrix}$$

Hence
$$A^{-1} = \frac{\begin{bmatrix} -3 & -5 \\ -10 & 2 \end{bmatrix}}{-56} = \frac{-1}{56}\begin{bmatrix} -3 & -5 \\ -10 & 2 \end{bmatrix}$$

(b) $$A = \begin{bmatrix} 1 & 3 & 4 \\ -1 & 1 & 0 \\ -1 & 0 & -1 \end{bmatrix}$$

$$\det A = |A| = 1[(1 \times -1 - 0 \times 0)] - (-1)[(3 \times -1) - (-4 \times 0)] + (-1)[(3 \times 0) - (4 \times 1)]$$

$$= -1 - 3 + 4 = 0$$

Since, the value of the determinant is zero, matrix A is singular. Hence, inverse of the matrix does not exist.

(c) $$A = \begin{bmatrix} 0 & 1 & 0 \\ 2 & -2 & 3 \\ 0 & 1 & 5 \end{bmatrix} = \begin{bmatrix} a_{11} & a_{12} & a_{13} \\ a_{21} & a_{22} & a_{23} \\ a_{31} & a_{32} & a_{33} \end{bmatrix}$$

The determinant

$$\det A = |A| = a_{11}(a_{22}a_{33} - a_{23}a_{32}) - a_{12}(a_{21}a_{33} - a_{23}a_{31})$$

$$= 0[(-2 \times 5 - 3 \times 1)] - 1[(-2 \times 5 - 3 \times 0)] + 0[(2 \times 1)(-2 \times 0)]$$

$$= 0 - 10 + 0 = -10$$

$$A_{11} = a_{22}a_{33} - a_{23}a_{32} \quad = -2 \times 5 - 3 \times 1 = -13$$

$$A_{12} = -(a_{12}a_{33} - a_{13}a_{32}) = -(1 \times 5 - 0 \times 1) = -5$$

$$A_{13} = a_{12}a_{23} - a_{13}a_{22} \quad = 1 \times 3 - 0 \times -2 = 3$$

$$A_{21} = -(a_{21}a_{33} - a_{23}a_{31}) = -(2 \times 5 - 3 \times 0) = -10$$

$$A_{22} = a_{11}a_{33} - a_{13}a_{31} \quad = 0 \times 5 - 0 \times 0 = 0$$

$$A_{23} = -a_{11}a_{23} - a_{21}a_{13} \quad = -(0 \times 3 - 2 \times 0) = 0$$

$$A_{31} = a_{21}a_{32} - a_{22}a_{31} \quad = (2 \times 1) - (-2 \times 0) = 2$$

$$A_{32} = -(a_{11}a_{32} - a_{12}a_{31}) = -[(0 \times 1) - (1 \times 0)] = 0$$

$$A_{33} = a_{11}a_{22} - a_{12}a_{21} \quad = (0 \times -2) - (1 \times 2) = -2$$

Hence
$$adj\ A = \begin{bmatrix} A_{11} & A_{12} & A_{13} \\ A_{21} & A_{22} & A_{23} \\ A_{31} & A_{32} & A_{33} \end{bmatrix} = \begin{bmatrix} -13 & -5 & 3 \\ -10 & 0 & 0 \\ 2 & 0 & -2 \end{bmatrix}$$

Therefore,
$$A^{-1} = \frac{1}{-10}\begin{bmatrix} -13 & -5 & 3 \\ -10 & 0 & 0 \\ 2 & 0 & -2 \end{bmatrix} = \begin{bmatrix} 1.3 & 0.5 & -0.3 \\ 1 & 0 & 0 \\ -0.2 & 0 & 0.2 \end{bmatrix}$$

SUMMARY

♦ Definitions of various types of matrices.

♦ Common matrix operations like addition, subtraction, multiplication and division.

♦ Calculations of determinant of a (2×2) and (3×3) matrix.

♦ Associative and distributive properties of the matrices.

♦ Transpose, adjoint and inverse of a matrix.

♦ Exponential of a matrix.

♦ Expressing given algebraic equations in matrix form.

REVIEW EXERCISE

1. Define matrix.
2. Has Null matrix any elements.
3. Which elements form the diagonal of Unity matrix.
4. Outline the procedure to compute determinant of a matrix. How can one find out whether a matrix is nonsingular.
5. Describe Associative and Distributive properties related to matrix operations.
6. Describe procedure to compute inverse of a matrix.
7. Define Determinant of a matrix. Let

$$A = \begin{bmatrix} 1 & 0 \\ 0 & 8 \end{bmatrix} \quad \text{and} \quad B = \begin{bmatrix} 2 & 4 \\ 0 & 4 \end{bmatrix};$$

   (a) Check if matrices A and B have same determinant.

   (b) If first row of matrix A is multiplied by 2 then what is the effect on the value of determinant.

(c) If the rows of matrices A and B are interchanged then what is the effect on the values of the determinants of the matrices.

(d) If a row matrix B is multiplied by a scaler k, then find out change in the value of determinant of matrix B.

8. How do you ascertain whether a matrix is singular or nonsingular.

9. Define inverse of a matrix. List out properties of inverse of a matrix. Evaluate inverse of matrix

$$A = \begin{bmatrix} 1 & 1 & -1 \\ 2 & 1 & 0 \\ 3 & 1 & 2 \end{bmatrix}$$

10. What is characteristic equation of a matrix. Evaluate inverse of matrix A given in Question No. 3 by using Cayley-Hamilton theorem. (Self study Question).

11. Find the adjoint and inverse of matrix

$$A = \begin{bmatrix} 1 & -2 & 3 \\ 0 & 2 & -1 \\ -4 & 5 & 2 \end{bmatrix}$$

12. Show that

(a) The inverse of the transpose is the transpose of the inverse $i.e.,$ $[A']^{-1} = [A^{-1}]'$.

(b) The inverse of matrix is unique.

(c) If A and B are two nonsingular matrices of the same order then $[AB]^{-1} = B^{-1}A^{-1}$. Verify this for matrices A and B given in Question No.1.

13. Ascertain the rank of matrix

(a) $\begin{bmatrix} 1 & 1 & -1 \\ 2 & 1 & 0 \\ 3 & 1 & 1 \end{bmatrix}$ (b) $\begin{bmatrix} 1 & 2 & 3 \\ -4 & 0 & 5 \end{bmatrix}$

14. Solve the following system of equations with the help of matrix inversion

$$x_1 + 2x_2 + 3x_3 = 1, \quad 2x_1 + 3x_2 + 2x_2 = 2 \quad \text{and} \quad 3x_1 + 3x_2 + 4x_3 = 1.$$

15. If $A = \begin{bmatrix} 3 & -3 & 4 \\ 2 & -3 & 4 \\ 0 & -1 & 1 \end{bmatrix}$; show that $A^3 = A^{-1}$.

16. If matrix A satisfies a relation $A^2 + A + I = 0$, prove that A^{-1} exists and that $A^{-1} = I + A$, I being an Identity matrix.

3

LAPLACE TRANSFORM

3.1 INTRODUCTION

The Laplace transform is a very useful mathematical tool utilised by control system engineers to solve simple linear differential equations. Laplace transform converts differential equations into algebraic equations in complex variable *s-form*. Exponential and trigonometric functions are also easily converted into algebraic complex variable functions. Mathematical operations such as differentiation and integration are also substituted in complex variable form. The solution of converted algebraic equation in complex variable s-form is easily obtained by solving for the dependent variable in *s*-domain. Application of inverse Laplace transform, finally gives the total solution i.e. the *complimentary* as well *as particular* solution is obtained in one operation.

LEARNING OBJECTIVES

- ◆ To introduce mathematical tool Laplace transform.
- ◆ To understand Laplace transform theorems and their use.
- ◆ To utilise Laplace transform operations and manipulations.

3.2 DEFINITION OF LAPLACE TRANSFORM

Laplace transform of a function $f(t)$ which satisfies the condition

$\int\limits_{0}^{\infty} f(t)\, e^{-\sigma t}\, dt < \infty$ for some finite, real value of σ, is defined as

$$Lf(t) \;=\; F(s) = \int\limits_{0}^{\infty} f(t)\, e^{-st}\, dt \tag{3.1}$$

where

$F(s) \;=\; Lf(t) =$ Laplace transform of $f(t)$

$s \;=\;$ Complex variable equal to $\sigma + j\omega =$ Laplace operator

$f(t) \;=\;$ A function of time such that $f(t) = 0$ for $t < 0$

The above definition ignores information pertaining to $f(t)$ prior to $t = 0$ and is considered as null. The analysis of the control systems in this book is based on this assumption. Students have not been burdened with the definition of the Laplace transform in which the limit of integration varies from 0^- to ∞. The retention of the lower limit implies that even if the function $f(t)$ is discontinuous at $t = 0$, integration process can commence prior to the discontinuity as long as the integral converges. This is particularly helpful in obtaining Laplace transform of the impulse functions.

SIGN CONVENTION

- The time function is depicted by lowercase letter; f, and
- The Laplace transform of the time function is represented by uppercase letter; F.

Nature of the Laplace Variable

The Laplace variable is a complex variable and is given by $s = \sigma + j\omega$. It can be easily represented on a s-plane as shown in Fig. 3.1. Students of control system engineering must have learnt by now that the system stability is easily ascertained from s-plane.

- The real part σ is associated with the system stability and depends upon the location σ in s-plane. LHS of the s-plane is the boundary of system stability.
- The imaginary part represents the frequency of the time signal $f(t)$.

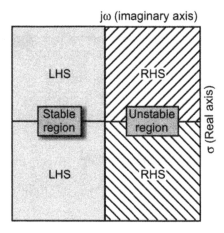

Fig. 3.1 *The Laplace variable s-plane*

3.3 LAPLACE THEOREMS

The two theorems which are widely used in the analysis of the control systems are *initial value* and *final value* theorems. *Initial value* theorem helps a control system engineer in ascertaining the initial value of $f(t)$ at $t \to 0^+$ from the Laplace transform of $f(t)$. Final value theorem provides the information about the final value or steady state value of the function $f(t)$.

3.3.1   Initial Value Theorem

It gives the initial value of the function $f(t)$ at $t \to 0^+$. The value obtained is thus not exactly at $t = 0$, but at a time slightly greater than zero. Initial value theorem states that *if f(t) is Laplace transformable, i.e. if F (s) is the Laplace transform of f(t), then*

$$f(0^+) \;=\; \lim_{t \to 0^+} f(t) = \lim_{s \to \infty} [sF(s)]$$

Proof

$$L\left[\frac{df(t)}{dt}\right] \;=\; \int_0^\infty \frac{df(t)}{dt} e^{-st} dt$$

Integrating by parts

$$\int_0^\infty \frac{df(t)}{dt} e^{-st} dt \;=\; \left[f(t)e^{-st}\Big|_0^\infty + s\int_0^\infty f(t)e^{-st} dt \right]$$
$$= -f(0^+) + sF(s)$$
$$= sF(s) - f(0^+)$$

Note : Refer Section 3.3.3.3 also.

Now, let $s \to \infty$, then

$$\lim_{s \to \infty} \int_{0^+}^\infty \left[\frac{df(t)}{dt}\right] e^{-st} dt \;=\; \lim_{s \to \infty} [sF(s) - f(0^+)]$$

On application of Limit

$$0 \;=\; \lim_{s \to \infty} [s\, F(s) - f(0^+)]$$

Hence $\qquad\qquad f(0^+) \;=\; \lim_{s \to \infty} [s(F(s)]$ (3.2)

3.3.2   Final Value Theorem

The ultimate aim of a control system engineer is to ascertain the steady state value of a control system so that the behaviour of the control system can be ascertained and analysed. Final value theorem states that *if f(t) is Laplace transformable i.e. if F (s) is the Laplace transform of f(t), and if F(s) is analytic in the right half of the s-plane and on the imaginary axis, then*

$$\lim_{t \to \infty} f(t) \;=\; \lim_{s \to 0} sF(s), \; \textit{if limit exists.}$$

The final value theorem is valid only if $sF(s)$ does not have any poles (refer section 3.7 for definition of Poles) in the right half of the s-plane and on imaginary ($j\omega$) axis.

Proof

A complex function is said to be analytic in a region if the function $F(s)$ and all its derivatives exist in that region of the s-plane. The points in the s-plane where the function $F(s)$ is not analytic are called *singularities*.

We know that $L\left[\dfrac{df(t)}{dt}\right] = \displaystyle\int_0^\infty \dfrac{df(t)}{dt} e^{-st} dt = sF(s) - f(0^+)$ (As found in Section 3.3.1)

Let $s \to 0$, then

$$\lim_{s \to 0} \int_0^\infty \frac{df(t)}{dt} e^{-st} dt = \lim_{s \to 0}[sF(s) - f(0^+)]$$

Since $\displaystyle\lim_{s \to 0} e^{-st} = 1$, we get on application of limit

$$\int_0^\infty \frac{df(t)}{dt} dt = f(t)\Big|_0^\infty = f(\infty) - f(0^+) = \lim_{s \to 0} [sF(s) - f(0^+)]$$

or $\qquad\qquad f(\infty) = \displaystyle\lim_{s \to 0} sF(s) = \lim_{t \to \infty} f(t)$ (3.3)

Keeping in view the analytic requirements, this theorem is not valid if the denominator of $sF(s)$ contains any pole whose real part is positive or zero i.e. poles lie on the imaginary axis or in the right half of s-plane. Such functions exhibit responses which are oscillating or exponentially increasing respectively. However, location of poles of $sF(s)$ pose no limitations while applying initial value theorem. Also, the steady state behaviour of $f(t)$ is similar to that of $sF(s)$ in the vicinity of $s = o$.

KEY POINTS LEARNT

- Laplace transform is an useful mathematical tool utilised for control system analysis
- Laplace variable s is a complex variable given by $s = \sigma + j\omega$. The real part σ is associated with system stability and the imaginary part is related to the time function $f(t)$. It can be conveniently represented on a *s-plane*.
- The Laplace transform is defined as

$$F(s) = L[f(t)] = \int_0^\infty f(t) e^{-st} dt$$

- Initial value theorem gives initial value of the time function *f(t)*

$$f(0^+) = \lim_{t \to 0} f(t) = \lim_{s \to \infty} sF(s)$$

- Final value theorem gives final or steady state value of the time function *f(t)*

$$f(\infty) = \lim_{t \to \infty} f(t) = \lim_{s \to 0} sF(s)$$

3.3.3 Other Laplace Theorems

Basic Laplace theorems which are utilised in the application of Laplace transforms are given in the succeeding paragraphs

3.3.3.1 Summation and difference It states that *the Laplace transform of the sum or difference of two functions is equal to the sum or difference of Laplace transforms of the individual functions.* If $F_1(s)$ and $F_2(s)$ are the Laplace transform of functions $f_1(t)$ and $f_2(t)$ respectively, then mathematically

$$L[f_1(t) \pm f_2(t)] = \int_0^\infty [f_1(t) \pm f_2(t)]e^{-st}\, dt = L[f_1(t)] \pm L[f_2(t)] = F_1(s) \pm F_2(s) \qquad (3.4)$$

3.3.3.2 Multiplication by a constant It states that *the Laplace transform of a function f(t) multiplied by a constant K is equal to K times the Laplace transform of the function f(t).* Mathematically it can be represented as

$$L[Kf(t)] = \int_0^\infty [Kf(t)e^{-st}\, dt = K\int_0^\infty f(t)e^{-st}dt = KF(s) \qquad (3.5)$$

3.3.3.3 Differentiation theorem It states that *if F(s) is the Laplace transform of a function f(t) and if f(t) and its derivatives are Laplace transformable, then the Laplace transform of the first derivative of the function f(t) is equal to s times the Laplace transform of f(t) minus the limit of f(t) as $t \to 0^+$.* Mathematically

$$L\left[\frac{df(t)}{dt}\right] = sF(s) - \lim_{t \to 0} f(t) = sF(s) - f(0^+)$$

Proof
$$Lf(t) = F(s) = \int_0^\infty f(t)e^{-st}dt = f(t)\frac{e^{-st}}{-s}\Big|_0^\infty - \int_0^\infty \left[\frac{d}{dt}f(t)\right]\frac{e^{-st}}{-s}\, dt$$

$$= -\frac{1}{s}f(t)e^{-st}\Big|_0^\infty + \frac{1}{s}\int_0^\infty \left[\frac{df(t)}{dt}\right]e^{-st}dt = \frac{f(0^+)}{s} + \frac{1}{s}L\left[\frac{df(t)}{dt}\right]$$

or
$$L\left[\frac{df(t)}{dt}\right] = sF(s) - f(0^+) \qquad (3.6)$$

Note: Integration has been done by parts.

In general, extending this to cover the higher order derivatives when they are Laplace transformable

$$L\left[\frac{d^n f(t)}{dt}\right] = s^n F(s) - \lim_{t \to 0^+}\left[s^{n-1}f(t) + s^{n-2}\frac{df(t)}{dt} + \ldots + \frac{d^{n-1}f(t)}{dt^{n-1}}\right] \qquad (3.7)$$

$$= s^n F(s) - s^{n-1}f(0^+) - s^{n-2}f'(0^+) - \cdots - f^{(n-1)}(0^+)$$

$f^i(0^+)$ denotes the *i th*-order derivative of $f(t)$ with respect to t at $t = 0$.

3.3.3.4 Integration It states that *if F(s) is the Laplace transform of a function f(t), then the Laplace transform of the first integral of the function f(t) with respect to time is the Laplace transform of f(t) divided by s.* Mathematically

$$L\int_0^t [f(t)dt] = \frac{F(s)}{s}$$

Proof $\quad L\int_0^t [f(t)dt] = \int_0^\infty \left[\int_0^t f(t)dt\right] e^{-st} dt$

Replace $\quad u = \int_0^t f(t)dt$ and $dv = e^{-st}dt$ which yields

$$du = f(t)dt \text{ and } v = -\frac{e^{-st}}{s}$$

Integrating by parts

$$L\left[\int_0^t f(t)dt\right] = \int_0^\infty udv = uv\Big|_0^\infty - \int_0^\infty vdu$$

Substituting values obtained above we get

$$L\left[\int_0^t f(t)dt\right] = \left[\frac{-e^{-st}}{s}\int_0^t f(t)dt\right]\Big|_0^\infty + \frac{1}{s}\int_0^\infty f(t)e^{-st}dt$$

$$= 0 - 0 + \frac{1}{s}\int_0^\infty f(t)e^{-st}dt = \frac{1}{s}L[f(t)] = \frac{F(s)}{s} \qquad (3.8)$$

In general for higher order i.e. n th-order integration

$$L\left[\int_0^{t_1}\int_0^{t_2}....\int_0^{t_n} f(t)dt_1 \ dt_2 \ .....dt_n\right] = \frac{F(s)}{s^n} \qquad (3.9)$$

3.3.3.5 *Shift in time* It states that *If F(s) is the Laplace transform of a function f(t), then the Laplace transform of the function f(t) delayed by time T is equal to the Laplace transform of f(t) multiplied by e^{-st}.* Mathematically

$$L[f(t-T)u(t-T)] = e^{-sT}F(s)$$

Where $u(t-T)$ is unit step time function that has shifted in time by T

Proof

$$L[f(t-T)u(t-T)] = \int_0^\infty [f(t-T)u(t-T)]e^{-st}dt$$

$$= \int_0^\infty f(t-T)e^{-st}dt = \int_0^\infty f(\tau)e^{-s(\tau+T)}d\tau$$

$$= e^{-sT} \int_0^\infty f(\tau)e^{-s\tau}d\tau = e^{-sT} F(s)$$

where $f(t-T)u(t-T)$ $= 0$ for $t < T$; and

$$= f(t-T) \text{ for } t > T$$

and $\qquad t - T = \tau$ or $t = \tau + T$

also $\qquad dt = d\tau$

with limits

$$t = T \Rightarrow \tau = 0 \text{ ; and } t = \infty \Rightarrow \tau = \infty, \text{ we get}$$

Therefore, $\quad L[f(t-T)u(t-T)] = e^{-sT} F(s)$ \hfill (3.10)

3.3.3.6 Multiplication by $e^{\pm at}$ *Laplace transform of function f(t) multiplied by $e^{\pm at}$ is equal to Laplace transform of that function in which s is replaced by $(s \mp a)$.* Mathematically

$$L[e^{\pm at} f(t)] = F(s \mp a)$$

Proof $\quad L[e^{\pm at} f(t)] = \int_0^\infty [e^{\pm at} f(t)] e^{-st} dt = \int_0^\infty f(t) e^{-(s \mp a)t} dt = F(s \mp a)$ \hfill (3.11)

3.4 LAPLACE TRANSFORM THEOREMS

Laplace transform theorems are tabutated in Table 2.1 for ready reference.

Table 2.1 Laplace Transform Theorems

Ser. No.	Theorem	Mathematical Expression
1.	Initial Value	$\lim\limits_{t \to 0} f(t) = \lim\limits_{s \to \infty} sF(s)$
2.	Final Value	$\lim\limits_{t \to \infty} f(t) = \lim\limits_{s \to 0} sF(s)$
3.	Summation & Difference	$L[f_1(t) \pm f_2(t)] = F_1(s) \pm F_2(s)$
4.	Multiplication by a Constant	$L[K f(t)] = KF(s)$
5.	Differentiation	$L\left[\dfrac{df(t)}{dt}\right] = sF(s) - f(0^+)$
6.	Integration	$L\int_0^t [f(t)dt] = \dfrac{F(s)}{s}$
7.	Shift in time	$L[f(t-T)u(t-T)] = e^{-sT}F(s)$
8.	Multiplication by $e^{\pm at}$	$L[e^{\pm at} f(t)] = F(s \mp a)$

3.5 IMPORTANT LAPLACE TRANSFORM PAIRS

Some important Laplace transform pairs widely used are given in Table 3.2

Table 3.2   Important Laplace Transform

Ser. No.	$f(t)$	$F(s)$
1.	Step function $u(t)$	$\dfrac{1}{s}$
2.	Unit Impulse $\delta(t)$	1
3.	t	$\dfrac{1}{s^2}$
4.	e^{-at}	$\dfrac{1}{s+a}$
5.	t^n	$\dfrac{n!}{s^{n+1}}$
6.	te^{-at}	$\dfrac{1}{(s+a)^2}$
7.	$\dfrac{t^{n-1}}{(n-1)!}$	$\dfrac{1}{s^n}$
8.	$\dfrac{t^{n-1}e^{-at}}{(n-1)!}$	$\dfrac{1}{(s+a)^n}$
9.	$\dfrac{e^{-at}-e^{-bt}}{b-a}$	$\dfrac{1}{(s+a)(s+b)}$
10.	$\sin at$	$\dfrac{a}{s^2+a^2}$
11.	$\cos at$	$\dfrac{s}{s^2+a^2}$
12.	$e^{-at}\sin \omega t$	$\dfrac{\omega}{(s+a)^2+\omega^2}$
13.	$e^{-at}\cos \omega t$	$\dfrac{s+a}{(s+a)^2+\omega^2}$

3.6 INVERSE LAPLACE TRANSFORM

The inverse Laplace transform is obtained by *inversion integral*

$$L^{-1}[F(s)] = f(t) = \frac{1}{2\pi j} \int_{\sigma-j\infty}^{\sigma+j\infty} F(s) e^{st} ds \qquad (3.12)$$

Inverse Laplace transformation designated as L^{-1} is the reverse process of obtaining the time function f(t) from the Laplace transform F(s). Obtaining inverse Laplace transform is a complicated process and is rarely used. Partial Fraction method which is a simpler method is generally employed to evaluate inverse Laplace transformation. The steps involved are:

◆ Find out inverse Laplace transform from the evaluated functions tabulated in Tables 3.1 & 3.2.

◆ For complex functions, obtain partial fractions and then use the functions tabulated in Tables 3.1 & 3.2.

3.7 PARTIAL-FRACTION EXPANSION METHOD

The analysis of control systems generally involves mathematical manipulation of Laplace transform

$$F(s) = \frac{P(s)}{Q(s)} \qquad (3.13)$$

where $P(s)$ and $Q(s)$ are polynomials in s and the order of $Q(s)$ in s is greater than that of $P(s)$. The polynomial $Q(s)$ is of the form

$$Q(s) = s^n + a_{n-1}s^{n-1} + \cdots + a_1 s + a_0 \qquad (3.14)$$

where $a_0, a_1, \cdots a_{n-1}$ are the real coefficients.

The denominator polynomial $Q(s)$ when equated to zero gives the *characteristic* equation. The name is derived from the fact that the roots of the characteristic equation indicates the character of the response of the control system in time-domain. The *roots* of the denominator polynomial or the characteristic equation are also referred to as *poles* of the system. **A pole is the most common type of singularity**. The roots of the numerator polynomial $P(s)$ gives the *zeros* of the systems.

3.7.1 Real and Distinct Roots of the Denominator Polynomial

When the roots of the denominator polynomial $Q(s)$ are distinct, it may be factored in the form

$$Q(s) = (s - p_1)(s - p_2)..........(s - p_n) \qquad (3.15)$$

The partial fraction expansion of equation (3.15) is of the form

$$F(s) = \frac{P(s)}{Q(s)} = \frac{K_1}{s - p_1} + \frac{K_2}{s - p_2} + ... + \frac{K_i}{s - p_i} + ... + \frac{K_n}{s - p_n} \qquad (3.16)$$

where K_1, K_2, $\cdots$ K_i, $\cdots$ K_n are constants. The value of the constants are ascertained by first multiplying both sides of equation (3.16) by $(s-p_i)$, i.e.

$$(s-p_i)F(s) = (s-p_i)\frac{P(s)}{Q(s)} = \frac{s-p_i}{s-p_1}K_1 + \cdots + \frac{s-p_i}{s-p_i}\ K_i + \cdots + \frac{s-p_i}{s-p_n}K_n \qquad (3.17)$$

Thereafter, we let $s = p_i$, which makes each term on the right hand side of equation (3.14) zero, except for K_i which remains and hence

$$K_i = \lim_{s\to p_i}\left[(s-p_i)\frac{P(s)}{Q(s)}\right] \qquad (3.18)$$

The procedure involves successive application of $i = 1, 2, \cdots, n.$ in equation (3.18), giving values of constants K_1, K_2, $\cdots$, K_n respectively. After substituting values in equation (3.16), inverse transform is obtained directly from the *Table 3.2* and is of the form

$$f(t) = K_1 e^{-p_1 t} + K_2 e^{-p_2 t} + \cdots + K_n e^{-p_n t} \qquad (3.19)$$

Example 3.1   Consider the Laplace transform function

$$F(s) = \frac{2s+3}{(s+1)(s+2)(s+3)}.$$

In the partial fraction expanded form, the function can be written as

$$F(s) = \frac{2s+3}{(s+1)(s+2)(s+3)} = \frac{A}{s+1} + \frac{B}{s+2} + \frac{C}{s+3}$$

where

$$A = (s+1)F(s)\Big|_{s=-1} = \frac{2s+3}{(s+2)(s+3)}\Bigg|_{s=-1} = \frac{2(-1)+3}{(-1+2)(-1+3)} = \frac{1}{2}$$

$$B = (s+2)F(s)\Big|_{s=-2} = \frac{2s+3}{(s+1)(s+3)}\Bigg|_{s=-2} = \frac{2\times(-2)+3}{(-2+1)(-2+3)} = 1$$

$$C = (s+3)F(s)\Big|_{s=-3} = \frac{2s+3}{(s+1)(s+2)}\Bigg|_{s=-3} = \frac{2\times(-3)+3}{(-3+1)(-3+2)} = -\frac{3}{2}$$

Therefore, $$F(s) = \frac{2s+3}{(s+1)(s+2)(s+3)} = \frac{1/2}{s+1} + \frac{1}{s+2} + \frac{-3/2}{s+3}$$

Thus $$f(t) = L^{-1}F(s) = \frac{1}{2}e^{-t} + e^{-2t} - \frac{3}{2}e^{-3t}$$

3.7.2   Real and Repeated Roots of the Denominator Polynomial

The roots of the denominator polynomial $Q(s)$ may have repeated roots p_i occurring r times, where $i \neq 1, 2, \cdots, n - r$. In such a case $Q(s)$ may be factored in the form

$$Q(s) = (s - p_i)^r (s - p_1)(s - p_2)......(s - p_{n-r}) \tag{3.20}$$

Partial Fraction expansion of $F(s) = \dfrac{P(s)}{Q(s)}$ is of the form

$$F(s) = \frac{P(s)}{Q(s)} = \frac{N_r}{(s - p_i)^r} + \cdots + \frac{N_2}{(s - p_i)^2} + \frac{N_1}{s - p_i} + \frac{K_1}{s - p_1} + \frac{K_2}{s - p_2}$$

$$+ \cdots + \frac{K_{n-r}}{s - p_{n-r}} \tag{3.21}$$

The coefficients $K_1, K_2, \cdots, K_{n-r}$ which correspond to simple poles may be determined as described in section 3.7.1. The evaluation of the coefficients $N_1, N_2, \cdots, N_r$ corresponding to repeated zeros is described below:

$$N_r = \left[(s - p_i)^r F(s) \right]\Big|_{s = p_i}$$

$$N_{r-1} = \frac{d}{ds} \left[(s - p_i)^r F(s) \right]\Big|_{s = p_i}$$

$$N_{r-2} = \frac{1}{2!} \frac{d^2}{ds^2} [(s - p_i)^r F(s)]\Big|_{s = p_i}$$

$$N_{r-k} = \frac{1}{K!} \frac{d^K}{ds^K} \left[(s - p_i)^r F(s) \right]\Big|_{s = p_i}$$

$$\vdots$$

$$N_1 = \frac{1}{(r-1)!} \frac{d^{r-1}}{ds^{r-1}} \left[(s - p_i)^r F(s) \right]\Big|_{s = p_i}$$

The inverse Laplace transform of $\dfrac{1}{(s + p_i)^n}$ is given by $\mathcal{L}^{-1}\left[\dfrac{1}{(s + p_i)^n} \right] = \dfrac{t^{n-1}}{(n-1)!} e^{-p_i t}$

After substituting values of the constants in equation (3.21), inverse Laplace transform yields

$$f(t) = \frac{N_r t^{r-1} e^{p_i t}}{(r-1)!} + \frac{N_{r-1} t^{r-2} e^{p_i t}}{(r-2)!} + \cdots + N_2 t e^{p_i t} + N_1 e^{p_i t} + K_1 e^{p_i t} + K_2 e^{p_2 t} + \cdots + K_{n-r} e^{p_{n-r} t} \tag{3.22}$$

Example 3.2 Consider the Laplace transformed function

$$F(s) = \frac{4}{(s+1)(s+2)^2}$$

In the partial fraction expansion form, the function can be written as

$$F(s) = \frac{4}{(s+1)(s+2)^2} = \frac{A}{s+1} + \frac{B}{(s+2)^2} + \frac{C}{s+2}$$

where

$$A = (s+2)^2 F(s)\Big|_{s=-1} = \frac{4}{(s+2)^2}\Big|_{s=-1} = \frac{4}{(-1+2)^2} = 4$$

$$B = (s+2)^2 F(s)\Big|_{s=-2} = \frac{4}{s+1}\Big|_{s=-2} = \frac{4}{-2+1} = -4$$

$$C = \frac{d}{ds}(s+2)^2 F(s)\Big|_{s=-2} = \frac{d}{ds}\left(\frac{4}{s+1}\right)\Big|_{s=-2} = \frac{(s+1)\times 0 - 4}{(s+1)^2}\Big|_{s=-2} = \frac{-4}{(-2+1)^2} = -4$$

Therefore,

$$F(s) = \frac{4}{(s+1)(s+2)^2} = \frac{4}{(s+1)} + \frac{-4}{(s+2)^2} + \frac{-4}{(s+2)}$$

Thus,

$$f(t) = L^{-1}F(s) = 4e^{-t} - 4te^{-2t} - 4e^{-2t}$$

3.7.3 Complex—Conjugate Roots of the Denominator Polynomial

Complex conjugate roots are always in pairs having same real parts but equal and opposite imaginary parts. Polynomial $Q(s)$ having a complex root $a + jb$ will have its complex conjugate $a - jb$. Let us assume denominator polynomial $Q(s)$ can be factored and is of the form

$$Q(s) = (s-a-jb)(s-a+jb)(s-p_1)\cdots(s-p_{n-2}) \qquad (3.23)$$

The partial fraction expansion is of the form

$$F(s) = \frac{P(s)}{Q(s)} = \frac{K_r}{s-a-jb} + \frac{K_r'}{s-a+jb} + \frac{K_1}{s-p_1} + \cdots + \frac{K_{n-2}}{s-p_{n-2}} \qquad (3.24)$$

Constant K_r is evaluated as given below:

$$K_r = \lim_{s\to a+jb}\left[(s-a\ jb)\frac{P(s)}{(s-a-jb)(s-a+jb)(s-p_1)\cdots(s-p_{n-2})}\right]$$

$$= \lim_{s\to a+jb}\left[\frac{1}{2jb}\frac{P(s)}{(s-p_1)\cdots(s-p_{n-2})}\right] \qquad (3.25)$$

$$= \frac{A}{2jb} \qquad (3.26)$$

where
$$A = \lim_{s \to a + jb} \left[\frac{P(s)}{(s - p_1)......(s - p_{n-2})} \right] = \left[(s^2 - 2as + a^2 + b^2) \frac{P(s)}{Q(s)} \right]_{s=a+jb} \quad (3.27)$$

Similarly
$$K_r' = -\frac{A'}{2jb} \quad \text{where}$$

$$A' = \lim_{s \to a - jb} \left[\frac{P(s)}{(s - p_1)......(s - p_{n-2})} \right] = \left[(s^2 - 2as + a^2 + b^2) \frac{P(s)}{Q(s)} \right]_{s=a-jb} \quad (3.28)$$

The constants A and A' are complex conjugate and can be represented on the complex plane as shown in Fig. 3.2

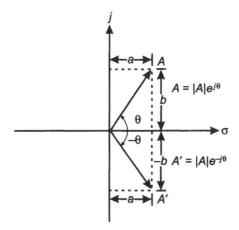

Fig. 3.1 *Representation of complex conjugate A and A'*

The length of vectors A and A', making angles θ and $-\theta$ is equal. Constant K_r and K_r', which are also complex conjugate can be written as

$$K_r = \frac{|A|}{2jb} e^{j\theta} \quad (3.29)$$

and
$$K_r' = -\frac{|A'|}{2jb} e^{-j\theta} = -\frac{|A|}{2jb} e^{-j\theta} \quad (3.30)$$

Substitution of values K_r and K_r' in equation (3.24) gives

$$f(t) = |A| e^{at} \frac{e^{j(bt+\theta)} - e^{-j(bt+\theta)}}{2jb} + K_1 e^{p_1 t} + \cdots + K_{n-2} e^{p_{n-2} t} \quad (3.31)$$

or
$$f(t) = \frac{1}{b} |A| e^{at} \sin(bt + \theta) + K_1 e^{p_1 t} + \cdots + K_{n-2} e^{p_{n-2} t} \quad (3.32)$$

Example 3.3 Consider

$$F(s) = \frac{10}{(s+6)(s^2+4s+13)}$$

$$(s^2+4s+13) \text{ yields } a \pm jb = -2 \pm j3.$$

Therefore $a = -2$ and $b = 3$.

Now

$$A = \left[(s^2 - 2as + a^2 + b^2)\frac{P(s)}{Q(s)} \right]_{s=a+jb}$$

$$= \frac{10}{s+6}\bigg|_{s=-2+j3} = \frac{10}{-2+j3+6} = \frac{10}{4+j3}$$

Magnitude and angle of A

$$A = \frac{10(4+j3)}{(4-j3)(4+j3)} = \frac{10 \times 5 \angle \tan^{-1}\left(\frac{3}{4}\right)}{16+9} = 2 \angle -36.8°$$

Therefore, $\qquad |A| = 2$ and $\theta = -36.8°$

The solution can be generalised as

$$f(t) = \frac{1}{b}|A|e^{at}\sin(bt+\theta) + K_1 e^{p_1 t}$$

$$K_1 = \frac{20}{s^2+4s+13}\bigg|_{s=-6} = 0.8$$

Therefore, $\qquad f(t) = \frac{2}{3}e^{-2t}\sin(3t-36.8°) + 0.8e^{-6t}$

3.8 CONVOLUTION INTEGRAL

If $F_1(s)$ and $F_2(s)$ are the Laplace transforms of functions $f_1(t)$ and $f_2(t)$ respectively and assuming that $f_1(t)$ and $f_2(t)$ are equal to zero for $t < 0$; then

$$F_1(s)F_2(s) = L[f_1(t) * f_2(t)] = L\left[\int_0^t f_1(t-\tau)f_2(\tau)d\tau\right] = L\left[\int_0^t f_1(\tau) \times f_2(t-\tau)d\tau\right] \text{(3.33)}$$

or $\qquad L^{-1}\left[F_1(s)F_2(s)\right] = f_1(t) * f_2(t)$

where the symbol '*' indicates *convolution* of the functions $f_1(t)$ and $f_2(t)$ in time domain.

Proof

By definition

$$L\left[\int_0^t f_1(t-\tau)f_2(\tau)d\tau\right] = \int_0^\infty\left[\int_0^t f_1(t-\tau)f_2(\tau)d\tau\right]e^{-st}dt \tag{3.34}$$

In order to show that equations (3.33) and (3.34) are same, equation (3.34) is rearranged as a product of two integrals ranging from 0 to ∞. Therefore

$$L\left[\int_0^t f_1(t-\tau)f_2(\tau)d\tau\right] = \int_0^\infty\left[\int_0^\infty f_1(t-\tau)f_2(\tau)u(t-\tau)d\tau\right]e^{-st}dt$$

where $u(t-\tau)$ is a unit step function which has been shifted, and

$$u(t-\tau) = 1 \text{ for } \tau \le t$$
$$= 0 \text{ for } \tau > t$$

Since as per the statement of the theorem, both $f_1(t)$ and $f_2(t)$ are Laplace transformable; the integrals are convergent, the sequence of integration operation with respect to variables t and τ can be reversed. Therefore

$$L\left[\int_0^t f_1(t-\tau)f_2(\tau)d\tau\right] = \int_0^\infty f_2(\tau)\left[\int_0^\infty f_1(t-\tau)u(t-\tau)e^{-st}dt\right]d\tau$$

because we know that $u(t-\tau) = 0$ for $t < \tau$. Putting $t-\tau = x$, $dt = dx$ and $t=0 \Rightarrow x = -\tau$ and $t = \infty \Rightarrow x = \infty$

$$\int_0^\infty\left[\int_0^t f_1(t-\tau)f_2(\tau)d\tau\right] = \int_0^\infty f_2(\tau)\left[\int_\tau^\infty e^{-s(T+x)}f_1(x)dx\right]d\tau$$

$$= \int_0^\infty f_2(\tau)\left[\int_0^\infty e^{-st}e^{-sx}f_1(x)dx\right]d\tau = \int_0^\infty f_2(\tau)\left[\int_0^\infty e^{-sx}f_1(x)dx\,e^{-st}d\tau\right]$$

$$= \int_0^\infty e^{-st}f_2(\tau)d\tau\int_0^\infty e^{-sx}f_1(x)dx = F_2(s)\,F_1(s) \tag{3.34}$$

Inverse Laplace transform of the function $F(s)$ which can be expressed as a product of two Laplace transformable functions is obtained as follows:

$$F(s) = F_1(s).F_2(s)$$

$$f(t) = L^{-1}F(s) = L^{-1}[F_1(s).F_2(s)]$$

$$= f_1(t) * f_2(t) = \int_0^t f_1(t-\tau)f_2(\tau)d\tau$$

Example 3.4 To illustrate usage of convolution integral for computing inverse Laplace transform, let us consider

$$F(s) = \frac{1}{s(s+6)}$$

Now
$$F(s) = F_1(s) \cdot F_2(s)$$

$$= \frac{1}{s} \cdot \frac{1}{s+6}$$

$$f_1(t) = L^{-1}F_1(s) = 1 \text{ ; and}$$

$$f_2(t) = L^{-1}F_2(s) = e^{-6t}$$

Therefore,
$$f(t) = \int_0^t f_1(t-\tau)f_2(\tau)d\tau$$

$$= \int_0^t 1 \times e^{-6\tau}d\tau = \left[\frac{e^{-6\tau}}{-6}\right]_0^t = \frac{e^{-6t}}{-6} - \frac{e^0}{-6} = \frac{e^{-6t}}{-6} + \frac{1}{6}$$

Hence,
$$f(t) = \frac{1}{6}(1 - e^{-6t})$$

3.9 SOLUTION OF DIFFERENTIAL EQUATIONS

Laplace transform method is very conveniently used in evaluation of linear time-invariant differential equations and yields the *complimentary* and *particular* solutions. The following steps are involved in obtaining the solution of the linear time-invariant differential equations:

- Convert the differential equation into Laplace transform form algebraic equation in *s*-domain.
- Rearrange the variables by manipulating the transformed algebraic equation.
- Carry out partial fraction expansion.
- Obtain the time solution by taking the inverse Laplace transform of the desired variable.

Example 3.5 Let us solve the differential equation

$$\frac{dy(t)}{dt} + 2y(t) = 8 \text{ where } y(0) = 0$$

To solve the differential equation, we first take the Laplace transform of both sides of the given differential equation.

$$sY(s) - Y(0) + 2Y(s) = \frac{8}{s}$$

Since $Y(0) = 0$, we have

$$Y(s)(s+2) = \frac{8}{s}$$

or

$$Y(s) = \frac{8}{s(s+2)}$$

or

$$Y(s) = \frac{8}{s(s+2)} = \frac{A}{s} + \frac{B}{s+2}$$

where

$$A = sY(s)\big|_{s=0} = \frac{8}{s+2}\bigg|_{s=0} = 4$$

$$B = (s+2)Y(s)\big|_{s=-2} = \frac{8}{s}\bigg|_{s=-2} = -\frac{8}{2} = -4$$

Therefore,

$$Y(s) = \frac{4}{s} + \frac{-4}{s+2}$$

Thus

$$Y(t) = L^{-1}Y(s) = 4 - 4e^{-2t} = 4(1 - e^{-2t})$$

3.10 CONCLUSION

In this chapter, fundamentals pertaining to Laplace transform have been explained. Laplace transform is widely utilised in design and analysis of control systems. The solution of differential equations is very conveniently obtained with the help of Laplace transform.

KEY POINTS LEARNT

- ◆ Definition of Laplace Transform.
- ◆ Initial & Final value theorems and application.
- ◆ Other important theorems of Laplace transform.
- ◆ Convolution Integral and its use of Laplace transform.
- ◆ Solution of differential equations by Laplace transform.
- ◆ Use of Partial-fraction expansion method to obtain inverse Laplace transform.
- ◆ Computation of Poles and Zeros.
- ◆ Use of Tables 3.1 and 3.2.

PROBLEMS AND SOLUTIONS

Problem 3.1

If Laplace transform of $e^{-at}f(t) = F(s + a)$, find the Laplace transform of $e^{-at}\cos \omega t$.

Solution

$L(e^{-at}\cos \omega t) = F(s+a)$ where $f(t) = \cos \omega t$; and

$$F(s) = L[\cos \omega t] = \frac{s}{s^2 + \omega^2}$$

Therefore, $F(s+a) = \dfrac{s+a}{(s+a)^2 + \omega^2}$

Problem 3.2

Find the poles and zeros of the following functions

(a) $F(s) = \dfrac{5s(s+1)}{(s+2)(s^2 + 3s + 1)}$

(b) $F(s) = \dfrac{5(s+3)}{s^2(s+1)(s+6)}$

(c) $F(s) = \dfrac{5(s+2)}{s(s^2 + 2s + 2)}$

(d) $F(s) = \dfrac{e^{-3s}}{5s(s+1)(s+3)}$

Solution

(a) Let $F(s) = \dfrac{5s(s+1)}{(s+2)(s^2 + 3s + 1)} = \dfrac{5s(s+1)}{(s+2)(s+2)(s+1)} = \dfrac{5s}{(s+2)(s+2)}$

Putting the numerator equal to zero gives

$5s = 0$ i.e. $s = 0$. Hence, Zeros are $s = 0$

Putting the denominator equal to zero gives

$(s + 2)(s + 2) = 0$. i.e. $s = -2, -2$. Hence, Poles are $s = -2$ and -2.

(b) Let $F(s) = \dfrac{5(s+3)}{s^2(s+1)(s+6)}$

Equating numerator and denominator to zero gives

Zeros at $s = -3$ and Poles at $s = 0, 0, -1$ and -6. Since, the function is a quotient of two polynomials of s, the total number of Poles equals Zeros. The function has thus three Zeros at infinity, since

$$\lim_{s \to \infty} F(s) = \lim_{s \to 0} \frac{5s(1+3/s)}{s^4(1+1/s)(s+6/s)}$$

Application of limit gives

$$= \lim_{s \to 0} \frac{5}{s^3} = 0 \text{ Hence, there are three Zeros at infinity also}$$

Therefore, Poles = 0, 0, –1 and –6 and Zeros = –3, ∞, ∞ and ∞

(c) $F(s) = \dfrac{5(s+2)}{s(s^2+2s+2)}$

$$= \frac{5(s+2)}{s(s+1-j)(s+1+j)}$$

Putting numerator and denominator equal to zero gives

Zeros at $s = -2$; and Poles at $s = 0$, $-1 + j$ and $-1 - j$

(d) $F(s) = \dfrac{e^{-3s}}{5s(s+1)(s+3)} = \dfrac{1}{5se^{3s}(s+1)(s+3)}$

Putting denominator equal to zero gives poles at

$s = 0$, –1, –3 and $e^{3s} = 0$ i.e. $s = \infty$. Hence; four poles are s = 0, –1, –3 and ∞.

Problem 3.3

Find the Laplace transform of the following functions

(a) $f_1(t) = 0$ for $t < 0$

    $= e^{-5t} \sin 10t$ for $t \le 0$

(b) $f_1(t) = 0$ for $t < 0$

    $= 3\sin(5t+45°)$ for $t \ge 0$

Solution

(a) $F(s+a) = L^{-1}[e^{-5t}\sin10\,t]$. Refer *Table 3.2*. We can write the Laplace transform straight away and is

$$F(s) = L[\sin10t] = \frac{10}{s^2+100}$$

Therefore, $F(s+a) = \dfrac{10}{(s+5)^2+100}$ where $a = 5$

(b) $f_1(t) = 0$ for $t < 0$

$\qquad = 3\sin(5t + 45°)$ for $t \geq 0$

Now $\qquad f_1(t) = 3\sin(5t + 45°) = 3[\sin 5t \cos 45° + \cos 5t \sin 45°]$

$\qquad\qquad = 2.121[\sin 5t + \cos 5t]$

$\qquad F_1(s) = [Lf_1(t)] = 2.121\left[\dfrac{5}{s^2 + 25} + \dfrac{s}{s^2 + 25}\right] = \dfrac{2.121(s + 5)}{s^2 + 25}$

Problem 3.4

Determine the Laplace transform of (a) $t - e^{-at}$ (b) At (c) $\cosh at$ (d) te^{-at} (e) $\sinh \omega t$ (f) $e^{-at}\sinh \omega t$ (g) $2e^{-2t}\cos 3t$.

Solution

(a) $f(t) = t - e^{-at}$

$\therefore \qquad\qquad F(s) = \dfrac{1}{s^2} - \dfrac{1}{s + a} = \dfrac{(s + a) - s^2}{s^2(s + a)} = \dfrac{-s^2 + s + a}{s^2(s + a)}$

(b) $f(t) = At \qquad \therefore \quad F(s) = \dfrac{A}{s^2}$

(c) $f(t) = \cosh at$

$\qquad F(s) = = \dfrac{1}{2}\int\limits_0^\infty (e^{at} + e^{-at})e^{-st}\,dt$

$\qquad\qquad = \dfrac{1}{2}\left[\int\limits_0^\infty e^{at}e^{-st}\,dt + \int\limits_0^\infty e^{-at}e^{-st}\,dt\right] = \dfrac{1}{2}\;[L(e^{at}) + L(e^{-at})]$

$\qquad\qquad = \dfrac{1}{2}\left[\dfrac{1}{s - a} + \dfrac{1}{s + a}\right] = \dfrac{1}{2}\left[\dfrac{2s}{s^2 - a^2}\right] = \dfrac{s}{s^2 - a^2}$

(d) $f(t) = te^{-at}$

$\qquad F(s) = L(t) = \dfrac{1}{s^2}$

$\qquad F(s + a) = \dfrac{1}{(s + a)^2}$

(e) $f(t) = \sinh at$

$\qquad F(s) = \dfrac{1}{2}\int\limits_0^\infty (e^{at} - e^{-at})e^{-st}\,dt$

$$= = \frac{1}{2}\left[\int_0^\infty e^{at}e^{-st}dt - \int_0^\infty e^{-at}e^{-st}dt\right]$$

$$= \frac{1}{2}[Le^{at} - Le^{-at}] = \frac{1}{2}\left[\frac{1}{(s-a)} - \frac{1}{(s+a)}\right] = \frac{1}{2}\left[\frac{s+a-s+a}{s^2-a^2}\right] = \frac{a}{s^2-a^2}$$

(f) $f(t) = e^{-at}\sinh \omega t$

$$F(s) = L\sinh \omega t = \frac{\omega}{s^2-\omega^2}$$

$$F(s+a) = \frac{\omega}{(s+a)^2-\omega^2}$$

$$F(s) = L\sinh \omega t = \frac{\omega}{s^2-\omega^2}$$

$$F(s+a) = \frac{\omega}{(s+a)^2-\omega^2}$$

(g) $f(t) = 2e^{-2t}\cos 3t$

$$F(s) = L\cos 3t = \frac{1}{2}L[e^{j3t} + e^{-j3t}]$$

$$= \frac{1}{2}\left(\frac{1}{s-3j} + \frac{1}{s+3j}\right) = \frac{s}{s^2+9}$$

Therefore, $\quad 2F(s+a) = \frac{2(s+2)}{(s+2)^2+9} = \frac{2s+4}{s^2+4s+13}$

Problem 3.5

Determine the initial and final value of the corresponding time function $y(t)$, for the given Laplace transform

$$Y(s) = \frac{17s^3 + 7s^2 + s + 6}{s^5 + 3s^4 + 5s^3 + 4s^2 + 2s}$$

Solution

Initial value

$$= \lim_{s\to\infty} sY(s) = \lim_{s\to\infty} \frac{s(17s^3 + 7s^2 + s + 6)}{s^5 + 3s^2 + 5s^3 + 4s^2 + 2s}$$

$$= \lim_{s\to\infty} \frac{17s^3 + 7s^2 + s + 6}{s^4 + 3s^3 + 5s^2 + 4s + 2}$$

$$= \lim_{s \to \infty} \frac{s^3\left(17 + \dfrac{7}{s} + \dfrac{1}{s^2} + \dfrac{6}{s^3}\right)}{s^4\left(1 + \dfrac{3}{s} + \dfrac{5}{s^2} + \dfrac{4}{s^3} + \dfrac{2}{s^4}\right)}$$

$$= \lim_{s \to \infty} \frac{1}{s} \times \left(\frac{14 + \dfrac{7}{s} + \dfrac{1}{s^2} + \dfrac{6}{s^3}}{1 + \dfrac{3}{s} + \dfrac{5}{s^2} + \dfrac{4}{s^3} + \dfrac{2}{s^4}}\right) = 0$$

Final value

$$= \lim_{s \to 0} sY(s) = \lim_{s \to 0} \frac{s(17s^3 + 7s^2 + s + 6)}{s^5 + 3s^4 + 5s^3 + 4s^2 + 2s} = \lim_{s \to 0} \frac{17s^3 + 7s^2 + s + 6}{s^4 + 3s^3 + 5s^2 + 4s + 2} = \frac{6}{2} = 3$$

Problem 3.6

Solve the differential equation $\dfrac{d^2 f(t)}{dt^2} + \dfrac{3 df(t)}{dt} + 2 f(t) = e^{-3t} u(t)$ by means of the Laplace transform. Assume zero initial conditions

---

Solution In the Laplace domain, the given differential equation can be written as

$$(s^2 + 3s + 2)F(s) = \frac{1}{s+3}$$

or $\qquad\qquad F(s) = \dfrac{1}{(s+1)(s+2)(s+3)}$

or $\qquad\qquad F(s) = \dfrac{1}{(s+1)(s+2)(s+3)} = \dfrac{A}{s+1} + \dfrac{B}{s+2} + \dfrac{C}{s+3}$

$$A = \frac{1}{(s+1)(s+3)}\bigg|_{s=-2} = \frac{1}{2}; \ B = \frac{1}{(s+2)(s+3)}\bigg|_{s=-1} = -1; C = \frac{1}{(s+1)(s+2)}\bigg|_{s=-3} = \frac{1}{2}$$

Therefore,

$$F(s) = \frac{1/2}{s+1} + \frac{-1}{s+2} + \frac{1/2}{s+3}.$$

Inverse Laplace transform

$$f(t) = \frac{1}{2}e^{-t} - e^{-2t} + \frac{1}{2}e^{-3t}$$

Problem 3.7

Solve the differential equation by means of Laplace transform

$$\frac{dx_1(t)}{dt} = x_2(t)$$

$$\frac{dx_2(t)}{dt} = -2x_1(t) - 3x_2(t) + u(t) \text{ when } x_1(0) = 1, \ x_2(0) = 0$$

Solution   Laplace transform of the differential equations give $sX_1(s) - x_1(0) = X_2(s)$ and

$$sX_2(s) - x_2(0) = -2X_1(s) - 3X_2(s) + \frac{1}{s}$$

Putting $\quad\quad\quad x_1(0) = 1$ and $x_2(0) = 0$, we get

$$sX_1(s) - 1 = X_2(s) \tag{1}$$

and $\quad\quad\quad sX_2(s) = -2X_1(s) - 3X_2(s) + \frac{1}{s} \tag{2}$

Putting value of $X_2(s)$ in equation (1), we get

$$s(s\,X_1(s) - 1) = -2X_1(s) - 3sX_1(s) + 3 + \frac{1}{s}$$

$$s^2 X_1(s) - s = -2X_1(s) - 3sX_1(s) + 3 + \frac{1}{s}$$

$$X_1(s)(s^2 + 3s + 2) = \frac{1}{s} + s + 3$$

$$X_1(s)(s^2 + 3s + 2) = \frac{1 + s^2 + 3s}{s}$$

$$X_1(s) = \frac{s^2 + 3s + 1}{s(s^2 + 3s + 2)} = \frac{s^2 + 3s + 1}{s(s+1)(s+2)} \tag{3}$$

or $\quad\quad\quad X_1(s) = \frac{A}{s} + \frac{B}{s+1} + \frac{C}{s+2}$

$$A = \frac{s^2 + 3s + 1}{(s+1)(s+2)}\bigg|_{s=0} = \frac{1}{2}$$

$$B = \frac{s^2 + 3s + 1}{s(s+2)}\bigg|_{s=-1} = 1$$

$$C = \frac{s^2 + 3s + 1}{s(s+1)}\bigg|_{s=-2} = \frac{-1}{2}$$

Therefore, $\quad X_1(s) = \dfrac{1/2}{s} + \dfrac{1}{s+1} + \dfrac{-1/2}{s+2}$

or $\quad x_1(t) = \dfrac{1}{2} + e^{-t} - \dfrac{1}{2}e^{-2t}$

Substituting the value of $X_1(s)$ from equation (3) in equation (1), we get

$$\frac{s(s^2 + 3s + 1)}{s(s+1)(s+2)} - 1 = X_2(s)$$

or $\quad X_2(s) = \dfrac{s^2 + 3s + 1 - (s+1)(s+2)}{(s+1)(s+2)}$

$$= \frac{s^2 + 3s + 1 - s^2 - 3s - 2}{(s+1)(s+2)}$$

$$= \frac{-1}{(s+1)(s+2)}$$

$$= \frac{-1}{s+1} + \frac{1}{s+2}$$

or $\quad x_2(t) = -e^{-t} + e^{-2t}$

Problem 3.8

Find the inverse Laplace transform of

(a) $F(s) = \dfrac{s^4 + 3s^3 + 3s^2 + 5s + 2}{s(s+1)}$ (b) $F(s) = \dfrac{6(s+1)}{s^2(s+2)(s+3)}$

Solution

(a) Numerator polynomial has a higher degree than that of denominator polynomial. Division of numerator by denominator yields

$$F(s) = \frac{s^4 + 3s^3 + 3s^2 + 5s + 2}{s(s+2)} = s^2 + s + 1 + \frac{3s+2}{s(s+2)}$$

$$= s^2 + s + 1 + \frac{A}{s} + \frac{B}{s+2}$$

$$A = \frac{3s+2}{s+2}\bigg|_{s=0} = 1, \quad \text{and } B = \frac{3s+2}{s}\bigg|_{s=-2} = 2$$

Hence $\qquad F(s) = s^2 + s + 1 + \frac{1}{s} + \frac{2}{s+2}$

Therefore, $f(t) = L^{-1}F(s) = \frac{d^2}{dt^2}\delta(t) + \frac{d}{dt}\delta(t) + \delta(t) + 1 + 2e^{-2t}$.

(b) $F(s) = \dfrac{6(s+1)}{s^2(s+2)(s+3)} = \dfrac{A}{s} + \dfrac{B}{s^2} + \dfrac{C}{s+2} + \dfrac{D}{s+3}$

$$B = \frac{6(s+1)}{(s+2)(s+3)}\bigg|_{s=0} = 1$$

$$C = \frac{6(s+1)}{s^2(s+3)}\bigg|_{s=-2} = \frac{-3}{2}$$

$$D = \frac{6(s+1)}{s^2(s+2)}\bigg|_{s=-3} = \frac{4}{3}$$

$$A = \frac{d}{ds}\left[\frac{6(s+1)}{(s+2)(s+3)}\right]_{s=0}$$

$$= \frac{6(s+2)(s+3) - 6(s+1)(2s+5)}{(s+2)^2(s+3)^2}\bigg|_{s=0}$$

$$= \frac{6 \times 2 \times 3 - 6 \times 1 \times 5}{4 \times 9}$$

$$= \frac{36-30}{36} = \frac{6}{36} = \frac{1}{6}$$

Therefore, $\qquad F(s) = \dfrac{1/6}{s} + \dfrac{1}{s^2} + \dfrac{-3/2}{s+2} + \dfrac{4/3}{s+3}$

$$f(t) = L^{-1}F(s) = \frac{1}{6} + t - \frac{3}{2}e^{-2t} + \frac{4}{3}e^{-3t} .$$

Problem 3.9

Find the Laplace transform of the following functions

(a) $u(t) = 0$ for $t < 0$ (b) $g(t) = A$ for $0 < t < t_0$

          1 for $t \geq 0$ 0 for $t < 0$ and $t \geq t_0$

Solution

(a) The Laplace transform of a unit step function $u(t)$ is

$$U(s) = L[u(t)] = \int_0^\infty u(t)e^{-st}\,dt = \frac{e^{-st}}{-s}\Big|_0^\infty = \frac{1}{s}$$

(b) The given function is shown in Fig. 3.3

$$G(s) = L[g(t)]$$

$$= \int_0^\infty e^{-st}g(t)\,dt = \int_0^\infty e^{-st}A\,dt + 0$$

$$= A\left(\frac{e^{-st}}{-s}\Big|_0^{t_0}\right) = \frac{A}{s}(1 - e^{-st_0})$$

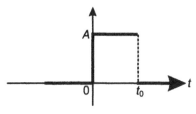

Fig. 3.3

Problem 3.10

Find the Laplace transform of the function shown in Fig. 3.4

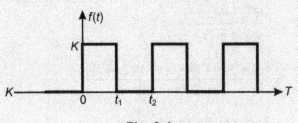

Fig. 3.4

Solution

The function $f(t)$ given in the Fig. 3.2 can be defined as

$$f_1(t) = Ku(t - t_1)$$
$$\text{where}$$
$$f_2(t) = -Ku(t - t_2)$$
$$f(t) = f_1(t) + f_2(t)$$

$u(t) = 0$ for $t < 0$

1 for $t \geq 0$

$$F(s) = Lf(t) = Lf_1(t) + Lf_2(t)$$

$$= K\left[\frac{e^{-t_1 s}}{s} - \frac{e^{-t_2 s}}{s}\right] = \frac{K}{s}[e^{-t_1 s} - e^{-t_2 s}]$$

Problem 3.11

Assuming that the switch S is closed at time $t = 0$, determine the current in the circuit shown in Fig. 3.5.

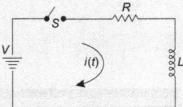

Fig. 3.5

Find the initial and final values.

Solution

Applying Kirchoff's Law, we get

$$V = Ri(t) + \frac{L\,di(t)}{dt}$$

Laplace transformation yields

$$L\left[Ri(t) + \frac{Ldi(t)}{dt}\right] = \frac{V}{s}$$

$$RI(s) + sLI(s) - i(0) = \frac{V}{s}$$

since $i(0) = 0$ at $t = 0$, we get

$$I(s) = \frac{V}{s(R + sL)} = \frac{V}{L}\left(\frac{1}{s(s + R/L)}\right) = \frac{V}{sR} - \frac{V}{R}\left(\frac{1}{s + R/L}\right)$$

Inverse Laplace transform yields

$$i(t) = \frac{V}{R} - \frac{V}{R}e^{-\frac{Rt}{L}} = \frac{V}{R}\left(1 - e^{-\frac{Rt}{L}}\right)$$

Initial Value Theorem

Initial value theorem yields

$$\lim_{t \to 0} f(t) = \lim_{s \to \infty} sF(s)$$

i.e.
$$\lim_{t \to 0} i(t) = \frac{V}{R}\left(1 - e^{\frac{-R \times 0}{L}}\right) = \frac{V}{R}(1-1) = 0$$

and similar result is obtained if apply

$$\lim_{s \to \infty} sI(s) = \lim_{s \to \infty} \frac{V}{R + sL} = 0$$

Final Value Theorem

$$\lim_{t \to \infty} f(t) = \lim_{s \to 0} sF(s)$$

Thus,
$$\lim_{t \to \infty} i(t) = \lim_{s \to 0} s\left[\frac{V}{s(R + sL)}\right] = \frac{V}{R}$$

REVIEW EXERCISE

1. What is analytic function and what are the analytic requirements of Initial and Final value theorems. Check the applicability of the theorems for the following time dependent functions

   (a) $\dfrac{d^2 y(t)}{dt^2} + \dfrac{5dy(t)}{dt} + 2y(t) = 7u(t)$ where $u(t)$ is the unit-step function.

   (b) $y(t) = \sin \omega(t)$
   (c) $y(t) = \cos \omega(t)$

2. If $F(s) = \dfrac{1}{s(s+1)(s+2)}$; find at what points in the s-plane the function is not analytic.

3. How do you define characteristic equation? What is the significance of the same?

4. What is the most common type of singularity. If $F(s) = \dfrac{5(s+3)}{s(s+1)(s+2)^3}$, find the number and values of poles and zeros of $F(s)$. [$Hint$: 5 poles and 5 zeros (one finite and four at infinity)].

5. Define Laplace transform. Is Laplace Operator a complex variable. If so, explain.

6. With reference to the differentiation theorem, obtain the second derivative of $f(t)$.

7. Find Laplace transform of $f(t) = \sin at$ from the Table 3.2. Apply final value theorem. Why is the result erroneous.

   (**Hint:** Two poles are located on the imaginary axis)

8. Write down the procedure for obtaining the solution of linear time-invariant differential equations.

4

MATLAB BASICS

4.1 INTRODUCTION

MATLAB is a registered software product of MATHWORKS Inc. It presents a platform for accelerated research analysis and development and production facilities. It provides a powerful technical computing environment by integrating the MATLAB technical language with mathematical computing and visualization of the output through specialized graphic features.

MATLAB is an abbreviation for *MATrix LABoratory*. It is essentially a matrix based computing environment. The variables manipulated are matrix based or array of numbers, making matrix as the basic computational unit. MATLAB has number of inbuilt functions which help in faster and accurate mathematical computation. It also provides to the developer, application specific toolbars for special application. *Control system toolbox* is of a special interest to control system engineer. It utilises MATLAB matrix functions and algorithms inbuilt in the form of M-files and helps in implementation of design, modelling and analysis of control systems. The functions not existing in the toolbox can be created by writing new M-files. The modelling of the control systems is possible both as *transfer function* and in *state-space* form. The analysis can be achieved by classical as well as modern techniques. Conversions from transfer function form to the state-space form can be easily done by using inbuilt functions of MATLAB.

LEARNING OBJECTIVES

- ◆ To introduce MATLAB environment.
- ◆ To understand MATLAB features.
- ◆ Practice and utilize inbuilt MATLAB functions, basic mathematical computation and manipulation associated with control systems : vectors, matrices, polynomials, partial-fraction, transfer function.
- ◆ Learn basic structure of MATLAB program, creation of M-files and evaluation.

> ◆ To utilize control system toolbox features.
> ◆ To use symbolic math toolbox to perform symbolic computations within the MATLAB numeric computing environment.

4.2 MATLAB ENVIRONMENT

MATLAB software can be easily loaded on pentium version computers. The computer should have adequate RAM for error free and faster graphic analysis of control systems. Recommended RAM memory is 128 MB.

Double-clicking on the MATLAB icon or by selecting MATLAB from the program menu, activates and opens the default MATLAB desktop shown in Fig. 4.1 consisting of the following windows:

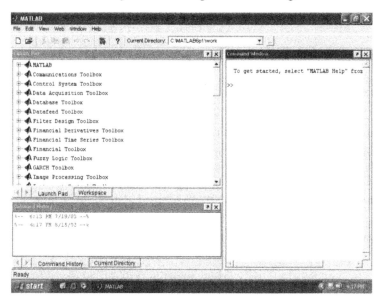

Fig. 4.1 *Matlab desktop*

◆ the *Command Window* appears on the right hand side.

◆ the *Command History* and *Current Directory Window* appear on the lower half of the left hand side.

◆ the *Launch Pad* and *Workspace Window* appear on the upper half of the left hand side.

The desktop apperance can be altered by manipulating the tools embedded in it. *View* menu provides facility for opening or closing the tools.It incorporates different tools for managing files and variables. We can choose to make any of the windows active by clicking anywhere inside the area of the window.

Command Window Command window can be viewed by selecting *view*, *desktop layout* and *command window only* as shown in Fig. 4.2.

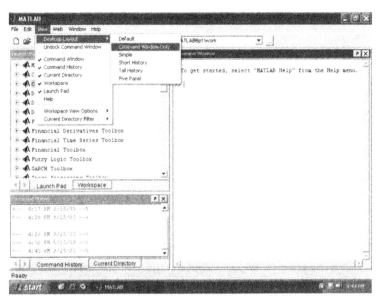

Fig. 4.2   *Procedure to activate command window only*

This results in opening of *command window* as shown in Fig. 4.3. This is the main window

Fig. 4.3   *Command window*

having MATLAB *command prompt* >>.This is the *workplace* where one interacts with MATLAB and performs numerical computation through calculator approach or by manipulating the variables by assigning values to them. The result of the operations are seen on pressing *return (↩)*. In Fig. 4.4, operation has been performed by assigning values to variables *a* and *b* and assigning *a + b* to variable *c*. On pressing *return (↩)*, the addition operation is executed and result is displayed.

Fig. 4.4   *Addition operation in command window.*

The computation tends to become tedious and slow as the number of commands increase and a large amount of data is required to be entered and manipulated. The problem is overcome by creating own M-files in which own created programs are written and saved and help us from rewriting the commands. Own programs can be written using MATLAB *editor/debugger*. The MATLAB files are created with MATLAB *editor* and errors found by using the *debugger*. MATLAB *editor/debugger* is activated by choosing *New* from *File* menu and then selecting the M-file as shown in Fig. 4.5. It opens the *editor* window as shown in Fig. 4.6

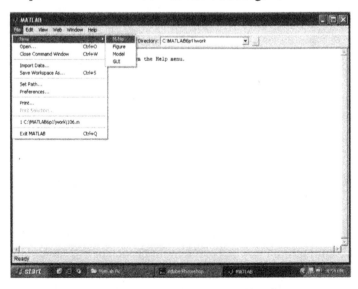

Fig. 4.5   *Opening M-file from the command window*

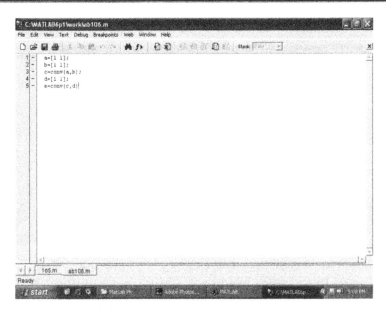

Fig. 4.6 *Editor window showing a user written program*

in which we can type our own program. After creating the program, the next step is to save the program. This is done by choosing *save* from the *File* of the *editor* window and by typing the name of the script file without extension *'m'*. This saves the created program in the *current directory*. The file can also be saved in a directory of own choice by changing the directory of MATLAB and saving in it.

The M-files, which are the programs created by the user also provide debugging and evaluating facility. This is achieved by selecting the *program* from *edit* menu and then selecting *run* from the *debug* menu as shown in Fig. 4.7.

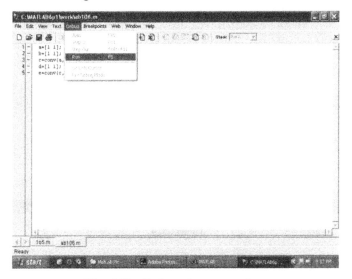

Fig. 4.7 *Debugging procedure*

This outputs the result in window shown in Fig. 4.8.

Fig. 4.8 *Output of user created program*

Alternatively, the selected program can be evaluated by clicking *evaluate* selection from *Text* menu as shown in Fig 4.9.

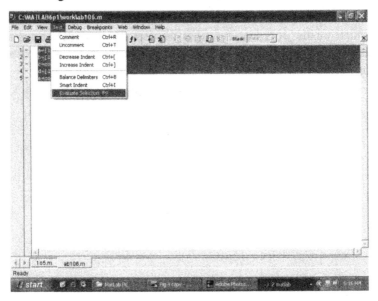

Fig. 4.9

The output obtained is shown in Fig. 4.10. Students should note that the output is accompanied with the user created program.

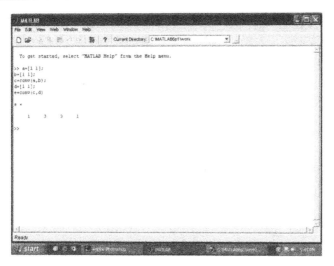

Fig. 4.10 *Output with user created program*

COMMAND HISTORY WINDOW

Command History Window maintains the log of all the statements/commands that are entered in the *command window.* By selecting any command and double-clicking on it executes the selected command. *Command history window* is shown in Fig. 4.11

Fig. 4.11 *Command history window*

Current Directory Window

The *Current Directory Window* (Fig. 4.12) indicates the directory in which the, user is operating. The program file can be created and saved. The default directory is MATLAB \ *Work directory.* User can also create his own directory. All created/existing MATLAB directories can be viewed, opened and altered by using the *current directory browser.*

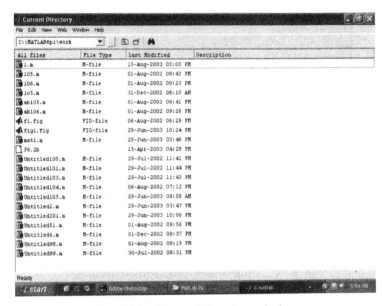

Fig. 4.12   *Current directory window*

Current directory can be easily viewed and changed with the help of current directory field in the desktop. Any previously saved file must be placed in the current directory field to run it.

Launch Pad

Launch Pad provides easy access to tools, demos and documentation in tree structure/view. Double-clicking on the selected icon executes the connected program. Fig. 4.13 depicts opening of MATLAB demo window by double-clicking *demo icon* in the *Launch Pad* window.

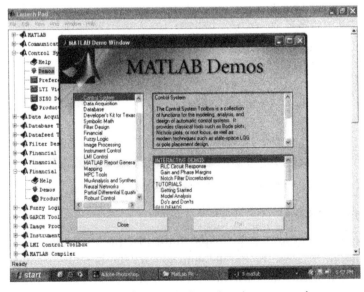

Fig. 4.13   *Launch pad window showing* MATLAB *demos*

Workspace

Workspace displays the variables created/defined and stored/saved during a MATLAB working. Fig. 4.14 illustrates workspace browser displaying details of the variables in *four* columns. Variables can be selected, deleted and modified. Variables can be deleted from the workspace by the selecting them with the mouse or otherwise with the help of the keyboard and pressing the *delete* key. This operation can also be performed by *right clicking* and then using the *delete* option. Double-clicking on a variable shows the *array editor* shown in Fig 4.15. The options depicted in *array editor* can be modified, if desired.

Fig. 4.14   *Workspace window*

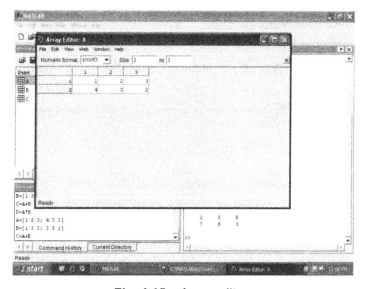

Fig. 4.15   *Array editor.*

Figure Window

Figure window displays two or three dimensional plots by execution of MATLAB commands.

Docking/Undocking of Windows

Undocking/docking can be done with the help of *dock arrow*. Following undocking operations can be performed

- ◆ undock tools from the desktop including all documents within it.
- ◆ undock documents from their tools in the desktop.
- ◆ undock documents from undocked tools.

The undocking operation can be performed by

- ◆ selecting the undock arrow for the item from the item's title bar;
- ◆ using the undock items from the desktop menu when the tool is active.

Docking is reverse of undocking. The operation is performed by use of dock arrow in the tool's menu bar. If the document is outside of the desktop and tool is inside the desktop, the document can be moved into the tool in the desktop by selecting desktop from the menu and then selecting the document name or by selecting the dock arrow.

KEY POINTS LEARNT

- ◆ The default MATLAB desktop contains the following windows
  - • Command Window.
  - • Command History and Current Directory Window.
  - • Launch Pad and Workspace Window.
- ◆ Other MATLAB desktop windows are
  - • Editor/Debugger Window.
  - • Array Editor.
  - • Help.
  - • Figure.
- ◆ Edit/Debug Window can
  - • Create a new M - file.
  - • Modify existing M - files.
  - • Debugs the error.
- ◆ Default desktop appearance can be altered from *View* menu.
- ◆ Workspace window displays the information displayed by *Whos* command and double - clicking on any variable in the window opens the Array Editor.
- ◆ Figure Window displays the graphic generated on execution of MATLAB commands.
- ◆ Windows can be docked and undocked based on the requirement by clicking the icon meant for this purpose.

4.3 MATLAB FEATURES

Computations in MATLAB can be done in the *Command Window* or by creating an *M-file*. MATLAB has many inherent capabilities which renders MATLAB a very powerful tool for numerical computing. Some of the important features are:

- ◆ MATLAB is fundamentally array-oriented. The basic data type is array and belongs to the class of double-precision. An array is a collection of values of data organised into rows and columns. Arrays can be classified as either *matrices* or *vectors*. The size of an array is defined by number of rows and number of columns.

- ◆ Real numbers, integers, matrices, cell arrays and structures, character strings, figures etc are the objects.

- ◆ Variables are case sensitive and can contain upto 31 characters. The variable name must commence with a *letter* and can be followed by any number of *letters*, *digits* or *underscores*. Punctuation characters are not permitted.

- ◆ There is no need to declare *data-type* or *data-object*. Variable assignment to data-type or object is automatic.

- ◆ MATLAB has a number of predefined variables called *special* variables. These are permanent variables. Few of the permanent variables store a value. They are listed in Table 4.1. These can assigned to a variable to obtain the desired output.

Table 4.1   Special Variables

ans, pi, beep, inf, eps, Nan, i or j, nargin, nargout, bitmax, realmax, realmin, varargin, varaout, date, clock, flops, computer, cputime.

- ◆ MATLAB has several *keywords* which cannot be declared as variables. These are listed in Table 4.2. Error will be reported if such *reserved* words are used as variables. However, by capitalizing one or more letters, the reserved list words may be used as variables.

Table 4.2   Keywords

If, elseif, else, for, break, end, while, function, global, catch, switch, continue, try, return, persistent, case, otherwise.

- ◆ The command *who* enables viewing of the variables created and the command *whos* gives more detailed information like size, bytes used and class of the variables. *Workpace* window displays information executed by *whos* command. *Clean* command cleans the variables created *e.g., clear var1 var2···*

- ◆ Dimensioning statements are not required for arrays. MATLAB does dimensioning automatically.

◆ Default output format is *format short*. Other output formats as given in Table 4.3 can be displayed by overriding the default output format from *preferences* item from the *File* menu as shown in Fig. 4.5 or by typing the required output format at the *prompt*. Output of every command is displayed on the screen. However, output display can be suppressed by typing a semicolon at the end of the command.

◆ On-line help can be requisitioned by clicking *help* on the *command window* or by selecting the icon assigned to *help*. It is one of the most useful commands available to the user on-line. If we type MATLAB command *help*, a list of the main topics appear on the users interface from which user can select and take further guidance. If the usage of a specific function say *polyval* is only required, then we can type *polyval*. This will give use of function polyval. Help can be requistioned by use of *lookfor* command. *Help* command searches and displays exact function matching the name of function requistioned while *lookfor* command displays the summary giving in each function to match. Comparatively *help* command responds faster than the *lookfor* command but *lookfor* command displays useful information connected with the help sought and hence improves the chances of extracting information.

Table 4.3   Output Formats

Type	Value of Pi
default (format short)	3.1416
format long	3.14159265358979
format short e	3.1416e+000
format long e	3.141592653589793 e+000
format long g	3.1416
format long g	3.14159265358979
format hex	400921fb5442d18
format rat	355/113
format bank	3.14

◆ The operators used for performing basic arithmetic operations are shown in Table 4.4

Table 4.4   Basic Arithmetic Operations

Operation	Operator	Example
Addition	+	x + y
Subtraction	−	x − y
Multiplication	*	x * y
Division	/	x / y
Power	^	x ^ 2

◆ MATLAB has predefined
- Special or Permanent variables
- Keywords or Reserved variables

◆ MATLAB is array oriented language in which
- Data-type or data-object need not be declared.
- Dimensioning statement not required
- Default output is format *short*

◆ Help can be requistioned in there ways
- By use of *Help* browser
- By typing *help* or *help* followed by function name in the command window
- By typing *lookfor* followed by function name in the command window

4.4 VECTORS

Vector is an array with only one dimension. Vectors, whether *row* or *column* can be manipulated easily.

◆ **Row Vector Input**   Elements of a *row vector* are entered in a line separated by a space and enclosed in square brackets ([ ]). A variable is assigned thereafter. e.g.

$$a = [2 \ 5 \ 3 \ 6]$$

◆ **Column Vector Input**   Elements of column vector are also enclosed in square brackets ([ ]) and assigned to a variable. However, in this case elements are separated by a semicolon. e.g.

$$a = [2; \ 5; \ 3; \ 6]$$

4.5 MATRICES

Matrix is an array with two or more dimensions. MATLAB has a number of predefined command functions which can be called based on the requirement to solve control engineering problems related to analysis and design of control systems. Few of the functions which are utilised for manipulation of matrices, polynomials, partial fractions etc are discussed in the succeeding paragraphs.

4.5.1 Matrix Input

MATLAB provides a very easy method to input any given matrix. Once the matrix is entered; it is automatically shown in the workspace. The following points are to be borne in the mind while proceeding with *matrix input operation*:-

MATRIX INPUT

- Elements of a matrix are entered against the command prompt in square brackets [ ].
- Elements are separated by *blanks* or *commas*.
- The rows are separated and distinguished by *semicolon*. i.e. row termination is indicated by *semicolon* (;)
- A small sized matrix may be input in a line. However, large sized matrix can be input across multiple lines by using the *carriage return*.
- Variable is generally assigned to a matrix to define its use subsequently in the program by use of *equal* to (=) sign.
- Dimension or type statements are not required while manipulating matrices.
- Memory allocation depending upon the size of the matrix is done automatically
- Elements of the matrix can contain real, complex, elementary mathematical functions or trigonometric funcitons.
- Matrices can be entered by the following methods:
  - Entering elements one by one as per the procedure explained above.
  - Exporting and loading from external files.
  - Creating matrices by using inbuilt functions explained in Section 4.5.2.

4.5.2 Matrix Functions

MATLAB has number of inbuilt matrix specific functions which are capable of generating elementary and special matrices without creating a M-file and performing frequently used matrix operations and manipulations. In this section we shall learn few basic functions.

- *eye.* It generates an *identity* matrix. e.g. *eye(3)* generates $\begin{bmatrix} 1 & 0 & 0 \\ 0 & 1 & 0 \\ 0 & 0 & 1 \end{bmatrix}$.

- *eye (m, n).* It generates a matrix of the order $(m \times n)$ with *ones* on the main diagonal and *zeros* at other places.

- *zeros.* It generates a matrix with all elements being zeros. e.g.. *zero(3, 2)* generates. $\begin{bmatrix} 0 & 0 \\ 0 & 0 \\ 0 & 0 \end{bmatrix}$

- *diag.* It generates a column vector whose elements are the diagonals of a matrix e.g. If matrix

  $A = \begin{bmatrix} 2 & 1 & 3 \\ 0 & 1 & 2 \\ 2 & 1 & 3 \end{bmatrix}$; then *diag(A)* produces a column vector $\begin{bmatrix} 2 \\ 1 \\ 3 \end{bmatrix}$. If we are interested in

  computing sum of the diagonal elements of matrix, then the function *sum (diag(A))* produces answer 6.

- ◆ **rank.**   Function *rank(A)* computes rank of matrix *A*.
- ◆ **det.**   Function *det(A)* computes determinant of matrix *A*. If the determinant of a matrix happens to be zero, then the matrix is singular.
- ◆ **inv.**   Function *inv(A)* computer inverse of matrix *A*. If the matrix happens to be singular a warning is generated.
- ◆ **trace.**   Function *trace(A)* produces trace of matrix *A* i.e. the sum of the diagonal elements of matrix *A*.
- ◆ **eig.**   Function *eig(A)* produces a column vector containing the eigenvalues of a square matrix *A*.
- ◆ **[V, D] = eig(A).**   It generates eigenvalues and eigenvectors. The result generated is a diagonal matrix *D* of eigenvalues and a *full* matrix *V* whose columns constitute the corresponding eigenvectors.
- ◆ **Transpose.** $B = A'$ generates transpose of matrix *A* and assigned to variable *B*.
- ◆ **Ones.**   Function *ones(m, n)* produces a $(m \times n)$ order matrix of *ones*.
- ◆ **Poly(A).**   Function *poly(A)* generates the coefficients of the characteristic polynomial of matrix *A* in a row. The coefficients are separated by blanks/ spaces e.g. 2 5 6 3 0 indicates that the characteristic polynomial is $2s^4 + 5s^3 + 6s^2 + 3s + 0$. The constant term is zero.

Example 4.1   Write a MATLAB program to input elements of 3×3 matrix and assign it variable *A*.

$$\begin{bmatrix} 1 & 4 & 2 \\ 3 & 0 & 1 \\ 5 & 2 & 3 \end{bmatrix}$$

MATLAB program is given below:

```
Solution

'Example 4.1'              %display enclosed text.
A= [1 4 2; 3 0 1; 5 2 3]   %input matrix A in one line

Output

Example 4.1
A =

     1     4     4
     3     0     1
     5     2     3
```

Note: The program illustrates matrix input in one line. Elements are separated by blanks and rows by semicolons. Semicolon is not used to depict end of row for the last row. Use of semicolon after the bracket closure ( ]) suppresses the printing though the command is executed.

Example 4.2   Write a MATLAB program to input (5×5) matrix A in multiple lines with a carriage return.

$$A = \begin{bmatrix} 2 & 5 & 2 & 4 & 1 \\ 1 & 5 & 3 & 3 & 2 \\ 7 & 1 & 3 & 5 & 1 \\ 4 & 2 & 0 & 8 & 6 \\ 5 & 3 & 1 & 0 & 2 \end{bmatrix}$$

MATLAB program is given below:

```
Solution

'Example 4.2'   %display enclosed text.
A= [2 5 2 4 1   %input matrix A in multiple
    1 5 3 3 2   %lines and output.
    7 1 3 5 1
    4 2 0 8 6
    5 3 1 0 2]

Output

Example 4.2
A =

    2   5   2   4   1
    1   5   3   3   2
    7   1   3   5   1
    4   2   0   8   6
    5   3   1   0   2
```

4.5.3   Matrix Operations

MATLAB has inbuilt command functions which help in performing basic matrix operations like addition, subtraction, multiplication, powers, transpose and the array operations. MATLAB recognises the mathematical operators and mathematical array operators given in Table 4.5.

Table 4.5					
Sl. No.	**Item**	**Symbols**			
		Addition	**Subtraction**	**Multiplication**	**Division**	**Power**
1	Basic Operation	+	−	*	/	^
2	Modified Array Operators	+	−	.*	./	.^

Basic matrix operations are executed element to element. Array operations are performed element to element modifying the basic operators by using a dot *preceding* the basic operators as shown in Table 4.5. Matrices operations require dimension compatibility of matrices. For performing matrices addition or subtraction, size of the matrices must be same. Matrix multiplication between matrices A and B can only be performed if number of rows of matrix A equal to number of columns of matrix B.

Example 4.3 If matrices A and B have elements

$$A = \begin{bmatrix} 1 & 3 & 2 \\ 4 & 0 & 1 \\ 2 & 1 & 2 \end{bmatrix}; \ B = \begin{bmatrix} 4 & 2 & 1 \\ 1 & 3 & 2 \\ 5 & 2 & 3 \end{bmatrix}$$ and C is a column vector having element $C = \begin{bmatrix} 1 \\ 3 \\ 2 \end{bmatrix}$, write a MATLAB

program for performing following operations:

   (a) $D = A + B$.   (b) $E = A * B$.   (c) $e = A * C$.   (d) F = Transpose of matrix A.

MATLAB program is given below:

```
Solution

'Example 4.3'            %display enclosed text.
A= [1 3 2;4 0 1;2 1 2];  %input matrix A and suppress print.
B= [4 2 1;1 3 2;5 2 3];  %input matrix B and suppress print.
C= [1;3;2];              %input matrix B and suppress print.
'Solution Example 4.3(a)'%display enclosed text.
D=A+B                    %add matrices A & B, assign to D &
                         %output.
'Solution Example 4.3(b)'%display enclosed text.
E=A*B                    %multiply matrices A & B, assign to
                         %E & output.
'Solution Example 4.3(c)'%display enclosed text.
e=A*C                    %multiply matrices A & C, assign to
                         %e & output.
'Solution Example 4.3(d)'%display enclosed text.
F=A'                     %transpose matrix A & assign to F

Output

Example 4.3
Solution Example 4.3(a)
D=
     5     5     3
     5     3     3
     7     3     5
Solution Example 4.3(b)
E=
    17    15    13
    21    10     7
    19    11    10
Solution Example 4.3(c)
e=
    14
     6
     9
Solution Example 4.3(d)
F=
     1     4     2
     3     0     1
     2     1     2
```

Example 4.4   Write a MATLAB program to input matrix

$$A = \begin{bmatrix} 0 & 2 & 0 \\ 0 & 0 & 2 \\ -6 & -5 & -4 \end{bmatrix}$$

(a) in single line.

(b) with carriage return.

MATLAB program is given below:

```
Solution

'Example 4.4'                          %display enclosed text.
'Input Matrix A in single line'        %display enclosed text.
A=[0 2 0;0 0 2;-6 -5 -4]               %input matrix A.
'Input matrix A with carriage return'  %display enclosed text.
A=[0 2 0                               %input matrix A with
   0 0 2                               %carriage return.
   -6 -5 -4]

Output

Example 4.4
Input Matrix A in single line
A=

    0     2     0
    0     0     2
   -6    -5    -4

Input matrix A with carriage return
A=

    0     2     0
    0     0     2
   -6    -5    -4
```

Example 4.5   Write a MATLAB program to input column vector matrix $A = \begin{bmatrix} 1 \\ 3 \\ 2 \end{bmatrix}$ by three different methods.

MATLAB program is given below:

```
Solution

'Example 4.5'       %display enclosed text.
'First method'      %display enclosed text.
A=[1;3;2]           %input matrix A.
'Second method'     %display enclosed text.
A=[1               %input matrix A with
   3;2]            %carriage return.
'Third method'     %display enclosed text.
A=[1 3 2]'         %input matrix A, transpose and
                   %output

Output

Example 4.5
First method
A=

      1
      3
      2

Second method
A=

      1
      3
      2

Third method
A=

      1
      3
      2
```

4.6 EIGENVALUES AND EIGENVECTORS

Often during the mathematical analysis of a control system, a control system engineer would have to deal with the terms *eigenvalue* and *eigenvector*. *Eigenvalues* of an $n \times n$ matrix A, also referred to as *characteristic roots*, are the roots of the characteristic equation $[\lambda I - A] = 0$

Eigenvalues and the *closed–loop poles* of a system are same. The location of eigenvalues in the s - plane indicates the system stability. For a system to be stable, eigenvalues cannot be located in the right hand side of the *s*-plane. Even a mere indication as to how many eigenvalues are located in the right hand side of s-plane points towards stability of the system. Eigenvalues can be directly obtained from state equations without converting the state equations into transfer functions. Readers would learn to assess the stability via eigenvalue approach in the succeeding chapters.

Eigenvectors x_i of matrix A are the nonzero vectors i.e. $x_i \neq 0$ associated with λ_i that satisfy the equations

$$(\lambda_i I - A) x_i = 0$$

where λ_i is the i th eigenvalue of matrix A,

Example 4.6 Find the eigenvalues and eigenvectors of the matrix

$$A = \begin{bmatrix} 0 & 1 \\ -2 & 3 \end{bmatrix}$$

Solution **Eigenvalues**

$$A = \begin{bmatrix} 0 & 1 \\ -2 & 3 \end{bmatrix}$$

$$| \lambda I - A | = \det \begin{vmatrix} \lambda & -1 \\ +2 & \lambda - 3 \end{vmatrix}$$

i.e. $\lambda(\lambda - 3) - (-1 \times 2) = 0$

or $\lambda^2 - 3\lambda + 2 = 0$

or $(\lambda - 1)(\lambda - 2) = 0$

Therefore, eigenvalues are $\lambda = 1, 2$

Eigenvectors

Let the eigenvectors be written as

$$x_1 = \begin{bmatrix} x_{11} \\ x_{21} \end{bmatrix}; \text{ and } x_2 = \begin{bmatrix} x_{12} \\ x_{22} \end{bmatrix}$$

$[\lambda_i I - A] x_i = 0$

Substituting $\lambda_1 = 1$ and x_1 we get

$$\left(1 \begin{bmatrix} 1 & 0 \\ 0 & 1 \end{bmatrix} - \begin{bmatrix} 0 & 1 \\ -2 & 3 \end{bmatrix} \right) \begin{bmatrix} x_{11} \\ x_{21} \end{bmatrix} = 0$$

or $\begin{bmatrix} 1 & -1 \\ 2 & -2 \end{bmatrix} \begin{bmatrix} x_{11} \\ x_{21} \end{bmatrix} = 0$

which gives

$x_{11} - x_{21} = 0$ and $2x_{11} - 2x_{21} = 0$

i.e. $x_{11} = x_{21}$ and $x_{11} = x_{21}$

If $x_{11} = 1$ then $x_{21} = x_{11} = 1$. Therefore, one choice of eigenvector is $x_1 = \begin{bmatrix} 1 \\ 1 \end{bmatrix}$

Similarly for $\lambda = 2$, gives

$$\left(2 \begin{bmatrix} 1 & 0 \\ 0 & 1 \end{bmatrix} - \begin{bmatrix} 0 & 1 \\ -2 & 3 \end{bmatrix} \right) \begin{bmatrix} x_{12} \\ x_{22} \end{bmatrix} = 0$$

$$\begin{bmatrix} 2 & -1 \\ 2 & -1 \end{bmatrix} = \begin{bmatrix} x_{12} \\ x_{22} \end{bmatrix} = 0$$

or

$2x_{12} - x_{22} = 0$ or $x_{22} = 2x_{12}$ and $2x_{12} - x_{22} = 0$ or $x_{22} = 2x_{12}$

Therefore, eigenvector $x_2 = \begin{bmatrix} x_{12} \\ x_{22} \end{bmatrix} = \begin{bmatrix} 1 \\ 2 \end{bmatrix}$

Example 4.7   Write a MATLAB program to find eigenvalues

$$A = \begin{bmatrix} 0 & 1 \\ -2 & 3 \end{bmatrix}$$

MATLAB program is given below

```
Solution

'Example 4.7'          %display enclosed text.
A=[0 1;-2 3]           %input matrix A.
'Eigenvalues matrix A' %display enclosed text.
eig(A)                 %compute eigenvalues
                       %of matrix A.

Output

Example 4.7
A=

      0      1
     -2      3

Eigenvalues matrix A
    = 1
      2
```

Example 4.8 Write a MATLAB program to find eigenvalues and eigenvectors of matrix A

$$A = \begin{bmatrix} 0 & 1 \\ -2 & 3 \end{bmatrix}$$

MATLAB program is given below

Solution

```
'Example 4.8'                %display enclosed text.
A=[0 1;-2 3]                 %input matrix A.
'Eigenvalues & Eigenvectors'%display enclosed text.
[X,D]=eig(A)                 %compute eigenvalues &
                             %eigenvectors of matrix
                             %A & assign to D & X.
```

Output
```
Example 4.8
A=
     0      1
    -2      3
Eigenvalues & Eigenvectors
X=
   -0.7071    -0.4472
   -0.7071    -0.8944

D=
     1      0
     0      2
```
Note: The eigenvalues have been represented in
diagonal matrix form.

4.7 TRANSFER FUNCTIONS

A *transfer function* depicting the ratio of output and input to a control system is generally expressed as a ratio of polynomials. In MATLAB, a transfer function is entered by using different MATLAB commands.

tf (num, den) command. num and den variables are defined as the row vectors of the coefficients of the numerator and denominator polynomial of a transfer function respectively. Let us consider the transfer function

$$G_1 = \frac{5s + 2}{s^2 + 4s + 1}$$

The coefficients of numerator are 5 and 2 and that of denominator are 1, 4 and 1. The command to represent G_1 in MATLAB is:

```
num      =      [5  2];↵
den      =      [1  4  2];↵
G1       =      tf(num,den)
```

The transfer function variable object G_1 created above can be utilised in the MATLAB program as an entity for further mathematical manipulations.

zpk command   This command is useful for representing a transfer function given in the factored form. The roots of the numerator and denominator are represented as row vectors. Let us consider a transfer function given as

$$G_2 = \frac{20(s+6)\ (s+2)}{(s+1)\ (s+3)\ (s-5)}$$

The MATLAB program to input transfer function G_2 is

```
Z  =  [-6   -2];↵
P  =  [-1   -3   5];↵
K  =  20;↵
G2 =  Zpk(Z,  P,  K)
or
G2 =  Zpk([-6  -2],[-1  -3  5],20)
```

The object G_2 so created can also be used in the MATLAB program for further mathematical manipulations. The expression *zpk* stands for *zeros, poles* and *gain* of the transfer function.

The transfer functions can be directly entered in the MATLAB program in the normal 'writing' fashion. This requires statements $s = tf\ ('s')$ or $s = zpk\ ('s')$ to precede the transfer function variable/object which is required to be created. The MATLAB program will be of the form:

$s = tf\ ('s')$ ↵
$G1 = (5 * s + 2) / (s \wedge 2 + 4 * s + 1)$

or

$s = zpk\ ('s')$ ↵
$G2 = 20 * (s + 6) * (s + 2)/(s + 1) * (s + 3) * (s - 5)$

4.8 POLYNOMIALS

Control system analysis and design involves manipulation of polynomials of higher orders generally in s-domain. Polynomial manipulation involves evaluation, factorization, finding roots, multiplications, division etc. MATLAB has the following in-built functions for performing mathematical computations on polynomials:

(a) **p = poly(A)** command converts the matrix A into characteristic equation and assigns to variable p.

(b) **r = roots(P)** command computes the roots of the characteristic equation p and assigns to r.

(c) **q = poly (r)** recreates original polynomial from the roots of the polynomial.

(d) **polyval(p, a)** evaluates the value of the polynomial p at a

(e) **x = conv(p, q)** computes the product of the polynomials p and q and assigns to x.

(f) **[q, r] = deconv (x, p)** computes division of polynomial x by p and assigns to q

(g) **polyvalm(p, A)**   It works in the same fashion as polynomial function. It evaluates the value of the polynomial p for each element of square matrix A and returns the matrix consisitng of values computed.

(h) **q = polyder (P)**   It computes the derivative of the given polynomial P and assigns to q

(i) **Polyder (a, b)**   It computes the derivative of the product of the polynomials a and b.

(j) **(n, q) = polyder(a, b)**   It computes the derivative of the polynomials a and b which is expressed as a ratio and assigns to n and q

(k) **[r, p. k] = residue(a, b)**   It computes and outputs the residues and poles in column vector and the constant term in row vector of the polynomials a and b expressed as a ratio. This depicts the partial fraction decomposition in MATLAB form.

(l) **(a, b) = residue(r, p, k)**   This matlab command converts the partial fraction decomposition into the polynomial ratio $a : b$. The coefficients of the polynomials a and b are returned in row vectors respectively, in the decreasing power order.

Polynomial is assigned by its coefficients in a closed bracket [ ] separated by space between the coefficients in decreasing power order. The zero coefficients are also to be considered e.g. $s^4 + 2s^2 + 1$ is represented as [1   0   2   1].

Example 4.9   If the two polynomials p and q are defined as

$$p = s^3 + 16s^2 + 48s + 96 = 0 \text{ and } q = s^2 + 2s + 4 = 0$$

Write a MATLAB program to find

(a) product of the polynomials.   (b) value of polynomials.

(c) roots of the polynomial *q*.   (d) result of the division of polynomial *p* by *q*.

MATLAB program is given below:

```
Solution

'Example 4.9'              %display enclosed text.
p=[1 16 48 96];           %input polynomial p
                          %& suppress output.
q=[1 2 4];                %input polynomial q
                          %& suppress output.
'Solution4.9(a)'          %display enclosed text.
'Product of polynomials'  %display enclosed text.
X=conv(p'q)               %compute product of p & q.
'solution4.9(b)           %display enclosed text.
'Value of polynomials'    %display enclosed text.
y=polyval(p,2)            %compute value of
                          %polynomial p at s=2
z=polyval(q,2)            %compute value of

'solution4.9(c)'          %display enclosed text.
'Roots of polynomials'    %display enclosed text.
r=roots(q)                %compute roots of
                          %polynomial q.
'Solution4.9(d)           %display enclosed text.
'Division of polynomials' %display enclosed text.
[s,t]=deconv(p,q)         %compute division of
                          %polynomial p by
                          %polynomial q
Output
Example 4.9
Solution4.9(a)
Product of polynomials
x=
     1    18    84    256    384    384
This is MATLAB representation or polynomial
```

$$s^5 + 18s^4 + 84s^3 + 256s^2 + 384s + 384$$

```
Solution4.9(b)
Value of polynomials
y=
    264
z=
     12
Solution4.9(c)
Roots of polynomial q
r=
 -1.0000 + 1.7321i
 -1.0000 - 1.7321i

Solution4.9(d)
Division of polynomials
s=
     1    14

t=
     0     0    16    40
```

Note: Here *s* denotes quotient and is given by *s+14* and *t* denotes remainder and is given by *16s+40*.

4.9 PARTIAL FRACTIONS

Partial Fraction method is an important mathematical tool for analysis of complex functions. Transfer function of a control system which represents the output and input ratio is generally expressed as ratio of polynomials in *s*. MATLAB has in-built function by which residues, poles and direct terms are easily obtained. Examples 4.10 & 4.11 illustrate the use of the in-built command functions. The residues and poles are computed in column vectors and the direct quotient is returned in a row vector.

Example 4.10 Write a MATLAB program to convert the function $F(s)$ into partial fraction

$$F(s) = \frac{2s^3 + 5s^2 + 3s + 6}{s^3 + 7s^2 + 20s + 8}$$

MATLAB program is given below:

```
Solution

'Example 4.10'              %display enclosed text.
'Partial fraction of F(s)'  %display enclosed text.
num=[2 5 3 6];              %input numerator polynomial
                            %& suppress output.
den=[1 7 20 8];            %input denominator polynomial
                            %& suppress output.
[r,p,k]=residue(num,den)    %compute residue, poles &direct
                            % quotient term & assign to
                            % r, p & k respectively.

Output

Example 4.10
Partial fraction of F(s)
r=
 -4.6953 + 0.8914i
 -4.6953 - 0.8914i
  0.3906
p=
 -3.2635 + 2.50.4i
 -3.2635 - 2.5024i
 -0.4730
k=
    2
```

Example 4.11 Partial Fraction expansion of a function F(s) is given by

$$F(s) = \frac{-5}{s+3} + \frac{-4}{s+2} + \frac{2}{s+1} + 3$$

Using MATLAB command funciton, convert the given partial fraction into polynomial representation.

MATLAB program is given below:

```
Solution

'Example 4.11'              %display enclosed text.
r=[-5;-4;2];                %input residues & suppress
                            %output.
P=[-3;-2;-1];               %input poles & suppress
                            %output.
k=3;                        %input direct term &
[num, den]=residue(r,p,k)   %compute numerator &
                            %denominator polynomials
                            %& assign to num & den.

Output

Example 4.11
num =
       3    11    12     8
den =
       1     6    11     6
```
Note: Here MATLAB representation of numerator & denominator polynomials is $3s^3+11s^2+12s+8$ and $s^3+6s^2+11s+6$ respectively.

4.10 SYMBOLIC MATH COMMANDS

Symbolic Math commands enormously enhances the computation power. These commands are used with MATLAB commands while creating MATLAB statements in M-files. The use of Symbolic Math commands require *Symbolic Math Toolbox* to append enhanced symbolic mathematics capabilities to the M-files.

Symbolic variables e.g. *syms* x_1 x_2 are declared at the commencement of MATLAB program along with MATLAB statements. Such declaration permit:

(a) Writing of equations and functions can be done symbolically. Alphabetical and numerical characters can be entered together in the M-files. For example, instead of writing $A = [2 \quad 5 \quad 7]$, one can express $A = 2 * s \wedge 2 + 5 * s + 7$

(b) Algebraic manipulation of symbolic expressions consisting of functions and equation in the form of alpha-numero characters gets simplified.

(c) Transfer functions can be entered in a normally written fashion. It enhances readable capability of the statement making it more user friendly.

(d) Time functions as well as Laplace transformed functions can be written as one would normally write it.

(e) Laplace and Inverse Laplace transforms can be obtained symbolically.

(f) Z-transform of a transfer function can be obtained symbolically.

(g) Integration capability is added by use of Symbolic Math Toolbox.

(h) 'Pretty' command allows distinguishable printing.

SUMMARY

♦ MATLAB is a matrix based computing environment having control system tool box which is of special interest to control engineers.

♦ MATLAB desktop has number of MATLAB tools for ease of working viz. command window, command history, current directory, workspace, array editor, editor/debugger, start button, launch pad, help, figure window.

♦ MATLAB is loaded with numerous features and tools.

♦ Matrices can be easily manipulated and loaded in several ways into the MATLAB desktop. It has *in-built* functions to create matrices and perform common operations and functions related to matrix algebra.

♦ MATLAB can manipulate transfer functions and polynomials with the help of *inbuilt* MATLAB functions.

♦ Symbolic Math tools features can be used to write MATLAB programs to enhance and simplify mathematical computations.

PROBLEMS AND SOLUTIONS

Problem 4.1

Write a MATLAB program to input matrix A

$$A = \begin{bmatrix} \cos(\pi/4) & 1 & \sqrt{3} \\ \sin(\pi/2) & -3 & \log(-2) \\ \tan^{-1}(0.6) & e^{0.5} & -2j \end{bmatrix}$$

Solution

```
'Problem 4.1'                    %display enclosed text.
A=[cos(pi/4),1,sqrt(3);          %input matrix A in multiple
   sin(pi/2),-3,log(-2)          %lines by using carriage
   atan(0.60),exp(0.5),-2*j]%return by inserting the
                                 %semi-colon in the first
                                 %line & use of carriage
                                 %return without semi-colon
                                 %in second line.
```

Output

```
Problem 4.1
A =
    0.7071      1.0000      1.7321
    1.0000     -3.0000      0.6931 + 3.1416i
    0.5404      1.6487      0      - 2.0000i
```

Problem 4.2

Consider two matrices

$$A = \begin{bmatrix} 4 & 3 \\ 1 & 2 \end{bmatrix}; \; B = \begin{bmatrix} 5 & 2 \\ 3 & 1 \end{bmatrix}$$

Using MATLAB, compute the following

(a) $A + B$ (b) AB (c) A^2 (d) $B^T A^T$ (e) $A^2 + B^2 - AB \cdot$

Solution

```
'Problem 4.2'      %display enclosed text.
A=[4 3;1 2]        %input matrix A and output.
B=[5 2;3 1]        %input matrix B and output.
'Solution 4.2(a)' %display enclosed text.
A+B                %compute addition of matrices A & B.
                   %and output.
'Solution 4.2(b)' %display enclosed text.
A*B                %compute multiplication of matrices
                   %A & B and output.
'Solution 4.2(c)' %display enclosed text.
A*A                %compute square of matrix A & output.
'Solution 4.2(d)' %display enclosed text.
```

```
B'*A'                    %compute product of transpose of
                         %matrices A & B and output.
'Solution 4.2(e)' %display enclosed text.
(A*A)+(B*B)-(A*B)%compute the expression.
```

Output

```
Problem 4.2
A =
     4       3
     1       2
B=
     5       2
     3       1
Solution 4.2(a)
     9       5
     4       3
Solution 4.2(b)
    29      11
    11       4
Solution 4.2(c)
    19      18
     6       7
Solution 4.2(d)
    29      11
    11       4
Solution 4.2(e)
    21      19
    13      10
```

Problem 4.3

Consider matrix A

$$A = \begin{bmatrix} 0 & 1 & 0 \\ 0 & 0 & 1 \\ -6 & -11 & -6 \end{bmatrix}$$

Write a MATLAB program to compute the characteristic polynomial and its evaluation for matrix

$$B = \begin{bmatrix} 0 & 1 \\ 2 & 3 \end{bmatrix}$$

Solution

```
'Problem 4.3'            %display enclosed text.
A=[0 1 0;0 0 1;          %input matrix A & output.
   -6 -11 -6]
B=[0 1;2 3]              %input matrix B & output.
p=poly(A)                %compute characteristic polynomial
                         %of matrix A & assign to p.
r=polyvalm(poly(A),B)%evaluate characteristic polynomial
                         %p for matrix B.
```

Output

```
Problem 4.3
A =
      0      1      0
      0      0      1
     -6    -11     -6
B=
      0      1
      2      3

p=
    1.0000    6.0000    11.0000    6.0000
Note: This represents polynomial s³+6s²+11s+6
r=
     24     40
     80    144
```

Problem 4.4

Consider the matrix

$$A = \begin{bmatrix} 1 & 2 & 1 \\ -1 & -1 & -3 \\ -1 & 1 & 3 \end{bmatrix}$$

Write a MATLAB program to compute determinant and inverse of matrix A.

Solution

```
'Problem 4.4'  %display enclosed text.
A=[1 2 1;      %input matrix A and output.
   -1 -1 -3
   -1 1 3]
'Determinant'  %display enclosed text..
d=det(A)       %compute determinant &
               %assign to d.
'Inverse'      %display enclosed text.
C=inv(A)       %compute inverse of matrix A &
               %assign to c.
'Adjoint'      %display enclosed text.
D=c*d          %compute adjoint & assign to D.
```

Output

```
Problem 4.4
A =
     1     2     1
    -1    -1    -3
    -1    -1     3
Determinant
d =
    10
Inverse
C =
        0   -0.5000   -0.5000
   0.6000    0.4000    0.2000
  -0.2000   -0.3000    0.1000

Adjoint
D=
        0   -5.0000   -5.0000
   6.0000    4.0000    2.0000
  -2.0000   -3.0000    1.0000
```

Problem 4.5

Write a MATLAB program to add and multiply complex numbers $-2 + j3$ and $-1 - j2$ and display the result in complex form. Find magnitude and phase angle of the resultant complex vectors.

Solution

```
'Problem4.5'      %display enclosed text.
a=(-2+3i)+(-1-2i)%add complex numbers,
                  %assign to a & display.
b=(-2+3i)*(-1-2i)%multiply complex numbers,
                  %assign to b & display.
r=abs(a)          %compute absolute value of
                  %complex vector a, assign to
                  %t & output.
theta=angle(a)    %compute phase angle of
                  %complex vector a, assign to
                  %theta & output.
t=abs(b)          %complex absolute value of
                  %complex vector b, assign to
                  %r & output.
beta=angle(b)     %compute phase angle of
                  %complex vector b, assign to
                  %beta & output.
```

Output

```
Problem4.5
a =
  -3.0000 + 1.0000i
b =
   8.0000 + 1.0000i
r=
    3.1623
theta =
    2.8198
t=
    8.0623
beta =
    0.1244
```

Problem 4.6

Write a MATLAB program to perform the following:

(a) Input polynomial $s^3 + 14s^2 + 63s + 90$ and find roots.

(b) Input polynomial $(s + 1)(s + 2)(s + 4)$ and display.

(c) Multiply polynomials $s^3 + 14s^2 + 63s + 90$ and $s^2 + 3s + 2$ and find roots.

<u>**Solution**</u>

```
'Problem 4.6'                    %display enclosed text.
'Solution 4.6(a)'                %display enclosed text.
A=[1 14 63 90]                   %assign polynomial s^3+14s^2+63s+90
                                 %to A & output.
rootsA=roots(A)                  %compute roots of polynomial A,
                                 %assign to roots A & output,
'Solution 4.6(b)'                %display enclosed text.
B=poly([-1 -2 -4])               %assign polynomial to B & output.
'Solution 4.6(c)'                %display enclosed text.
D=conv([1 14 63 90],[1 3 2])     %multiply polynomials s^3+14s^2+63s+90
                                 %& s^2+3^s+2, assign to D & output.
rootsD=roots(D)                  %compute roots of polynomial D,
                                 %formed above, assign to roots D
                                 %& output.
```

<u>**Output**</u>

```
Problem 4.6
Solution 4.6(a)
A =
    1      14      63      90
rootsA=
   -6.0000
   -5.0000
   -3.0000
Solution 4.6(b)
B=
1      7      14      8
Note: This is MATLAB representation of polynomial
```
$s^3+7s^2+14s+8$.
```
Solution 4.6(c)
D=    1      17      107      307      396      180
Note: This is MATLAB representation of polynomial
```
$s^5+17s^4+107s^3+307s^2+396s+180$.
```
rootsD=
   -6.0000
   -5.0000
   -3.0000
   -2.0000
   -1.0000
```

Problem 4.7

Write a MATLAB program to perform the following

(a) Multiply polynomials $s(s + 2)$ $(s^2 +10s + 24)$ and $s^2 +14s^2 + 63s + 90$

(b) Multiply polynomials $(s + 3) (s + 5) (s + 8)$ and $(s + 1) (s + 2) (s + 4)$.

Solution

```
'Problem 4.7'                %display enclosed text.
'Solution 4.7(a)'            %display enclosed text.
A=conv(poly([0 -2]),[1 10 24]); %represent polynomial
                             %s(s+2)(s^3+10s+24),assign
                             %to A & suppress output.
B=[1 14 63 90];              %represent polynomial
                             %s^3+14s^2+63s+90, assign
                             %to B & suppress output.
C=conv(A,B)                  %multiply polynomials
                             % A & B formed above,
                             %assign to C & output.
'Solution 4.7(b)'            %display enclosed text.
A1=poly([-3 -5 -8]);         %represent polynomial
                             %(s+3)(s+5)(s+8),assign
                             %to A1 & suppress output.
B1=poly([-1 -2 -4])          %represent polynomial
                             %(s+1)(s+2)(s+4), assign
                             %to B1 & suppress output.
C1=conv(A1,B1)               %multiply polynomials
                             %A1 & B1 formed above,
                             %assign to C1 & output.
```

Output

```
Problem 4.7
Solution 4.7(a)
C =
    1   26   275   1510   4524   6984   4320   0
```
Note: This is MATLAB representation of polynomial
$s^7+26s^6+275s^5+1510s^4+4524s^3+6984s^2+4320s$
```
solution 4.7(b)
C1 =
    1   23   205   905   2074   2312   960
```
Note: This is MATLAB representation of polynomial
$s^6+23s^5+205s^4+905s^3+2074s^2+2312s+960$

Problem 4.8

Use MATLAB program to ascertain the Laplace transform of the function
$y(t)=2t^2 e^{-5t} sin(3t+30°)$

Solution   We will utilize the Symbolic Math toolbox features to write the MATLAB program. The given function $y(t)$ is a function of time 't', therefore, we will commence writing the MATLAB program by declaring t as symbolic variable by use of command *syms t*. Command *laplace* (*y*), computes the

Y(*s*), in partial fraction form. Command *simple* (*Y*) yields simplest expression for *Y* with minimum number of terms. Command *pretty* (*y*) and *pretty* (*Y*); outputs pretty prints of expressions assigned to *y* & *Y* respectively.

Solution

```
'Problem 4.8'                  %display enclosed text.
syms t;                        %declare t as symbolic
                               %variable.
y=2*t^2*exp(-5*t)*sin(3*t+30)  %input y(t)& suppress
                               %output.
'Mathematical expression of Y' %display enclosed text.
pretty(y)                      %print y in format that
                               %resembles type-set
                               %mathematics.
'Laplace Transform y(s)'       %display enclosed text.
Y=laplace(y);                  %compute laplace of y,
                               %assign to Y.
'Simple symbolic expression of Y'%display enclosed text.
Y=simple(y)                    %display simplest form of
                               %symbolic expression of Y.
'Mathematical expression of Y' %display enclosed test.
pretty(Y)                      %print Y in format that
                               %resembles type-set
                               %mathematics.
```

Output
```
Problem 4.8
y =
2*t^2*exp(-5*t)*sin (3*t+30)
Mathematical expression of y
```

$$2\ t^2 \exp(-5\ t)\ \sin(3\ t + 30)$$

```
Laplace Transform Y(s)
Simple symbolic expression of Y
Y =
4*(sin(30)*s^3+15*sin(30)*s^2+48*s*sin(30)-10*sin(30)
+9*cos(30)*s^2+90*cos(30)*s+198*cos(30)/(s^2+10*s+34)^3
```

```
Mathematical expression of Y
```

$$4(\sin(30)s^3 + 15\sin(30)s^2 + 48s\ \sin(30) - 10\sin(30)$$
$$+ 9\cos(30)s^2 + 90\cos(30)s + 198\cos(30))/\ (s^2 + 10\ s + 34)^3$$

Problem 4.9

Write MATLAB program to ascertain the Laplace transform of the time function.

$y(t) = y(t) = \dfrac{1}{2}te^{-t} - \dfrac{3}{2}t^2e^{-3t} + \dfrac{3}{16}\sin(t + 45°)$. The result be displayed upto two decimal places for the fractions.

Solution   We will utilize the features of *Symbolic Math toolbox* to enhance and simplify mathematical computations. The use of following command functions would be demonstrated to achieve the desired results:

(a) **syms t:**   Since the given function $f(t)$ is a function of time t, command *syms t* declares t as symbolic variable at the commencements of MATLAB program.

(b) **laplace(y):**   The command *laplace* (y) computes Laplace transform $Y(s)$ in partial fractions form.

(c) **pretty(y) and pretty(Y):**   Expressions y and Y are pretty printed.

(d) **simplify(Y):**   This command combines the partial fractions of $Y(s)$ obtained by the command *laplace* (y).

(e) **vpa(Y, 2):**   This command converts the symbolic fractions upto two decimal places.

MATLAB program using above mentioned commands can be written as:

```
'Problem 4.9'                          %display enclosed text.
syms t;                                %declare t as symbolic
                                       %variable.
y=1/2*t*exp(-t)-3/2*t^2*exp(-3*t       %input y(t)& suppress
+3/16*sin(t+45);                       %output.
'Mathematical expression of Y'         %display enclosed text.
pretty(y)                              %print y in format that
                                       %resembles type-set
                                       %mathematics.
'Laplace Transform Y(s)'               %display enclosed text.
Y=laplace(y);                          %compute laplace of y,
                                       %assign to Y.
'Mathematical expression of Y'         %display enclosed text.
pretty(y)                              %print Y computed above in format
                                       %that resembles type-set
                                       %mathematics.
Y=simplify(y)                          %display after combining partial
                                       %fractions of Y
                                       %symbolic expression of Y.
'Mathematical expression of Y'         %display enclosed test.
pretty(Y)                              %print Y  format above
                                       %in format that resembles type-set
                                       %mathematics.
'Conversion fo symbolic expression'%display enclosed test
'up to two decimal places'
Y=vpa(Y,2)                             %convert symbolic fractions up to
                                       %two decimal places.
```

```
'Mathematical expression of Y'        %display enclosed text.
 pretty(Y)                            %print Y found above
                                      %in format that resembles type-set
                                      %mathematics.
```

Output

```
Problem 4.9
Mathematical expression of y

                           2
    1/2 t exp(-t) - 3/2 t  exp(-3t) + 3/16 sin(t + 45)


Laplace Transform Y(s)
Mathematical expression of Y




           1            3              s sin(45) + cos(45)
    1/2 --------- - --------- + 3/16 ---------------------
              2           3                    2
         (s + 1)     (s + 3)                 s + 1

Y =

1/16*(8*s^5+128*s^3+24*s^4+192*s^2+120*s+168+3*s^6*sin(45)
+33*s^5*sin(45)+138*s^4*sin(45)+270*s^3*sin(45)
+243*s^2*sin(45)+81*s*sin(45)+3*cos(45)*s^5+33*cos(45)*s^4
+138*cos(45)*s^3+270*cos(45)*s^2+243*cos(45)*s+81*cos(45))
/(s^7+47*s^5+11*s^6+101*s^4+127*s^3+117*s^2+81*s+27)

Mathematical expression of Y
          5       3      4       2                    6
1/16(8s + 128s + 24s + 192s + 120s + 168 + 3s sin (45)
      5             4              3              2
+ 33s sin (45) + 138s sin (45) + 270s sin (45) + 243s sin (45)
                           5              4              3
+ 81s sin (45) + 3cos (45) s + 33cos (45) s + 138cos (45) s
             2                                    /
+ 270cos (45) s + 243cos (45) s + 81cos (45)    /
                                               /

   7      5      6       4       3       2
(s + 47 s + 11 s + 101 s + 127 s + 117 s + 81s + 27)

Conversion of symbolic expression
up to two decimal places

Y =.63e-1*(38.*s^5+.43e3*s^3+016e3*s^4+.53e3*s^2+.32e3*s+.21e3+2.6*s^6)
/(s^7+47.*s^5+11.*s^6+.10e3*s^4+.13e3*s^3+.12e3*s^2+81.*s+27.)

Mathematical expression of Y

             5       3       4       2                    6
     38. s +430.s +160.s + 530.s +320.s+210.+2.6s
     ------------------------------------------------
      7       5      4       4       3       2
     s + 47.s +11.s +100.s +130.s +120.s +81.s+27.
```

Problem 4.10

Write a MATLAB program to yield the transfer function $G(s)$ into factored and polynomial forms.

$$G(s) = \frac{20(s+1)(s+3)(s+5)}{s(s+2)(s^2+10s+24)(s+7)(s^2+17s+72)}$$

Solution Transform function G(s) has been given as a ratio of polynomials. The MATLAB program can be written as

```
'Problem 4.10'                                        %display enclosed text.
num={-1 -3 -5}                                        %input numerator.
den=[0 -2 -7(roots([1 10 24]))'(roots([1 17 72]))']%input denominator.
'Polynomial form'                                     %display enclosed text.
[num,den]=zp2tf(num',den',20)                         %form transfer function
                                                      %in co-efficient form in
                                                      %descending power. num &
                                                      %den are to be column
                                                      %vectors as shown.
G=tf(num,den)                                         %convert transfer function
                                                      %as a ratio of num and den.
'Factored form'                                       %display enclosed text.
Gfac=zpk(G)                                           %display transfer function
                                                      %as a ratio of num and den.
'Alternatively'                                       %display enclosed text.
'Factored form'                                       %display enclosed text.
Gfac=zpk([-1 -3 -5],[0 -2 -7 roots([1 10 24])'roots([1 17 72]),20)%input
                                                      %num & den & display transfer
                                                      %function as a ratio in
                                                      %factored form.
'Polynomial form                                      %display enclosed text.
Gpoly=tf(Gfac)                                        %convert & display transfer.
                                                      %function in polynomial form.


Output
Problem 4.10
num =
   -1    -3    -5
den =
    0    -2    -7    -6    -4    -9    -8
Polynomial form
num =
    0     0     0     0    20   180   460   300
den =
  Columns 1 through 6
    1    36   523  3900  15604  31344
  Columns 7 through 8
   24192     0
```

```
Note: This represents numerator & denominator polynomials
as 20s³ + 180s² +460s+300 & s⁷ + 36s⁶ + 523s⁵ + 3900s⁴ +15604s³ +31344s²+2419²s
respetively.

Transfer function:
                20 s^3 + 180 s^2+ 460 s + 300
    ------------------------------------------------------------------------
    s^7 +36 s^6 + 523 s^5 + 3900 s^4 + 15604 s^3 + 31344 s^2 + 24192 s

Factored form
zero/pole/gain:

            20 (s+5) (s+3) (s+1)
    ---------------------------------------
    s (s+9) (s+8) (s+7) (s+6) (s+4) (s+2)

Alternatively

Factored form
zero/pole/gain:
            20 (s+1) (s+3) (s+5)
    ---------------------------------------
    s (s+2) (s+4) (s+6) (s+7) (s+8) (s+9)

Polynomial form
Transfer function:
                20 s^3 + 180 s^2 + 460 s + 300
    ---------------------------------------------------------------------
    s^7 + 36 s^6 + 523 s65 + 3900 s^4 + 15604 s^3 + 31344 s^2 + 24191 s
```

Problem 4.11

Solve problem 4.10 using MATLAB and Symbolic math toolbox features.

Solution Complicated transfer functions can be easily manipulated by use of *symbolic command* functions. The steps involved are:

(a) Declare symbolic variable for the given function symbolically by use of command *syms s*.

(b) Enter the expression for the given transfer function symbolically and assign to a variable G. Construct the given transfer function in symbolic form.

(c) Command *[N, D] = numden(G)* helps in separating symbolic numerator and denominator from symbolic G and assigning to N and D respectively.

(d) *sym2poly(N)* and *sym2poly(D)* converts symbolic numerator and denominator to vectors.

(e) Transfer function in polynomial form is obtained by MATLAB command *tf(N, D)* and is converted to factored form by executing command *zpk(poly)*.

Computer program using symbolic math toolbox commands is given below:

Solution

```
'Problem 4.11'              %display enclosed text.
syms s                      %declare s as symbolic variable.
G=20*[s+1)*(s+3)*(s+5)]...%input G symbolically.
   /[s*(s+2)*(s+3)*(s+7)*(s^2+10*s+24)*(s^2+17*s+72)];
'Transfer function'         %display enclosed text.
Prety(G)                    %output G in mathematical form.
[N,D]=numden(G);            %separate symbolic num & den.
N=sym2poly(N);              %form num vector.
D=sym2poly(D);              %form den vector.
'Transfer function polynomial form'%display enclosed text.
Gpoly=tf(N,D)               %construct & output transfer
                            %function in polynomial form.
'Transfer function factored form''%display enclosed text.
Gfac=zpk(Gpoly)             %construct & output polynomial
                            %transfer function in factored
                            %form.
```

Output
```
Problem 4.11
Transfer function
                (s + 1)(s + 3)(s + 5)
   20-----------------------------------------------------
                     2                2
      s(s + 2)(s + 7)(s + 10s + 24)(s + 17s + 72)

Transfer function polynomial form

Transfer function:

20 s^3 + 180 s^2 + 460 s + 300
------------------------------------------------------------
s^7 + 36 s^6 + 523 s^5 + 3900 s^4 + 15604 s^3 + 31344 s^2 + 24192 s

Transfer function factored form

zero/pole/gain:
          20(s+0.6407)(s^2 - 0.6407s + 23.41)
------------------------------------------------------------
s(s^2 + 9.315s + 28.92)(s^2 - 9.315 + 28.92)(s^2 + 28.92)
```

Problem 4.12

Solve the following set of linear algebraic equations
$$5x_1 + 6x_2 + 10x_3 = 4$$
$$-3x_1 + 14x_3 = 10$$
$$-7x_2 + 21x_3 = 0$$

by using MATLAB.

In the matrix form, the given equations can be written as

$$\begin{bmatrix} 5 & 6 & 10 \\ -3 & 0 & 14 \\ 0 & -7 & 21 \end{bmatrix} \begin{bmatrix} x_1 \\ x_2 \\ x_3 \end{bmatrix} = \begin{bmatrix} 1 \\ 40 \\ 0 \end{bmatrix}$$

The MATLAB program is given below

Solution

```
'Problem 4.12'                  %display enclosed text.
A=[5 6 10; -3 0 14; 0 -7 21]    %input matrix A & output.
B=[4;10;0]                      %input matrix B & output.
x=A\B                           %solve equation A*x=B by
                                %Gaussian elimination.
'Alternatively'                 %display enclosed text.
x=inv(A)*B                      %solve equation A*x=B.
```

Output
```
Problem 4.12
A =
     5       6      10
    -3       0      14
     0      -7      21
B =
     4
    10
     0
x =
   -1.4545
    1.2078
    0.4026
Alternatively
x=
   -1.4545
    1.2078
    0.4026
```

Problem 4.13

Use MATLAB program to represent the transfer function

$$G(s) = \frac{200(s^3 + 2s^2 + 4s + 1)}{s(2s^3 + 3s^2 + 2s + 1)}$$

Solution The given transfer function is a ratio of polynomials made up of numerator and denominator. In the MATLAB program, numerator and denominator polynomials are input separately and assigned variables N and D respectively. The transfer function is constructed by using the command $tf(N, D)$. Denominator polynomial has been expanded by multiplying term $2s^3 + 3s^2 + 2s + 1$ by s for easy representation. The MATLAB program is given below:

```
Solution

'Problem 4.13'      %display enclosed text.
N=200*[1 2 4 1];    %input numerator polynomial,assign to
                    %N & suppress output.
D=[2 3 2 1 0];      %input expanded denominator polynomial,
                    %assign to D & suppress output.
G=tf(N,D)           %construct transfer function & output.

Output

Problem 4.13
Transfer function:

200 s^3 + 400 s^2 + 800 s + 200
---------------------------------
  2 s^4 + 3 s^3 + 2 s^2 + s
```

Problem 4.14

Use MATLAB program to form transfer function $G(s) = \dfrac{5s(s+1)(s+3)(s+5)}{(s+2)(s+4)(s+6)}$

Solution The given transfer function G(s) represented by ratio of numerator and denominator in factored form is constructed by defining them separately and assigning to variables N and D respectively. MATLAB command *zpk (N, D, K)* forms the transfer function in the factored form. K is value of the constant which is equal to 5, in the problem under consideration.

```
Solution

'Problem 4.14'          %display enclosed text.
N=[0 -1 -3 -5];         %input numerator polynomial,assign to
                        %N & suppress output.
D=[-2 -4 -6];           %input  denominator polynomial,
                        %assign to D & suppress output.
'Transfer function'     %display enclosed text.
G=zpk(N,D,5)            %construct transfer function & output.
```

```
Output
Problem 4.14
Transfer function:

Zero/pole/gain:
5 s (s+1) (s+3) (s+5)
----------------------
  (s+2) (s+4) (s+6)
```

Problem 4.15

Use MATLAB program to find the poles and zeros of the transfer function

$$G(s) = \frac{s^2(s^2 + 3s + 2)}{s^4 + 5s^3 + 6s^2 + 5s + 1}$$

Solution Transfer function G(s) is a ratio of polynomials. The roots of the numerator yield the zeros and that of the denominator yield the *poles*. MATLAB program is given below:

```
Solution

'Problem 4.15'                  %display enclosed text.
N=conv(poly([0 0]),[1 3 2]);%input numerator,assign to
                                %to N & suppress output.
D=[1 5 6 5 1]                   %input  denominator polynomial,
                                %to D & suppress output.
'zeros of transfer function'%display enclosed text.
z=roots(N)                      %compute zeros of transfer
                                %function, assign to Z & output.
'Poles of transfer function'%display enclosed text.
P=roots(D)                      %compute poles of transfer
                                %function, assign to p & output.

Note: Numerator N can also be written as
N=conv([1 0 0],[1 3 2]);

Output
Problem 4.15
zeros of transfer function
z =
     0
     0
    -2
    -1
```

```
Note: There are four zeros.
Poles of transfer function
P =
   -3.7321
   -0.5000 + 0.8660i
   -0.5000 - 0.8660i
   -0.2679
Note: There are four poles.
```

Problem 4.16

Use MATLAB program to find out partial fraction of

$$F(s) = \frac{6(s+1)}{s^2(s+2)(s+3)}$$

Solution $F(s)$ is expressed as a ratio of polynomials. The numerator and denominator polynomials would be input separately and assigned to variables *num* and *den* respectively. Partial fraction expansion is obtained by executing MATLAB command [*r*, *p*, *k*] = *residue(num, den)*, which gives residues, poles and direct quotient and in the output are assigned to *r*, *p* & *k* respectively. MATLAB program is given below:

```
Solution

'Problem 4.16'              %display enclosed text.
N=6*[1 1];                  %input numerator,assign
                            %to N & suppress output.
D=poly([0 0 -2 -3]);        %input  denominator, assign
                            %to D & suppress output.
'Partial fraction Expansion'%display enclosed text.
[r,p,k]=residue(N,D)        %compute Partial fraction expansion.

Output

Note: Denominator can also be written as
D=conv(poly([-2 -3]),[1 0 0]);

Output
Problem 4.16
Partial fraction Expansion
r =
    1.3333
   -1.5000
    0.1667
    1.0000
```

```
P =
    -3
    -2
     0
     0
k=
    []
```

Problem 4.17

Use matlab to represent the transfer function

$$G(s) = \frac{s^2 + 11s^2 + 3}{s^2 + 11s^2 + 38s + 40}$$

Find the zeros and poles and plot them.

Solution

```
'Problem 4.17'              %display enclosed text.
num=[1 4 3];                %enter numerator.
den=[1 11 38 40];           %enter denominator.
g=tf(num,den);              %construct transfer function.
z=zero(g)                   %find zeros & output.
p=pole(g)                   %find poles & output.
pzmap(g)                    %map poles & zeros.
```

Output

```
Transfer function:
s^2 + 4 s + 3
-------------------------
s^3 + 11 s^2 + 38 s + 40


z =
    -3
    -1
P =
    -5.0000
    -4.0000
    -2.0000
The map showing poles and zeros is depicted in Fig. 4.16.
```

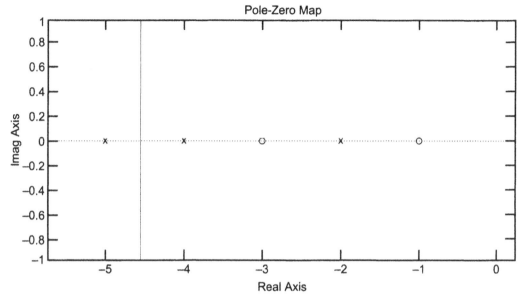

Fig. 4.16

1. Which Windows are opened normally in the default MATLAB desktop.

2. What is the purpose of:
   (a) MATLAB Command Window.
   (b) Edit/Debugger Window.
   (c) Command History Window.

3. Help can be requistioned in how many ways. What is the basic difference when *help* is asked by the commands *help* and *lookfor*.

4. What is the use of Launch Pad Window. How can you launch a required Demo.

5. What is displayed in a Workspace Window. How can Array editor be activated and contents of a Workspace cleared.

6. What are the rules governing the variables. If a variable is named from the reserved work list, what will MATLAB report.

7. If printing is to be suppressed in the output, what has to be placed after the command.

8. What is an array. What is the difference between a matrix and a vector.

9. What is the current directory when MATLAB is started. How can the current directory be changed easily.

10. What are the different ways of entering matrices into MATLAB.

11. Enter a matrix

$$A = \begin{bmatrix} 8 & 5 & 6 \\ 7 & 4 & 8 \\ 9 & 6 & 4 \end{bmatrix}$$

   and perform the following operations.

   (a) Sum of the rows.

   (b) Sum of the columns by transposing the matrix.

   (c) Display the diagonal and find its sum.

12. Create following matrices by use of functions:

   (a) Matrix having three rows and three columns with elements equal to zero.

   (b) An Identity matrix of three rows and three columns. Manipulate the function to create a similar matrix but all elements equal to four (4).

13. List the operators related to basic matrix operations.

14. Find out which of the following expressions are legal and what arrays are created

   (a) [1,2]   (b) [1.0 2.0 3.0 4.0]
   (c) [1 2 3; 1 2]   (d) [1; 2; 3]
   (e) [1.0; 2.0; 3.0]   (f) [1,2,3; 1,2, 3]
   (g) [1, 2, 3; 1, 2, 3]

15. What expressions will be generated by the following

   (a) a = [ ]   (b) a(1,2) = 3
   (c) a = [0 1+2]   (d) a = [1:5]′
   (e) a = zeros(3)   (f) a = zeros (3,4)
   (g) a = eye(3,3)

5

CONTROL SYSTEM: BASIC THEORY

5.1 INTRODUCTION

Ever since the evolution of life on the earth, control systems have been associated with the mankind. The desire of man has always been to control the nature and the visible/invisible forces; in order to utilize them in performing; first the basic task and later on as the requirements become more specific and complex, use them to perform such tasks which were beyond the capabilities of a man. As the requirements turned difficult, the reliance on his own strength for accomplishing the task was slowly replaced by devices/systems such as wheels, levers, windmills and use of sail for sailing vessels, etc. The invention of steam engine revolutionized the world and provided a tool by which enormous power could be produced and utilized. The world wars in twentieth century generated the military requirements and saw the invention of automatic control systems such as fire control, auto pilots and aeroplanes. This opened the door for further invention of complex systems such as radars, guided missiles, telecommunication etc.

LEARNING OBJECTIVES

- ◆ To formulate generalised defination of control systems.
- ◆ To classify control systems and formulate a family tree.
- ◆ To describe basic elements of a control system.
- ◆ To learn representation of control systems:
  - ◆ Block diagram.
  - ◆ Signal flow graph.

The requirement of economies in production of goods required by man, created devices which could be effectively utilized in industry to produce everything which man desired. Today, use of control system has transformed man's hope and dreams into reality. Control system science is not restricted to a particular field. It finds wide applications in the fields of mechanical, electrical, civil, electronics, aeronautical, chemical, biosciences, nuclear field management, economics, telecommunication, computers and other related fields and in future its use will be seen in many more undiscovered fields. Control systems thus *posses* great potential in days to come. Having seen the potential of control systems, we must now try to evolve a generalized definition of a control system, which suits the interdisciplinary subject and relates to all fields.

5.2 DEFINITION

Control system is a science, which deals with mechanisms, devices or objects joined together for interaction with the aim to achieve the desired results. The entire unit or system consisting of an element or number of elements work in cohesion to perform desired controlled function. The arrangement of devices or elements, their appearance and design vary with the objectives to be achieved. The main *criterion* is that the entire system works as a cohesive unit when actuated with any predetermined input signal, takes control over the disturbances and exhibit a controlled and stable output.

5.3 CLASSIFICATION

There are numerous ways by which a control system can be classified. The classification can be based on *state, principle of superposition, homogeneity, nature of signal flow, nature mathematical equations* governing the system and also on number of input/output signals. When classified on *states*, a system can be *static*, or *dynamic.* In a static system, the steady state values are reached instantly and remain for a long time. *Dynamic* systems exhibit transients when actuated by an input signal because of certain energy storage elements. A purely resistive circuit can be termed as a *static* system and RLC circuit is an example of a *dynamic* system.

Based on the *principle of superposition & homogeneity*, control systems can be classified, as *linear* and *nonlinear* systems. *Linear* systems satisfy principles of *superposition* and *homogeneity*. Simple mathematical tools help analyze such systems easily as the input-output relationship can be represented by a straight line. *Non-linear* systems do not obey the principles of superposition and homogeneity. Non-linearities are caused by hysteresis, frictional forces, saturation effect, backlash and dead play in the components. In practice linear systems are difficult to realize. However, if the deviation from linearity is small and considered to be insignificant for the problem under consideration, such nonlinear systems can be approximately linearised about the operating point.

Control systems are also classified as *single-input-single-output* (SISO) and *multi-input-multi-output* (MIMO) systems based on the number of input and output signals.

Another method of classification of control systems is based on the parameters and the mathematical equations which describe the functioning of a control system.

Distributed and Lumped Parameter Control Systems

In *lumped parameter* systems, the parameters of the devices/elements that help in analysis can be lumped or concentrated at a point and utilize ordinary differential or difference equations for description with time as the only independent variable. *Distributed-parameter* control systems are described by partial differential equations having space and time as independent variables. In such systems, parameters may be considered distributed at more them one point or location.

Random and Non-Random control systems

Random systems are also called *stochastic* as *probabilistic* systems. The parameter of such systems cannot be described with certainty. Statistical terms and properties are often used to describe the system parameters. *Non-Random* systems also called *deterministic* systems posses systems parameters which can be described at all times and precisely. Also, in certain practical systems in which uncertainties can be neglected and ignored and a practical model can be evolved, *deterministic* model can be adopted for analysis.

Continuous and Discrete Time Control Systems

Continuous time systems are those systems in which the system variables/parameters are the functions of time. In *discrete* system the systems parameters/variables are known at discrete intervals of time. Signal appears as a train of pulses at one or more points rather than as a continuous function.

Time-Variant and Time-Invariant Control Systems

In *time-varying* systems one or more system parameters vary with time irrespective of functional dependence of input and output on time parameter. In time-invariant system the systems parameters are constant and independent of time.

Swisher [1] has illustrated the classification of control systems by forming a family tree shown is Fig. 5.1

5.4 GENERAL CLASSIFICATION

In general, control systems are classified as *open-loop* and *closed-loop* control systems. The basic difference between them is the presence of feedback element in the closed-loop control system.

Open-loop Control System   In an open-loop system output depends upon the input. The changes in the output can be affected by altering the input. The output on the other hand cannot

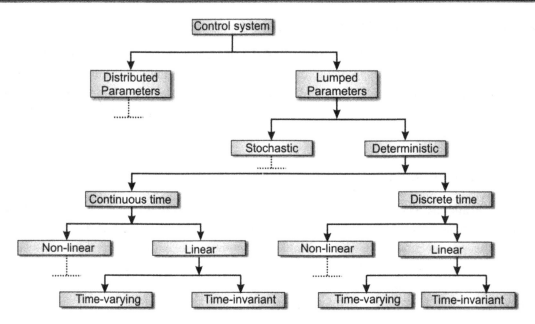

Fig. 5.1 *Control system classification based on nature of mathematical equations describing a system. Dashed lines signify the existence of sub-divisions of similiar type to the other at same level.*

influence the input. It is also called *control system without feedback*. An open-loop system is represented is Fig. 5.2. Open-loop systems are simple, economical, easily designed and assembled. However, the exact behavior of the output cannot be determined and the system does not alter by itself on experiencing external disturbances. Human operator is required to exercise control on the system to achieve desired results. The performance of an open-loop system is dependent on the calibration and is prone to instability.

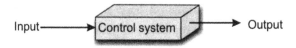

Fig. 5.2 *Open-loop control system*

Closed-loop System In a *closed-loop* system the input is made to depend on the output or changes in the output. Such systems are also termed as *control system with feedback*. Changes in the output are fed back with the help of feedback elements and generates error signal on comparison with input. A closed-loop system is represented in Fig 5.3. Such systems are complex, difficult to design, costly but accurate and self-correcting. However, tendency to overcorrect may cause oscillations in the system.

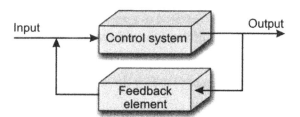

Fig. 5.3 *Closed-loop system*

Feedback is the essential characteristic of the closed-loop systems, which distinguishes it from open-loop systems. It is the presence of feedback which renders accuracy and reduces effects of nonlinearities.

5.5 ELEMENTS OF A FEEDBACK CONTROL SYSTEM

A feedback control system with all its essential elements is shown in Fig. 5.4. The definition, nomenclature/symbols used are widely accepted and employed globally by control system engineering books and engineers.

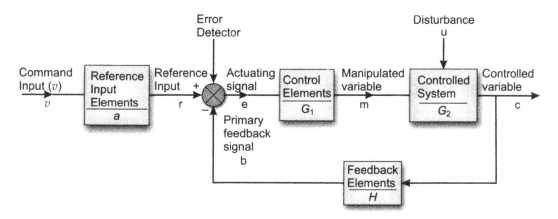

Fig. 5.4

- ◆ **Command Input (v)** It is the input externally produced independent of feedback control system.
- ◆ **Reference Input Elements (a)** It produces a signal (r) proportional to the command input (v).
- ◆ **Reference Input (r)** It is the actual input signal proportional to the command input (v)
- ◆ **Error Detector** It is an element which sums or compares the signal (b) obtained from feedback elements (H) with the reference input signal (r).

- **Actuating Signal (e)**   It is also termed as *error* signal. It is the signal which is produced by *error detector* and is the difference between the reference input signal (r) and feedback signal (b).

- **Control Elements (G₁)**   Elements which develop the manipulated variable (m) from the actuating signal (e).

- **Manipulated Variable (m)**   It is the signal developed by control elements (G_1) and applied to the controlled system (G_2).

- **Controlled System (G₂)**   It is also termed as plant and is the device which needs to be controlled.

- **Controlled Variable (c)**   It is the output signal which is desired to be controlled and fed to the feedback elements (H).

- **Disturbance (u)**   It is undesired disturbing signal which tends to affect the controlled output variable signal (c).

- **Feedback Elements (H)**   It is a device which receives the controlled output (c) and after acting on it produces the primary feedback signal (b).

- **Primary Feedback Signal (b)**   It is the signal developed by the feedback elements (H) and is compared with the reference input signal (r) with the help of error detector. The primary feedback signal may either have positive ($+$) or negative ($-$) polarity.

5.6 REPRESENTATION OF CONTROL SYSTEMS

Block diagrams and signal flow graphs are the commonly used pictorial representations of control systems and show interrelationship in the form of cause and effect with physical devices which constitute a control system.

5.6.1 Block Diagrams

A control system consisting of a number of physical components or devices can be easily represented with the help of block diagram representation. A simple open-loop control system having one input and one output is shown in Fig. 5.5

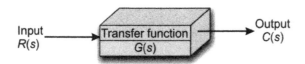

Fig. 5.5 *Block diagram of a simple open-loop system*

The transfer function relating the input and output signal is:

$$G(\text{s}) = \frac{C(s)}{R(s)}$$

The general block diagram representation of a control system with negative feedback is shown in Fig. 5.6.

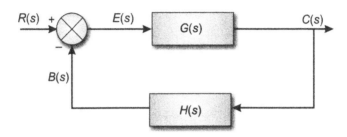

Fig. 5.6 *Closed-loop system with negative feedback*

$R(s)$, $C(s)$, $B(s)$, $E(s)$, $G(s)$ and $H(s)$ are the Laplace transforms of the respective signals and transfer functions. From the definition of transfer function, we can write:

$$G(s) = \frac{C(s)}{E(s)} \quad \text{or} \quad C(\text{s}) = E(s)G(s) \tag{5.2}$$

The transfer function $T(s)$ of the closed-loop control system is defined as

$$T(s) = \frac{C(s)}{R(s)} \tag{5.3}$$

Also

$$H(s) = \frac{B(s)}{C(s)}$$

or

$$B(s) = H(s)\ C(s) \tag{5.4}$$

From the block diagram, we can write the following relationship

$$C(s) = G(s)E(s) \tag{5.5}$$

$$E(\text{s}) = R(\text{s}) - B(s)$$

$$E(s) = R(\text{s}) - H(s)C(s) \tag{5.6}$$

Substituting the value of $E(s)$ in eqn (5.5), we get

$$C(s) = G(s)\ [R(s) - H(s)C(s)]$$

$$C(s) + G(s)\ H(s)C(s) = G(s)R(s)$$

$$C(s)\ [1 + G(s)H(s)] = G(s)R(s)$$

or
$$\frac{C(s)}{R(s)} = \frac{G(s)}{1+G(s)H(s)} = T(s) \tag{5.7}$$

Eqn (5.7) can be represented by a block diagram as shown in Fig. 5.7

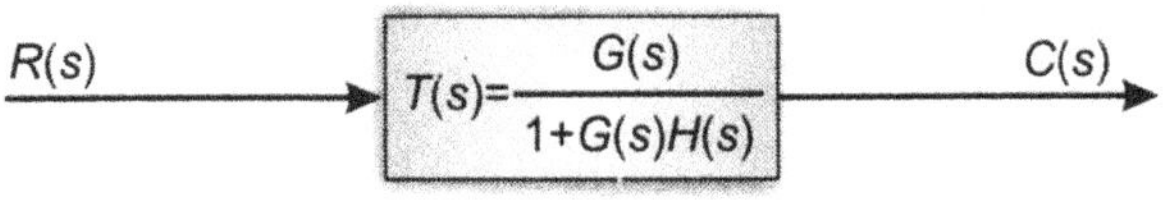

Fig. 5.7 *Representation of closed - loop system depicted in Fig.5.6*

Any complex control system consisting of number of paths and loops can be reduced to a very simple block diagram by the use of block diagram reduction algebra/rules illustrated in Table 5.1.

Table 5.1   Block Diagram Reduction Rules

Transformation	*Original Block Diagram*	*Equivalent Block Diagram*
1. Combining blocks in Cascade	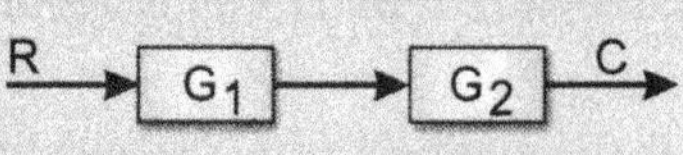	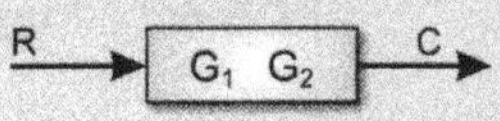 $C=(G_1 G_2)R$
2. Combining blocks in parallel or eliminating a forward loop	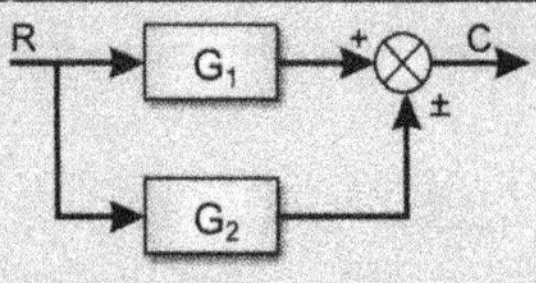	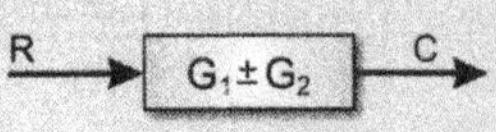 $C=(G_1\pm G_2)R$
3. Removing a block from a forward path	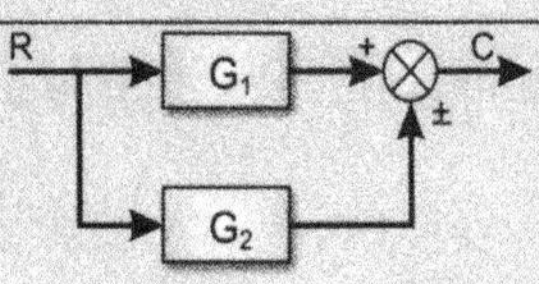	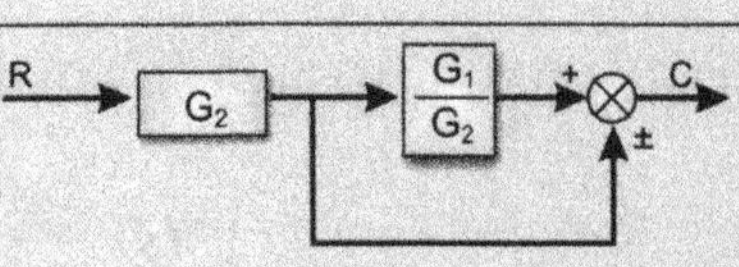 $C = G_2\left(1\pm\dfrac{G_1}{G_2}\pm 1\right)R$
4. Eliminating a feedback loop	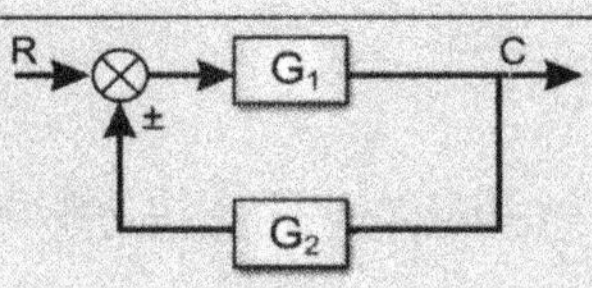	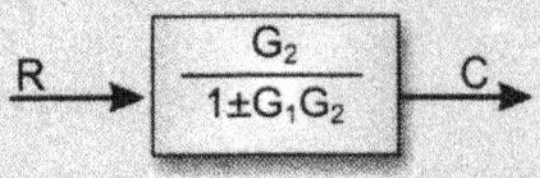 $C =\left(\dfrac{G_2}{1\mp G_1 G_2}\right)R$
5. Removing a block from a feedback loop	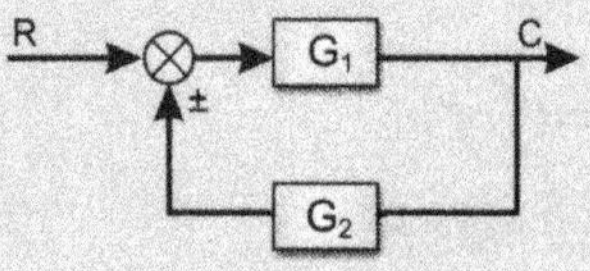	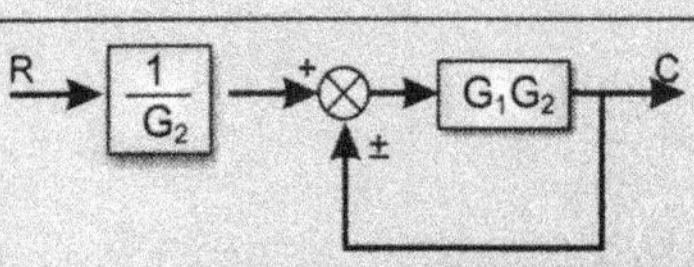 $C = \dfrac{1}{G_2}\left(\dfrac{G_1 G_2}{1\mp G_1 G_2}\right)R$

(Table 5.1 Contd...)

Transformation	Original Block Diagram	Equivalent Block Diagram
6. Rearranging summing points	$C = R \pm X \pm Y$	$C = R \pm X \pm Y$
7. Rearranging summing points	$C = (R \pm X) \pm Y$ $= R + (\pm X \pm Y)$	$C = R + (\pm X \pm Y)$
8. Moving a take-off point ahead of a block	$C = GR$	$C = GR$
9. Moving a take-off point beyond a block	$C = GR$	$C = GR$
10. Moving a summing point ahead of a block	$C = RG \pm X$	$C = G\left(R \pm \dfrac{1}{G}\right) X = RG \pm X$
11. Moving a summing point behind a block	$C = RG \pm XG = G(R \pm X)$	$C = RG \pm GX = G(R \pm X)$
12. Moving a take-off point ahead of a summing point	$C = R \pm X$	
13. Moving a take-off point beyond a summing point	$C = R \pm X$	

5.6.2 Signal Flow Graph

A linear control system can be described by a set of linear equations having the form

$$Y_i = \sum_{j=i}^{n} a_{ij} Y_j \quad \text{where } i = 1, 2, \text{------------------}, n. \tag{5.8}$$

Y_i and Y_j are the variables and a_{ij} may be a constant. The eqn (5.8) expresses each of the n variables in terms of the others and themselves. Assuming that

$i = i$ and $j = 1$ to 4 then

$$Y_i = \sum_{j=1}^{4} a_{ij} Y_j = a_{i1} Y_1 + a_{i2} Y_2 + a_{i3} Y_3 + a_{i4} Y_4 \tag{5.9}$$

Eqn. (5.9) can be represented by signal flow diagram shown in Fig. 5.8

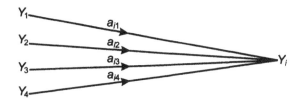

Fig 5.8 *Representation of eqn (5.9) by signal flow diagram*

Let us consider a simple equation

$$x_i = a_{ij} y_j \tag{5.10}$$

The signal flow diagram of eqn (5.10) is shown in Fig. 5.9

Fig 5.9 *Signal flow diagram of eqn (5.10)*

The variables x_i and y_j are represented by small dots (.) called a node and a_{ij} is written on the arrow on the branch connecting nodes x_i and y_j and is called the *transmittance*. Let us now understand the commonly used terms connected with the signal flow diagrams.

5.6.2.1 Terminology

A signal flow diagram is shown in Fig. 5.10

- ◆ **Branch** It represents the transmission of signal from one node to the other and is represented by a line. It also represents dependence of one node upon the other e.g. x_2 is dependent upon x_1.
- ◆ **Arrow** It represents the direction of transmission between the nodes.

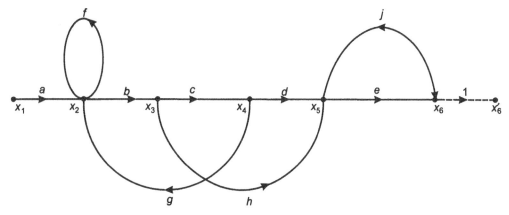

Fig 5.10 *Signal flow diagram*

- ◆ **Node** Nodes represent the variables and are represented by dots (.). Nodes are arranged generally from left to right depicting input-output relationship of the elements of a control system. A **source** node has outgoing branches. x_1 is a source node. It is also called **input** node. **Sink** node has only one incoming branch. It is also called **output** node. This condition may not always be met. In such cases; an additional node connected with dash lines is introduced to meet the condition. In Fig 5.11 node x_6' has been introduced and can be termed as **sink or output** node.

  - $x_6' = 1 \times x_6 = x_6$

- ◆ **Forward path** The paths originating form *source* node to *sink* node with no node traversed twice, is called a *forward* path. The two forward paths which satisfy the conditions in Fig. 5.10 are

  - $x_1 - x_2 - x_3 - x_4 - x_5 - x_6 - x_6'$ having transmittance/ gain of '*abcde*'; and

  - $x_1 - x_2 - x_3 - x_5 - x_6 - x_6'$ having transmittance/ gain of '*abhe*'.

- ◆ **Transmittance/Gain** It is associated with paths and loops. **Path gain** is the product of the gains of all the branches encountered in a path and the **loop gain** is the product gain of the branches forming a loop.

- ◆ **Loop** It is a closed path which originates from a node and terminates at the same node and along which no intermediate path is traversed twice. In Fig. 5.10, there are three loops.

  - $x_2 - x_3 - x_4 - x_2$ having loop gain of '*bcg*'
  - $x_5 - x_6 - x_5$ having loop gain of '*ej*'
  - $x_2 - x_2$ having loop gain of '*f*'

- ◆ **Self or feedback loop** is a path originating form a node and terminating at the same node. In Fig. 5.10, $x_2 - x_2$ is the *feedback* loop. *Non-touching* loops are such loops having no paths or branches in common. In Fig. 5.10 $x_2 - x_3 - x_4 - x_2$ and $x_5 - x_6 - x_5$ and $x_2 - x_2$ and $x_5 - x_6 - x_5$ are the two possible combinations of non-touching loops.

5.6. 2.2 Signal Flow Graph Algebra

Complex signal flow diagrams can be simplified by use of signal flow graph reduction algebra shown is Table 5.2

Table 5.2 Signal Flow Fraph Reduction Algebra

Operation	Original Signal Flow Graph	Equivalent Signal Flow Graph or Diagram
1. Addition		$x_2=(a+b)x_1$
2. Multiplication	$x_2 = ax_1$ $x_3 = abx_1$	$x_3 = abx_1$
3. Elimination of feedback loop		$a/(1\pm ab)$ $x_2 = \dfrac{a}{1\pm ab}$
		$a/(1+b)$ $x_2 = \dfrac{a}{1+ab}$
4. Miscellaneous		ac bc $x_4 = c(ax_1 + bx_2)$
		$ab/(1-bc)$

5.7 MASON'S GAIN FORMULA

A complex signal flow diagram can be simplified by repeatedly applying the signal flow graph reduction algebra and reducing it to a signal flow diagram containing a source and sink node. This is a time consuming and tedious process. Mason was the first to identify the inherent property of the signal flow diagrams and evolved a theorem popularly known as *Mason's gain formula*. The result can be obtained by inspection by the use of the Mason's theorem. The general expression for signal flow diagram as given by Mason is:

$$T = \frac{1}{\Delta} \sum P_k \Delta_k \qquad (5.10)$$

T = Overall transmittance of the system

Δ = 1 − (sum of gain of all individual loops) + (sum of the gain product of all possible combination of two non-touching loops) − (sum of the gain product of all possible combination of three non-touching loops) + (-----) − (------) + ------

P_k = Gain of the kth forward path

Δ_k = Same as Δ but formed by loops not touching the kth forward path.

Δ is know as the *determinant* of the signal flow diagram and Δ_k is called the *cofactor* of the forward path k.

5.8 CONCLUSION

In this chapter we described linear control systems with the help of transfer functions, block diagrams and signal flow diagrams. The dynamics of the control systems are described by input-output relationship called the transfer function. The block-diagram reduction rules, signal-flow graph reduction algebra and Mason's gain theorem are the means to achieve input-output relationship. The use is restricted to linear time-invariant systems. The initial conditions are not considered and intermediate variables get eliminated. The block-diagram and signal flow-graph representation technique will be utilized for representing the control systems in state-space in succeeding chapters.

KEY POINTS LEARNT

- Generalised definition of a control system.
- Classification of control systems
  - Based on state.
  - Based on the equations describing any control system.
  - Based on nature of signal flow.
  - Based on number of input-output signals.
- General classification
  - Open-loop control system.
  - Closed-loop control system. Also called feedback control system.
- Advantages/disadvantages of open and closed loop control systems.
- General representation of a closed-loop system.
- Elements of closed-loop system: *command input, reference input, reference input elements, error detector, actuating signal,* controlled *system or plant, controlled variable, disturbance, feedback elements, primary feedback signal.*
- Block diagram representation and block diagram reduction rules.
- Signal flow diagram and signal flow graph reduction rules, basic terminology-*node (source, sink), branch, transmittance, arrow, forward path, path gain, loop (self, non-touching), loop gain.*

◆ Definition of transfer function–depicts input-output relationship of control systems.
◆ Mason's Gain theorem

$$T = \frac{1}{\Delta}\sum P_k \Delta_k$$

PROBLEMS AND SOLUTIONS

Problem 5.1

Reduce the block diagram shown in Fig. 5.11 and find the input-output relationship ratio (C/R).

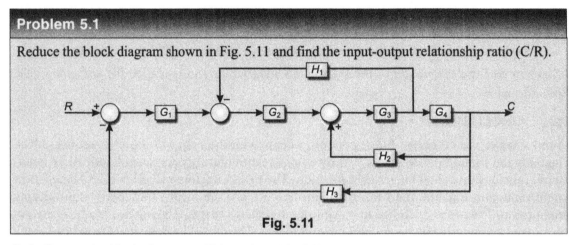

Fig. 5.11

Solution   The block diagram will be reduced in following steps:

(a) Moving take off point of block H_1 behind block G_4 (Rule 9)

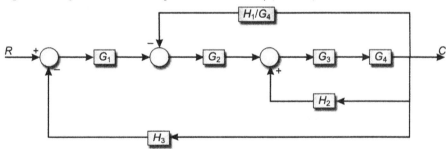

Fig. 5.12

(b) Combining blocks G_3 and G_4 (Rule 1) and eliminating the feedback loop formed by block H_2 (Rule 4).

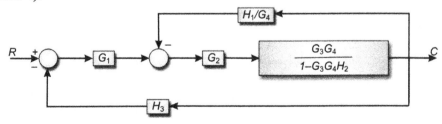

Fig. 5.13

(c) Combining blocks G_2 and $\dfrac{G_3G_4}{1-G_3G_4H_2}$ (Rule 1) and eliminating feedback loop formed by block H_1/G_4 (Rule 4).

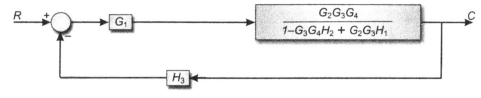

Fig. 5.14

(d) Combining blocks G_1 and $\dfrac{G_2G_3G_4}{1-G_3G_4H_2+G_2G_3H_1}$ (Rule 1) and eliminating feedback loop formed by block H_3 (Rule 4).

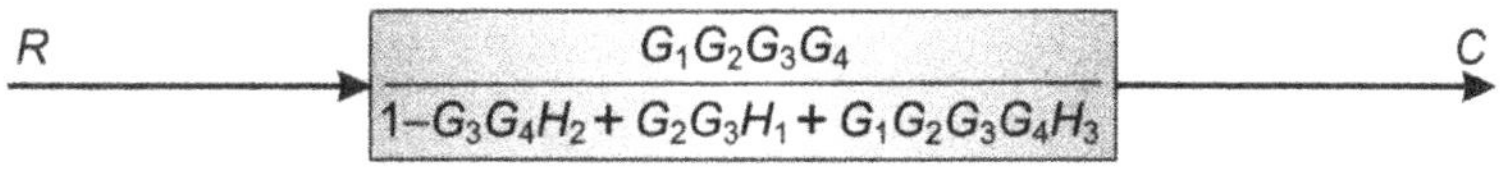

$$R \longrightarrow \boxed{\dfrac{G_1G_2G_3G_4}{1-G_3G_4H_2+G_2G_3H_1+G_1G_2G_3G_4H_3}} \longrightarrow C$$

Fig. 5.15

The input-output relationship C/R is

$$\frac{C}{R}=\frac{G_1G_2G_3G_4}{1-G_3G_4H_2+G_2G_3H_1+G_1G_2G_3G_4H_3}$$ **Ans.**

Problem 5.2

Convert the block diagram shown is Fig 5.16 to equivalent signal flow graph and reduce the signal flow graph using Mason's gain theorem and obtain C/R.

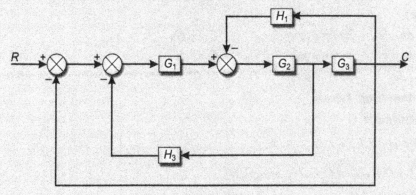

Fig. 5.16

Solution   Mark nodes a, b, c, d and e on the given block diagram as shown in Fig. 5.17

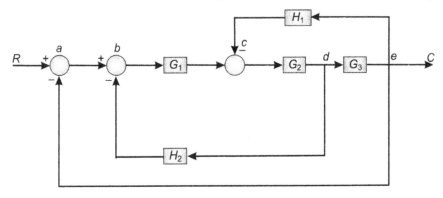

Fig. 5.17

Replace by signal flow diagram as shown is Fig. 5.18

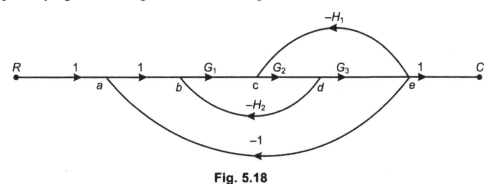

Fig. 5.18

(a) **Forward Paths**   There is only one forward path:

   • $R\,a\,b\,c\,d\,e\,C$ having gain $P_1 = G_1 G_2 G_3$

(b) **Loops**

   • $abcdea$ having gain $L_1 = -G_1 G_2 G_3$
   • $bcdb$ having gain $L_2 = -G_1 G_2 H_2$
   • $cdec$ having gain $L_3 = -G_2 G_3 H_1$

(c) **Non-touching Loops**   Nil

(d) **Determinant**

$$\Delta = 1 - (L_1 + L_2 + L_3)$$

$$= 1 + G_1 G_2 G_3 + G_1 G_2 H_2 + G_2 G_3 H_1$$

(e) **Cofactor Δ_1**

   The forward path P_1 touches all the three loops, hence

   $\Delta_1 = 1$

$$\frac{C}{R} = T = \frac{P_1\Delta_1}{\Delta} = \frac{G_1G_2G_3}{1 + G_1G_2G_3 + G_1G_2H_2 + G_2G_3H_1}$$ **Ans.**

Problem 5.3

Obtain the input-output relationship C/R for the signal flow diagram shown is Fig. 5.19

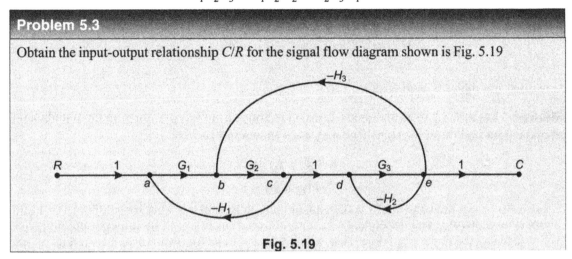

Fig. 5.19

Solution:

(a) Forward Path   There is only one forward path.

- $RabcdeC$ having gain $P_1 = G_1G_2G_3$

(b) Loops
- $abca$ having gain $L_1 = -G_1G_2H_1$
- ded having gain $L_2 = -G_3H_2$
- $bcdeb$ having gain $L_3 = -G_2G_3H_3$

(c) Non-touching loops   There is one combination of non-touching loop formed by L_1 and L_2 having gain product.

$$L_1 L_2 = (-G_1G_2H_1)(-G_3H_2) = G_1G_2G_3H_1H_2$$

(d) Determinant

$$\Delta = 1 - (L_1 + L_2 + L_3) + (L_1 L_2)$$
$$= 1 + G_1G_2H_1 + G_3H_2 + G_2G_3H_3 + G_1G_2G_3H_1H_2$$

(e) Cofactor

Since all the three loops touch the forward path P_1, hence
$$\Delta_1 = 1$$

$$T = \frac{C}{R} = \frac{P_1\Delta_1}{\Delta} = \frac{G_1G_2G_3}{1 + G_1G_2H_1 + G_3H_2 + G_2G_3H_3 + G_1G_2G_3H_1H_2}$$ **Ans.**

Problem 5.4

A system having input and output represented by u and x is described by the following equations

$$x = x_1 + t_3 u$$
$$\dot{x}_1 = q_1 x_1 + x_2 + t_2 u$$
$$\dot{x}_2 = -q_2 x_1 + t_1 u$$

Construct the signal flow diagram.

Solution Let u, x, x_1, x_2 be the nodes. Locate the nodes from left to right with input node u on left and output node x on the right. The nodes are shown in Fig 5.20

$$\dot{u} \quad \dot{x}_2 \quad \dot{x}_1 \quad \dot{x}$$

Fig. 5.20

(a) $x = x_1 + t_3 u$ indicates that x is dependent on node x_1 and u having transmittance of 1 and t_3 respectively. The dependence is superimposed on Fig 5.20 by drawing the branches between the node x and dependent nodes u and x_1. The transmittances are marked on the arrows as shown in Fig 5.21

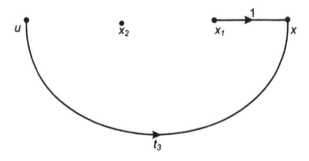

Fig. 5.21

(b) $x_1 = q_1 x_1 + x_2 + t_2 u$ indicates dependence of x_1 on nodes x_1, x_2 and u. The dependence is shown by connecting node x_1 with nodes x_1, x_2 and u as shown in Fig. 5.22(a). The transmittance q_1, 1 and t_2 are then marked on the respective branches as shown in Fig. 5.22 (b)

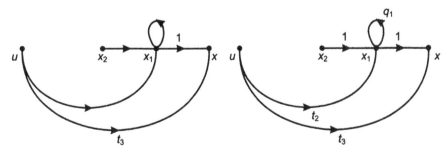

Fig. 5.22(a)   **Fig. 5.22(b)**

(c) $x_2 = -q_2 x_1 + t_1 u$ indicates dependence of node x_2 on nodes x_1 and u with transmittance of $-q_2$ and t_1 respectively. this is achieved by connecting node x_1 with x_2 with a branch and transmittance $-q_2$ marked on it and then connect node u with x_2 with transmittance t_1 marked on it. The final signal flow diagram is shown in Fig. 5.23.

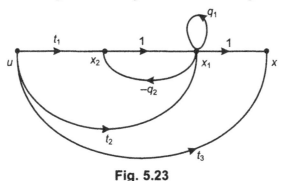

Fig. 5.23

Problem 5.5

Block diagram of an open-loop transfer function is shown in Fig. 5.24 where

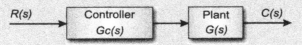

Fig. 5.24

The transfer function of the controller and plant is $\dfrac{s+2}{s+3}$ and $\dfrac{s+1}{s^2+6s+8}$ respectively.

Write a MATLAB program to find the transfer function.

```
'Problem 5.5'                          %display enclosed text.
'Controller'                           %display enclosed text.
Nc=[1 2]                                %input numerator of controller.
Dc=[1 3]                                %input denominator of controller.
G1=tf(Nc,Dc)                           %create transfer function of
                                        %controller in G1.
'plant'                                 %display enclosed text.
Np=[1 1]                                %Input numerator of plant.
Dp=[1 6 8]                              %form denominator of plant.
G2=tf(Np,Dp)                           %create transfer function of plant
                                        %in G2.
'Open-Loop Transfer Function'          %display enclosed text.
G=series(G1,G2)                        %combine G1 and G2.
G=minreal(G)                           %cancel common poles and zeros.
'Alternatively'                        %display enclosed txt.
Galt=G1*G2                             %multiply G1 and G2 being in
                                        %series
Galt=mineral(Galt)                     %cancel common poles and zeros.
```

```
Output

Problem 5.5
Controller
Nc=
   1 2
Pc=
   1 3
Transfer function:
s+2
───
s+3
Plant
Np=
   1 1
Dp=
   1 6 8
Transfer function:
s+1
─────────
s^2+6s+8
Open-Loop Transfer Function
Transfer function:
s^2+3s+2
──────────────────
s^3+9s^2+26s+24
Transfer function:
s+1
──────────
s^2+7s+12
Alternatively
Transfer function:
s^2+3s+2
──────────────────
s^3+9s^2+26s+24
Transfer function
s+1
──────────
s^2+7s+12
```

Problem 5.6

Find the transfer function of the control system represented by block diagram in Fig. 5.25 by using MATLAB.

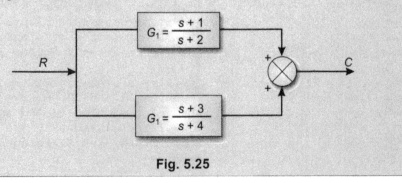

Fig. 5.25

Solution MATLAB program is given below:

```
'Problem 5.6'                              %display enclosed text.
'Input Transfer Function G1'               %display enclosed text.
N1=[1 1]                                   %input numerator of G1
D1=[1 2]                                   %input denominator of G2
G1=tf(N1,N2)                               %create transfer function G1
'Input Transfer Function G2'               %display enclosed text.
N2=[1 3]                                   %input denominator of G2
D2=[1 4]                                   %input numerator of controller.
G2=tf(N2,D2)                               %create transfer function G2
'Transfer Function Control System G=C/R/P  %display enclosed text.
G=Parllel(G1,G2)                           %combine G1&G2.
G=minreal(G)                               %cancel common poles & zeros.
'Alternatively'                            %display enclosed text.
Galt= G1+G2                                %cancel G1&G2
Galt=minreal(Galt)                         %cancel common poles &
                                           %zeros.


Output

Problem 5.6
Input Transfer Function G1
N1=
   1 1
D1=
   1 2

Transfer function
s+1
───
s+2
Input Transfer Function G2
N2=
   1 3
D2=
   1 4
Transfer function
s+3
───
s+4
Transfer Function Control System G=C/R
Transfer function:
s^2+4s+3
────────
s^2+6s+8
Transfer function:
s^2+4s+3
────────
s^3+6s+8
Alternatively
Transfer function:
s^2+4s+3
────────
s^2+6s+8
Transfer function:
s^2+4s+3
────────
s^2+6s+8
```

Problem 5.7

Find the transfer function of the closed-loop system shown in Fig. 5.26 by using MATLAB.

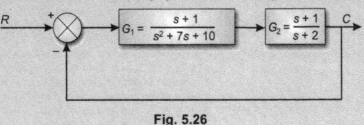

Fig. 5.26

Solution MATLAB program is given below:

```
'Problem 5.7'                       %display enclosed text.
'Input Transfer Function G1'        %display enclosed text.
N1=[1 2]                            %input numerator
D1=[1 7 10]                         %input denominator of
G1=tf(N1,D1)                        %create transfer function
                                    %in G1.

'Input Transfer Function G2'        %display enclosed text.
N2=[1 1]                            %input numerator
D2=[1 2]                            %input denominator
G2=tf(N2,D2)                        %create transfer function
                                    %in G2.

'Forward Path Transfer function     %display enclosed text.
G12=series(G1,G2)                   %combine G1 and G2.
G=feedback(G12,1)                   %combine the feedback loop.
G=minreal(G)                        %cancel common poles and
                                    %zeros.

'Alternatively'                     %display enclosed text.
Galt=G1*G2                          %multiply G1 and G2 being
                                    %in series
Galt1=feedback(Galt,1)             %combine the feedback loop.
Galt1=minreal(Galt1)               %cancel common poles and
                                    %zeros.
```

Output

```
Proble 5.7
Input Transfer Function G1
N1=
   1 2
D1=
   1 7 10

Transfer function
s+2
s^2+7s+10
Input Transfer function G2
```

```
N2=
   1 1
D2=
   1 2
Transfer function
s+1
s+2
Forward Path Transfer function
Transfer function:
s^2+3s+2
s^3+9s^2+24s+20

  Transfer function:
  s^2+3s+2
  s^3+10s^2+27s+22
  Transfer function:
  s+1
  s^2+8s+11
  Alternatively
  Transfer function
  s^2+3s+2
  s^3+9s^2+24s+20
  Transfer function:
  s^2+3s+2
  s^3+10s^2+27s+22
  Transfer function:
  s+1
  s^2+8s+11
```

REVIEW EXERCISE

1. Define Control system. Is its use restricted to one field of operation.

2. Cite three applications of control system.

3. What are the methods by which control system can be classified.

4. How are the control systems classified in general.

5. Define open-loop control system. What are its advantages/disadvantages.

6. Define closed-loop control system. What are its advantages/ disadvantages.

7. What are the reasons for using feedback control system.

8. Draw a feedback control system and label its elements on it.

9. How are the control systems represented pictorially.

10. Draw block-diagram of an open-loop control system.

11. Draw block-diagram of a closed-loop system

- Define transfer function
- Describe transfer function relationship

12. What are the advantages of using Masons gain formula over block-diagram approach. Explain the formula.

13. What is a node. Name different types of nodes.

14. Define loop. Name different types of loops.

15. Using block diagram reduction technique, find the input-output relationship:

(a)

(b)

(c)

(d)

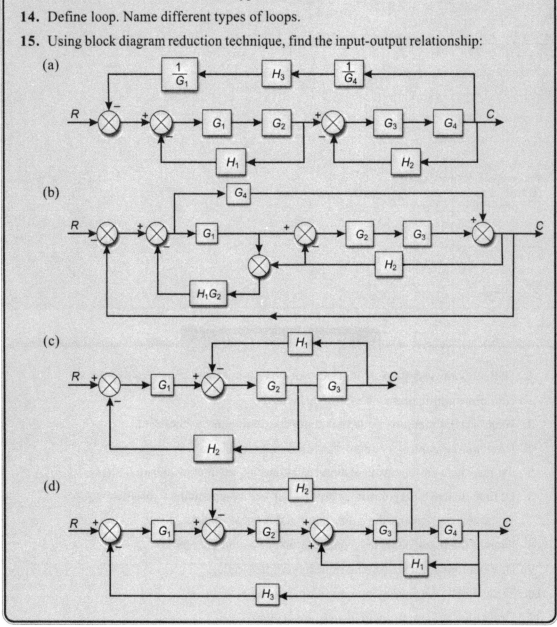

16. Use Mason's gain formula to determine the input-output relationship

(a)

(b)

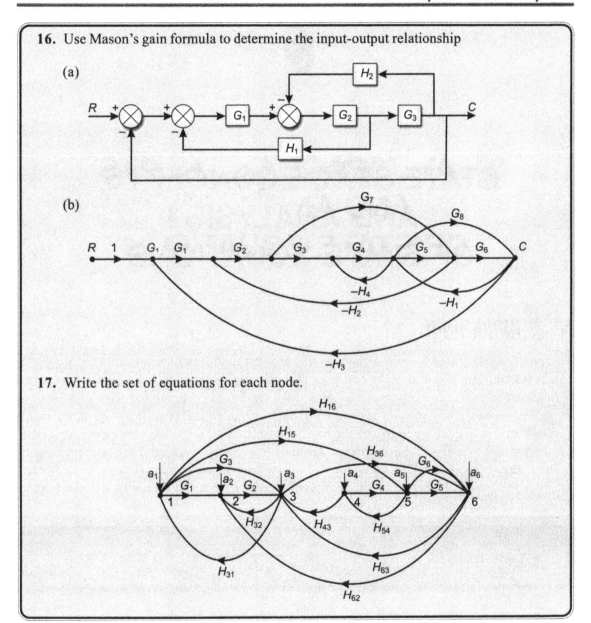

17. Write the set of equations for each node.

6

STATE SPACE CONCEPTS
AND ANALYSIS
OF STATE-EQUATIONS

6.1 INTRODUCTION

In chapter 1, we discussed various methods available to the control engineer to study and analyse a control system. The classical technique uses the *convolution integral* and *Laplace transform* approach to analyse control systems under zero initial conditions i.e. in relaxed state. Convolution integral technique is a *time-domain approach* and Laplace transform method is a *frequency domain approach*. In this chapter, we introduce the basic concepts linked with state-space method; a time domain approach to analyse control systems. Advantages using this technique have been spelt out in chapter 1. This approach simplifies the problem of solving multivariable control systems. Higher-order systems represented by higher-order differential equations is analysed by forming a set of first-order differential equations and then solving them.

LEARNING OBJECTIVES

- ◆ Understand basic concepts associated with state-space analysis-state, state-variables, state-vector, state-trajectories, state-equation, state-transition matrix and state-transfer function.
- ◆ Learn the importance and properties of state-transition matrix.
- ◆ Obtain the solution of state-equations-homogenous and non-homogenous and learn the meaning of free and forced response.
- ◆ Understand the significance of characteristic equation, eigenvalues and eigenvectors.

6.2 BASIC TERMS

In this section, we will commence discussion on the various terms associated with state variable analysis of control systems.

State *The state of a control system at time t = t_0 is the smallest set of variables called state variables such that the information of these state variables along with the information of the input at t = t_0 is sufficient to determine the output dynamics of the system at $t \geq 0$.*

State Variables The state variables of a dynamic control system *are the variables which constitute the smallest set of variables that determine the state of the system.* The state variables describe the output response of a control system for specified inputs and the existing state. Number of variables selected should be sufficient enough to describe the dynamics of the system. The variables chosen need not have any dimension for measurement and may not be observable. However, measurable and observable variables, present added advantage in the analysis of a control system.

Let us consider a spring-mass-damper system shown in Fig. 6.1. Application of force $f(t)$ produces

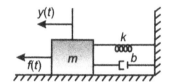

Fig. 6.1 *Spring-mass-damper system*

displacement $y(t)$ of mass m. The mass is assumed to be stationary before application of the force $f(t)$. Application of Newton's law give:

$$\frac{md^2 y(t)}{dt^2} + \frac{bdy(t)}{dt} + Ky(t) = f(t) \tag{6.1}$$

or
$$m\ddot{y}(t) + b\dot{y}(t) + ky(t) = f(t) \tag{6.2}$$

This second-order differential equation can be written as a set of two first-order differential equations. To achieve this we define

$$x_1(t) = y(t) \quad \text{and} \quad x_2(t) = \dot{y}(t) \tag{6.3}$$

resulting into
$$\dot{x}_1(t) = \dot{y}(t) = x_2(t) \tag{6.4}$$

and
$$\dot{x}_2(t) = \ddot{y}(t) = \frac{-b}{m}\dot{y}(t) + \frac{-k}{m}y(t) + \frac{1}{m}f(t) \tag{6.5}$$

or
$$\dot{x}_2(t) = \frac{-b}{m}x_2(t) + \frac{-k}{m}x_1(t) + \frac{1}{m}f(t) \tag{6.6}$$

Equations
$$\dot{x}_1(t) = x_2(t) \quad \text{and} \quad \dot{x}_2(t) = \frac{-b}{m}x_2(t) - \frac{k}{m}x_1(t) + \frac{1}{m}f(t) \tag{6.7}$$

are a set of first-order differential equations which replace the second-order differential equation (6.1) for describing the dynamics of mass-spring-damper system shown in Fig. 6.1. These equations, together with the initial conditions of the system represented by $[x_1(t_0), x_2(t_0)]$ can be utilised for determining the output response of the system and its future behaviour. Hence, *the set of two numbers $x_1(t_0)$ and $x_2(t_0)$, are defined as the state of the system at time t_0 and the variables $x_1(t)$ and $x_2(t)$ qualify as state variables* of the system.

State Vector   The *state vector* of a system is *column vector* that represents the *state* of the system. If a system needs n state variables to describe the dynamics of a system, then these n state variables are made to constitute a column vector $x(t)$ that represents the state of the system. A *state vector* is thus defined as *a column vector $x(t)$ that describes the state of system for $t \geq t_0$*. In the example considered above,

$$x(t) = \begin{bmatrix} x_1(t) \\ x_2(t) \end{bmatrix} \tag{6.8}$$

is the *state vector*. A state vector is *composed* of state variables. The state variables are not unique and with the knowledge of the input, output of the system can be computed.

State Space   *If $x_1(t)$, $x_2(t), \cdots, x_n(t)$ are the minimum number of state variables which are necessary to describe the dynamics of a control system, then the n-dimensional space whose coordinate axis consist of x_1-axis, x_2-axis $\cdots x_n$-axis is called a state-space.* For the example under consideration, the state-space would consist of $x_1(t)$ axis and $x_2(t)$ axis. The state-space for the system is shown in Fig. 6.2

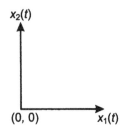

Fig. 6.2   *State-space of mass-spring-damper system*

6.3 TRAJECTORIES IN STATE SPACE

A trajectory is a plot of the n state variables in the n-dimensional state-space with time being the implicit variable. The trajectory starts at time $t = t_0$ and is traced till the final time $t > t_0$. The plot tracing the trajectory can be easily visualised geometrically for lower-order systems. Plotting of the trajectory for higher-order systems is difficult.

Example 6.1   Consider a second-order system represented by a set of first-order differential equations as given below.

$$x_1(t) = 2 - \frac{1}{2} e^{-t} + \frac{1}{4} e^{-3t}$$

$$x_2(t) = \frac{1}{2} e^{-t} + \frac{1}{2} e^{-3t}$$

where $x_1(t)$ and $x_2(t)$ are the state variables. Plot the trajectory in the state-space starting from $t = 0$ to $t = 3$ seconds

Solution

Since there are two state variables, the state-space would be two dimensional having $x_1(t)$ axis and $x_2(t)$ axis. Let us tabulate the values of state variables.

time t	$x_1(t)$	$x_2(t)$
0	1.75	0.25
1	1.8	0.2
2	1.9	0.07
3	2.0	0.02

The state trajectory is plotted in Fig. 6.3.

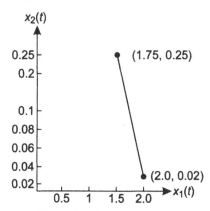

Fig. 6.3   *State trajectory*

6.4 NONLINEARITIES AND LINEARISATION

A system is considered to be *linear* if it follows the *principle of superposition*. A *relaxed system* when subjected to two inputs separately i.e. $u_1(t)$ and $u_2(t)$ records the outputs $c_1(t)$ and $c_2(t)$ respectively, then the system would be considered to obey the principle of superposition if the combined input $u(t) = a_1 u_1(t) + a_2 u_2(t)$ results in the output $c(t) = a_1 c_1(t) + a_2 c_2(t)$. In fact, principle of superposition encompasses the properties of *additivity* and *homogeneity*.

Any system which does not follow the principle of superposition is considered as a *nonlinear system*. A linear system is difficult to be realised practically. All systems exhibit nonlinearities due to inherent physical characteristics and properties of the elements forming part of the system. A servo-amplifier exhibits *saturation effect* at high input voltages. Dampers have *stiction* and *square law nonlinearity*. Gears exhibit *backlash*, motors possess *dead zone* nonlinearity as the frictional forces do not let a motor respond to lower input voltages. An iron-cored coil in electrical circuits exhibit linearity over a range of input voltage and current. However, *hysteresis* and *saturation* result in nonlinear behaviour. Nonlinear behaviours of different types is plotted in Fig. 6.4. Nonlinearities may not be considered as an evil by control engineers and may be introduced during the design of the control system, intentionally, to achieve the desired goals *e.g.*

Damping is many times introduced to obtain desired amplitude dependent time response.

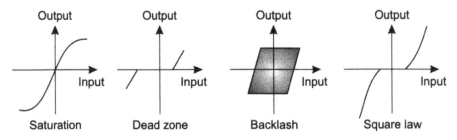

Fig. 6.4   *Nonlinear behaviour curves due to various nonlinearities*

Linearisation is approximation of a nonlinear system by a linear model within a desired or operating range so that complicated solution of a nonlinear system can be realised mathematically. The method involves formulation of a *nonlinear equation* for the control system and then selecting an *operating point*. The next step is to linearise the nonlinear differential equation formulated earlier at or around the operating point. This yields an approximate but a functional model which helps the control engineer to study, experiment, analyse and design a satisfactory control system. The range around the operating point should be so chosen that the desired performance of the control system is achieved. The ultimate aim is to be obtain a stable control system.

A Taylor series expansion about the operating point x_0 brings about the linear approximation to the nonlinear function. The tangent line to the curve at the operating point makes the nonlinear system practically linear in its neighborhood. The first two terms of the Taylor series expansion of $y = f(x)$ sufficiently gives linear approximation about x_0.

$$y = f(x_0) + \left[\frac{df}{dx}\bigg|_{x=x_0}\right] x - x_0$$

6.5   STATE EQUATION

State-equation is a set of n first-order simultaneous differential equations having n-state variables. Let us consider a linear, time-invariant control system. It is also assumed that u and y are the time functions depicting input and output respectively.

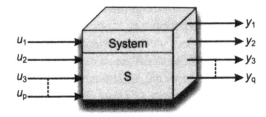

Fig. 6.5   *Control system with multivariable input and output*

Considering that the starting time is t_o, the initial state of the system refers to its initial condition at t_0. The system is subjected to inputs u from t_0 to T. The outputs y of the system, hence depends upon the inputs u and the initial state of the system at t_0.

The system can be represented by n^{th} order differential equation

$$a_n \frac{d^n y}{dt^n} + a_{n-1} \frac{d^{n-1}}{dt^{n-1}} y^{n-1} + \cdots + a_o y = u \tag{6.9}$$

where all coefficients are constants and $a_n \neq 0$.

It is assumed that details of input u are known and from the initial conditions at $t_0 = 0$, the values of $y(0), \dfrac{dy(0)}{dt}, \cdots \dfrac{d^{n-1}y(0)}{dt^{n-1}}$ can be ascertained, then the dynamic behaviour of the system can be completely determined by selecting a set of n state-variables as given below:

$$\left.\begin{aligned} x_1 &= y \\ x_2 &= \dot{y} \\ &\cdots \\ x_n &= \overset{(n-1)}{y} \end{aligned}\right| \tag{6.10}$$

Equation. (6.9) can then be represented by a set of first order differential equations as given below

$$\dot{x}_1 = x_2 = \dot{y}$$

$$\dot{x}_2 = x_3 = \ddot{y}$$

$$\cdots$$

$$\dot{x}_{n-1} = x_n = \overset{(n-1)}{y}$$

$$\overset{(n)}{y} = \dot{x}_n = \frac{1}{a_n}(u - a_0 x_1 - \cdots - a_{n-1} x_n)$$

or

$$\dot{x}_n = \frac{-a_0}{a_n} x_1 - \cdots - \frac{a_{n-1}}{a_n} x_n + \frac{1}{a_n} u \tag{6.11}$$

Expressing equation (6.11) in matrix form, we get state-equation

$$\dot{x} = Ax + Bu \quad \text{and} \quad y = Cx \tag{6.12}$$

For a nth order single-input single-output system

$$x = \text{state vector } (n \times 1) = \begin{bmatrix} x_1 \\ x_2 \\ \vdots \\ x_{n-1} \\ x_n \end{bmatrix}$$

$$A = \text{system matrix } (n \times n) = \begin{bmatrix} 0 & 1 & 0 & \cdots & 0 \\ 0 & 0 & 1 & \cdots & 0 \\ \vdots & \vdots & \vdots & \cdots & \vdots \\ 0 & 0 & 0 & \cdots & 1 \\ -a_0 & -a_1 & -a_2 & \cdots & -a_{n-1} \\ \overline{a_n} & \overline{a_n} & \overline{a_n} & & \overline{a_n} \end{bmatrix} \tag{6.13}$$

$$B = \text{driving matrix } (n \times 1) = \begin{bmatrix} 0 \\ 0 \\ \vdots \\ \dfrac{1}{a_n} \end{bmatrix}$$

$$\dot{x} = \text{state variable vector } (n \times 1) = \begin{bmatrix} \dot{x}_1 \\ \dot{x}_2 \\ \vdots \\ \dot{x}_{n-1} \\ \dot{x}_n \end{bmatrix}$$

$u = \text{scaler input } (p \times 1)$

The output is given by

$$y = \begin{bmatrix} 1 & 0 & 0 & \cdots & 0 \end{bmatrix} \begin{bmatrix} x_1 \\ x_2 \\ x_3 \\ \vdots \\ x_n \end{bmatrix} \tag{6.14}$$

where $\qquad y = Cx$

and $\qquad C = \text{output matrix } (1 \times n) = \begin{bmatrix} 1 & 0 & 0 & \cdots & 0 \end{bmatrix}$; and

$y = \text{scaler output } (1 \times 1)$

Thus, the solution of the equation (6.12) is uniquely determined by the knowledge of initial state x_0 together with the input $u = (t_0, T)$ and hence $x(t)$ qualifies to be termed as state vector for A.

The general state-equation and output-equation for a nth order multivariable control system depicted in Fig. 6.6 with p inputs and q outputs are given by

$$\dot{x} = Ax + Bu \qquad\qquad 6.15\,(a)$$
$$y = Cx + Du \qquad\qquad 6.15\,(b)$$

The output-equation has another term Du. This comes into existence because of the fact that each output may interact with the other inputs.

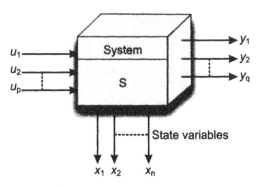

Fig. 6.6 *Multivariable control system*

x = state vector of n elements $(n \times 1)$

y = output vector of q elements $(q \times 1)$

u = input vector of p elements $(p \times 1)$

A = *system* matrix or *companion* matrix or *coefficient* matrix or *evolution* matrix $(n \times n)$

B = *driving* matrix or *control* matrix $(n \times p)$

C = *output* matrix or *observation* matrix $(q \times n)$

D = *transmission* matrix $(q \times p)$

$\dot{x}$ = state-variable vector $(n \times 1)$

6.6 STATE EQUATION: HOMOGENOUS AND NON-HOMOGENOUS PARTS

The general form of the state-equation derived in Section (6.5) is

$$\dot{x}(t) = Ax(t) + Bu(t) \quad \text{or} \quad \dot{x} = Ax + Bu \tag{6.16}$$

$Ax(t)$ is termed as the *homogenous* part of the state-equation and $Bu(t)$ is called the *forcing function* or the *non-homogenous* part. If the input $u(t) = 0$, the state-equation can be expressed as

$$\dot{x}(t) = Ax(t) \quad \text{or} \quad \dot{x} = Ax \tag{6.17}$$

The analysis of the state-equation (6.17) yields the *unforced response* of the control system and is governed by the initial conditions only. If the input $u(t)$ is specified and is not equal to zero, the response obtained is termed as *forced response*

6.7 SOLUTION OF NON-HOMOGENEOUS STATE EQUATIONS

Let us consider the scaler case

$$\dot{x}(t) = ax(t) + bu(t) \tag{6.18}$$

where $x(t)$ and $u(t)$ are scaler variables and a and b are scaler coefficients respectively. Assuming a and b to be constants and initial conditions at $t = t_0$, equation (6.18) can be written as

$$\dot{x}(t) - ax(t) = bu(t) \tag{6.19}$$

Premultiplying both sides by e^{-at}, we get

$$e^{-at}[\dot{x}(t) - ax(t)] = e^{-at}bu(t) \tag{6.20}$$

Now $\quad \dfrac{d}{dt}[e^{-at}x(t)] = e^{-at}\dfrac{d}{dt}x(t) - ax(t)e^{-at} \tag{6.21}$

$$= e^{-at}\dot{x}(t) - ax(t)e^{-at} \tag{6.22}$$

$$= e^{-at}[\dot{x}(t) - ax(t)] \tag{6.23}$$

Substituting the value of $e^{-at}[\dot{x}(t) - ax(t)]$ from equation (6.23) in equation (6.20), we get

$$\frac{d}{dt}[e^{-at}x(t)] = e^{-at}bu(t) \tag{6.24}$$

or $\qquad e^{-at}x(t) = \displaystyle\int_{-\infty}^{t} be^{-at}u(t) + C_1 \tag{6.25}$

where C_1 is the constant of integration. Considering initial conditions defined for time $t = t_0$, equation (6.25) can be written as

$$e^{-at}x(t) = \int_{-\infty}^{t_0} e^{-a\tau}bu(\tau)d\tau + \int_{t_0}^{t} e^{-a\tau}bu(\tau)d\tau + C_1 \tag{6.26}$$

or $\qquad e^{-at}x(t) = C_2 + \displaystyle\int_{t_0}^{t} e^{a\tau}bu(\tau)d\tau + C_1 \tag{6.27}$

Where C_2 is a definite intergral between the limits $-\infty$ to t_0 is a function of initial condition. Combining the constants C_1 and C_2 as

$$C = C_1 + C_2 \tag{6.28}$$

Equation (6.27) can be written as

$$e^{-at}x(t) = C + \int_{t_0}^{t} e^{-a\tau}bu(\tau)d\tau \tag{6.29}$$

or $\qquad x(t) = e^{at}C + \displaystyle\int_{t_0}^{t} e^{a(t-\tau)}bu(\tau)d\tau \tag{6.30}$

At $t = t_0$ we have $x(t) = x(0)$. Substituting in equation (6.30), we get

$$x(0) = e^{at_0}C + \int_{t_0}^{t_0} e^{a(t-\tau)}bu(\tau)d\tau \tag{6.31}$$

or $\qquad C = e^{-at_0}x(0) \tag{6.32}$

Substituting the value of C from equation (6.32) in equation (6.30), we get

$$x(t) = e^{at}e^{-at_0}x(0) + \int_{t_0}^{t} e^{a(t-\tau)}bu(\tau)d\tau \tag{6.33}$$

$$x(t) = \underbrace{e^{a(t-t_0)}x(0)}_{\substack{\text{Free Response} \\ \text{(due to initial conditions)}}} + \underbrace{\int_{t_0}^{t} e^{a(t-\tau)}bu(\tau)d\tau}_{\substack{\text{Forced Response} \\ \text{(due to input } u(t))}} \tag{6.34}$$

The response obtained in equation (6.34) consists of response due to initial conditions (*free response*) and the response due to the input $u(t)$ (*forced response*).

The analysis presented in this Section shall be the basis for analysis of the first-order matrix differential equation. The aim is to solve for $y(t)$ for time $t > t_0$, considering that input $u(t)$ is known and state $x(t_0)$ is specified. The solution for state $x(t)$ yields solution of output $y(t)$ by application of simple matrix algebra.

6.8 SOLUTION OF NON-HOMOGENOUS TIME-INVARIANT MATRIX DIFFERENTIAL EQUATION

Let us consider the equation

$$\dot{x}(t) = Ax(t) + Bu(t) \text{ with } x(t_0) = x(0) \tag{6.35}$$

where

$x(t)$ = column vector $(n \times 1)$

$u(t)$ = input column vector $(m \times 1)$

A = matrix $(n \times n)$

B = matrix $(n \times m)$

Rearranging equation (6.35) as

$$\dot{x}(t) - Ax(t) = Bu(t) \tag{6.36}$$

Premultiplying equation. (6.36) by e^{-At}, we get

$$e^{-At}(\dot{x}(t) - Ax(t)) = e^{-At}Bu(t) \tag{6.37}$$

The right hand side of equation (6.37)

$$= \frac{d}{dt}(e^{-At}x(t)) \tag{6.38}$$

and is obtained as given below. We known that

$$e^{At} = I + At + \frac{1}{2!}A^2t^2 + \frac{1}{3!}A^3t^3 + \dots \tag{6.39}$$

$$\frac{d}{dt}e^{At} = A + A^2 t + \frac{1}{2!}A^3 t^2 + \frac{1}{3!}A^4 t^3 + ... \tag{6.40}$$

$$= A(I + At + \frac{1}{2!}A^2 t^2 + \frac{1}{3!}A^3 t^3 + ...) \tag{6.41}$$

$$= Ae^{At} \tag{6.42}$$

or $$\frac{d}{dt}e^{At} = (I + At + \frac{1}{2!}A^2 t^2 + \frac{1}{3!}A^3 t^3 + ...)A \tag{6.43}$$

$$= e^{At}A \tag{6.44}$$

Now $$\frac{d}{dt}[e^{At}x(t)] = e^{At}\left(\frac{d}{dt}x(t)\right) + \left(\frac{d}{dt}e^{At}\right)x(t) \tag{6.45}$$

$$= e^{At}\dot{x}(t) + e^{At}Ax(t) \tag{6.46}$$

Similarly $$\frac{d}{dt}[e^{-At}x(t)] = e^{-At}\dot{x}(t) - e^{-At}Ax(t) \tag{6.47}$$

Substituting the value of $e^{-At}\dot{x}(t) - e^{-At}Ax(t)$ from equation (6.47) in equation (6.37) (6.38) we get

$$\frac{d}{dt}[e^{-At}x(t)] = e^{-At}Bu(t) \tag{6.48}$$

or $$e^{-At}x(t)\Big|_0^t = \int_0^t e^{-A\tau}Bu(\tau)d\tau \tag{6.49}$$

or $$e^{-At}x(t) - x(0) = \int_0^t e^{-A\tau}Bu(\tau)d\tau \tag{6.50}$$

Premultiplying both sides by e^{At}, we get

$$x(t) = e^{At}x(0) + \int_{t_0}^t e^{A(t-\tau)}Bu(\tau)d\tau \tag{6.51}$$

If the initial state is known at $t = t_0$ equation (6.49) can be written as

$$e^{-At}x(t)\Big|_{t_0}^t = \int_{t_0}^t e^{-A\tau}Bu(\tau)d\tau \tag{6.52}$$

$$e^{-At}x(t) - e^{-At}x(t_0) = \int_{t_0}^t e^{-A\tau}Bu(\tau)d\tau \tag{6.53}$$

or $$\boxed{x(t) = e^{A(t-t_0)}x(t_0) + \int_{t_0}^t e^{A(t-\tau)}Bu(\tau)d\tau} \tag{6.54}$$

Output Response

The output equation is given as

$$y(t) = Cx(t) + Du(t) \tag{6.55}$$

where $y(t)$ = Output

C = Constant matrix

D = Constant matrix

Substituting the value of $x(t)$ from equation. (6.54) in equation. (6.55), we have

$$y(t) = C\left[e^{A(t-t_0)}x(t_0) + \int_{t_0}^{t} e^{A(t-\tau)} Bu(\tau)d\tau\right] + Du(t) \tag{6.56}$$

$$= Ce^{A(t-t_0)}x(t_0) + \int_{t_0}^{t} Ce^{A(t-\tau)} Bu(\tau)d\tau + Du(t) \tag{6.57}$$

If the initial state $x(0)$ is at $t_0 = 0$, then the output response $y(t)$, obtained from equation. (6.57) is given as

$$\boxed{y(t) = Ce^{At}x(0) + \int_{0}^{t} Ce^{A(t-\tau)} Bu(\tau)d\tau + Du(t)} \tag{6.58}$$

6.9 SOLUTION OF HOMOGENOUS TIME-INVARIANT MATRIX DIFFERENTIAL EQUATION

The general form of homogenous time-invariant equation is given as

$$\dot{x}(t) = Ax(t) \tag{6.59}$$

The solution to the equation (6.59) is obtained by putting $u(t) = 0$ in equation. (6.54)

$$\dot{x}(t) = e^{A(t-t_0)}x(t_0) \tag{6.60}$$

If initial time is specified as $t_0 = 0$ and $x(t_0) = x(0)$, then

$$\dot{x}(t) = e^{At}x(0) \tag{6.61}$$

6.10 STATE-TRANSITION MATRIX

We have seen in the preceding section that response of the system is obtained in terms of the input function and the initial state of the system. The transition matrix relates the initial state of the system to a state after time t considering that the input is zero. The *state transition matrix is defined as a matrix that satisfies the homogenous part of the state-equation represented by equation. (6.62)*

$$\dot{x}(t) = Ax(t) \tag{6.62}$$

It is denoted as $\phi(t)$ and is also defined as a $n \times n$ matrix such that $x(t) = \phi(t, t_0)$ yields the solution of the state-equation $\dot{x}(t) = Ax(t)$ where $\phi(t, t_0)$ is the denotation used for state-transition matrix and $x(t_0)$ is the initial condition. The state-transition matrix describes the transition of the state from time t_0 to t with input $u(t) = 0$. It represents the *free* response and hence is governed by the initial condition only. It is also called the *unforced* response.

Mathematically

$$\dot{x}(t) = Ax(t) \tag{6.63}$$

or

$$\frac{dx(t)}{dt} = Ax(t) \tag{6.64}$$

since state-transitions matrix has to satisfy the above equation

$$\frac{d\phi(t)}{dt} = A\phi(t) \tag{6.65}$$

Laplace transform of equation. (6.64) yields

$$s X(s) - x(0) = AX(s) \tag{6.66}$$

or

$$X(s) = (sI - A)^{-1} x(0)$$

where $(sI - A)$ is a *nonsingular* matrix. Inverse Laplace transform of equation. (6.66) gives

$$x(t) = L^{-1}[(sI - A)^{-1}] x(0) \quad t \geq 0 \tag{6.67}$$

If $x(0)$ denotes the initial state at $t = 0$, then $\phi(t)$ is defined as

$$x(t) = \phi(t) x(0) \tag{6.68}$$

where

$$\phi(t) = L^{-1}[(sI - A)^{-1}]$$

Alternatively, state-transition matrix is also defined as

$$\phi(t) = e^{At} \tag{6.69}$$

This is derived as given below

$$\dot{x}(t) = Ax(t)$$

or

$$\frac{dx(t)}{dt} = Ax(t) \tag{6.70}$$

where

$$x(t) = n \text{ order vector and}$$
$$A = n \times n \text{ constant matrix}$$

Assuming a Taylor series expansion as a power series in t.

$$x(t) = a_0 + a_1 t + a_2 t^2 + \cdots + a_k t^k + \cdots \tag{6.71}$$

substituting in equation. (6.70) we get

$$a_1 + 2a_2 + 3a_3 t^2 + \ldots \ldots = A(a_0 + a_1 t + a_2 t^2 + \ldots \ldots) \tag{6.72}$$

Equating the coefficients for equal power of t on both sides of equation (6.72),

$$a_1 = Aa_0$$
$$2a_2 = Aa_1 \qquad (6.73)$$
$$3a_3 = Aa_2$$

Equation. (6.73) yields

$$a_1 = A a_0 \qquad (6.74)$$

$$a_2 = \frac{1}{2}Aa_1 = \frac{1}{2}AAa_0 = \frac{1}{2}A^2 a_0$$

$$a_3 = \frac{1}{3}Aa_2 = \frac{1}{3}A\frac{1}{2}A^2 a_0 = \frac{1}{6}A^3 a_0$$

$$\cdots\cdots\cdots$$

$$a_k = \frac{1}{k!}A^k a_0$$

Rewriting equation (6.71)

$$x(t) = a_0 + a_1 t + a_2 t^2 + \cdots + a_k t^k + \cdots$$

Substituting $t = 0$, we get

$$x(0) = a_0 \qquad (6.75)$$

Substituting $a_0 = x(0)$ is equation (6.74), we get

$$a_1 = Ax(0)$$

$$a_2 = \frac{1}{2}A^2 x(0)$$

$$a_3 = \frac{1}{6}A^3 x(0)$$

$$\cdots\cdots\cdots \qquad (6.76)$$

$$a_k = \frac{1}{k!}A^k x(0)$$

Substituting in equation (6.71), value from equation (6.75) & equation (6.76), we get

$$x(t) = x(0) + Ax(0)t + \frac{1}{2}A^2 x(0)t^2 + \frac{1}{6}A^3 x(0) t^3 + \cdots + \frac{1}{k!}A^k x(0) t^k + \cdots \qquad (6.77)$$

or

$$x(t) = \left(I + At + \frac{1}{2!}A^2 t^2 + \frac{1}{3!}A^3 t^3 + \cdots + \frac{1}{k!}A^k t^k + \cdots\right)x(0) \qquad (6.78)$$

or

$$x(t) = e^{At}x(0) \qquad (6.79)$$

Thus, *matrix exponential is also termed as transition matrix. It is also called fundamental matrix. It is a n × n matrix and signifies linear transformation of the initial state x(0) at time t_o to another state x(t) at time t.*

6.11 PROPERTIES OF STATE TRANSITION MATRIX

The state-transition matrix possesses the following properties.

1. $\dfrac{d\,\phi(t)}{dt} = A\,\phi(t)$

We know that

$$\dot{x}(t) = Ax(t)$$

Substituting $x(t) = \phi(t)x(0)$ in the above equation gives

$$\frac{d\,\phi(t)}{dt}\,x(0) = A\phi(t)\,x(0)$$

or

$$\frac{d\,\phi(t)}{dt} = A\phi(t) \tag{6.80}$$

2. $\phi(0)=$ Identity matrix $= I$

Proof We know that

$$e^{At} = \phi(t) = I + At + \frac{1}{2!}A^2t^2 + \frac{1}{3!}A^3t^3 + \cdots$$

Pulling $t = 0$ gives

$$\phi(t) = I \tag{6.81}$$

3. $\phi^{-1}(t) = \phi(-t)$

Proof We know that

$$\phi(t) = e^{At}$$

Multiplying both sides by e^{-At} gives

$$\phi(t)e^{-At} = e^{At}e^{-At}$$

or

$$\phi(t)e^{-At} = I$$

Multiplying both sides by $\phi^{-1}(t)$, we get

$$\phi^{-1}(t)\,\phi(t)e^{-At} = \phi^{-1}(t)I$$

or

$$e^{-At} = \phi^{-1}(t)$$

or

$$\phi^{-1}(t) = e^{-At} \tag{6.82}$$

or

$$\phi^{-1}(t) = \phi(-t) \tag{6.83}$$

4. $x(0) = \phi(-t)\,x(t)$

Proof We know that

$$x(t) = e^{At}x(0)$$

or

$$e^{At}x(0) = x(t)$$

or

$$x(0) = e^{-At}x(t)$$

Using equation (6.82), we get

$$x(0) = \phi(-t)x(t) \tag{6.84}$$

5. $\phi(t_2 - t_1)\,\phi(t_1 - t_2) = \phi(t_2 - t_1)$

Proof We know that

$$\phi(t) = e^{At}$$

Therefore, $\phi(t_2 - t_1) = e^{A(t_2 - t_1)}$

and $\phi(t_1 - t_0) = e^{A(t_1 - t_0)}$

Hence

$$\phi(t_2 - t_1)\,\phi(t_1 - t_0) = e^{A(t_2 - t_1)}e^{A(t_1 - t_0)}$$
$$= e^{A(t_2 - t_1 + t_1 - t_0)}$$

$$= e^{A(t_2 - t_0)} \tag{6.85}$$

$$= \phi(t_2 - t_0)$$

It is called the *transition property* and implies process of state transition consisting of number of sequential transitions. The property illustrates that transition from t_2 to t_0 can be achieved in two parts i.e. from t_2 to t_1 and then from t_1 to t_0.

The process of state transition in parts is illustrated is Fig. 6.7.

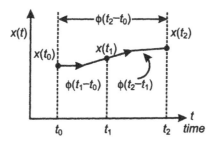

Fig. 6.7 *Process of State Transition in Parts*

6. $[\phi(t)]^K = \phi(Kt)$ *where K is positive integer*

Proof

$$[\phi(t)]^K = e^{At}e^{At}\ldots\ldots e^{At}\quad (K\text{ terms})$$

$$= e^{KAt}$$

$$= e^{AKt} \tag{6.86}$$

$$= \phi(Kt)$$

State Transition Matrix Properties
• $\dfrac{d\phi(t)}{dt} = A\,\phi(t)$
• $\phi(0) = I$
• $\phi^{-1}(t) = \phi(-t)$
• $x(0) = \phi(-t)\,x(t)$
• $\phi(t_2 - t_1)\,\phi(t_1 - t_2) = \phi(t_2 - t_1)$
• $[\phi(t)]^K = \phi(Kt)$

6.12 COMPUTATION OF STATE TRANSITION MATRIX

Various methods are available to compute state-transition matrix.

1. **Laplace transform approach.**

   Equation (6.68) and (6.69) represent

   $$\phi(t) = e^{At} = L^{-1}[(sI - A)^{-1}]$$

   $$= L^{-1}\left[\frac{\text{adjoint}[sI - A]}{|sI - A|}\right]$$

2. **By use of $\wedge$ matrix**

   Let
   $$\Lambda = P^{-1} AP$$

   or
   $$A = P \wedge P^{-1}$$

   or
   $$A^2 = (P \wedge P^{-1})(P \wedge P^{-1})$$
   $$= P \wedge^2 P^{-1}$$

   or
   $$A^m = P \wedge^m P^{-1}$$

   $$e^{At} = I + At + \frac{A^2 t^2}{2!} + \cdots$$

   $$= I + P\Lambda P^{-1} t + \frac{P\Lambda^2 P^{-1} t^2}{2!} + \ldots = Pe^{\wedge t} P^{-1}$$

The steps involved in computing state transition matrix are

- Find eigenvalues. Let $\lambda_1, \lambda_2, \ldots \lambda_n$
- Find $\wedge$ matrix

$$\wedge = \begin{bmatrix} \lambda_1 & \cdots & \cdots \\ \cdots & \lambda_2 & \cdots \\ \cdots & \cdots & \lambda_3 \end{bmatrix}$$

- Find e^{At}

$$e^{At} = Pe^{\wedge t} P^{-1}$$

where P = modal matrix formed by eigenvectors.

3. By Cayley-Hamilton Method

The steps involved for computation of state-transition matrix are

- Use the result $e^{At} = \sum_{K=0}^{n-1} a_K A^K$ where, A is the matrix of order $n \times n$

- Find eigenvalues of matrix A.

- Write down the equation in terms of aK by substituting eigenvalues in place of A one by one.

- Find the values of aK from the equations formulated above.

- Expand equation given above to compute e^{At}.

6.13 LAPLACE TRANSFORM SOLUTION OF STATE-EQUATIONS (FREQUENCY-DOMAIN ANALYSIS)

In this section Laplace transform solution of the state-equations

$$\dot{x}(t) = Ax(t) + Bu(t) \tag{6.87}$$

$$y(t) = Cx(t) + Du(t) \tag{6.88}$$

is discussed. Let us assume that the initial conditions at $t = 0$ is $x(0)$. Laplace transform of equation (6.87) yields

$$sX(s) - X(0) = AX(s) + BU(s) \tag{6.89}$$

solving for $X(s)$ gives

$$X(s) = (sI - A)^{-1}x(0) + (sI - A)^{-1}BU(s) \tag{6.90}$$

Inverse Laplace transform of equation (6.90) yields

$$x(t) = L^{-1}[(sI - A)^{-1}]x(0) + L^{-1}[(sI - A)^{-1}BU(s)] \tag{6.91}$$

Similarly, Laplace transform of equation (6.88) gives

$$Y(s) = CX(s) + DU(s) \tag{6.92}$$

Substituting value of $X(s)$ from equation (6.90) in equation (6.92) gives

$$Y(s) = C(sI - A)^{-1}x(0) + [(sI - A)^{-1}B + D]U(s) \tag{6.93}$$

Equations (6.90) and (6.93) are the Laplace transforms of the time-domain solutions to the state-equations. The solution consists of two parts i.e. free and forced response parts as explained in Section 6.7.

6.14 TRANSFER FUNCTION MATRIX

Transfer function matrix relates the output vector $Y(s)$ to the input vector $U(s)$. Let us consider a control system whose state-equation and output equation are given as

$$\dot{x} = Ax(t) + Bu(t) \tag{6.94}$$

$$y = Cx(t) + Du(t) \tag{6.95}$$

The Laplace transform of the state-equation and output equation yields

$$sX(s) - x(0) = AX(s) + BU(s) \tag{6.96}$$

$$Y(s) = CX(s) + DU(s) \tag{6.97}$$

or $\qquad (sI - A)X(s) = x(0) + BU(s)$

or $\qquad X(s) = (sI - A)^{-1}x(0) + ((sI - A)^{-1}B + D)\,U(s) \tag{6.98}$

or $\qquad Y(s) = C(sI - A)^{-1}x(0) + (C(sI - A)^{-1}B + D)\,U(s) \tag{6.99}$

If the initial condition $x(0)$ and input $U(s)$ are known, $X(s)$ and $Y(s)$ can be calculated from equations (6.98) and (6.99). For zero initial condition i.e. $x(0) = 0$.

$$Y(s) = 0 + (C(sI - A)^{-1}B + D)U(s) \tag{6.100}$$

or $\qquad \dfrac{Y(s)}{U(s)} = G(s) = C(sI - A)^{-1}B + D \tag{6.101}$

We call $G(s)$ as the *transfer function matrix*. $(sI - A)^{-1}$ can be computed as given below:

$$C(sI - A)^{-1} = \frac{C\,\text{adjoint}\,[sI - A]}{\det\,[sI - A]} \tag{6.102}$$

or $\qquad \dfrac{Y(s)}{U(s)} = G(s) = \dfrac{C\,\text{adjoint}\,[sI - A]}{\det[sI - A]}B + D \tag{6.103}$

6.15 REPRESENTATION OF STATE-EQUATION SOLUTION IN TERMS OF $\phi(t)$

The solution of state-equations for a non-homogenous time invariant control system was obtained in Section 6.13 as

$$x(t) = L^{-1}[(sI - A)^{-1}]\,x(0) + L^{-1}[(sI - A)^{-1}BU(s)] \tag{6.104}$$

or $\qquad x(t) = \phi(t)x(0) + \int_0^t \phi(t - \tau)\,Bu(\tau)d\tau \quad t \geq 0 \tag{6.105}$

Equation (6.105) is called the *state-transition equation*. It should be noted that initial time has been considered to be zero. This imposes, restrictions in the study of control systems, especially discrete-data control systems, where it is desirable to specify a more general initial time to achieve the desired flexibility in study and analysis of control systems. Let

$$\text{initial time} = t_0$$
$$\text{initial state} = x(t_0)$$

Let us also assume that input $u(t)$ is applied at $t \geq 0$. Substituting the assumptions in equation (6.105), we get

$$x(t_0) = \phi(t_0)x(0) + \int_0^{t_0} \phi(t_0 - \tau)Bu(\tau)d\tau \qquad (6.106)$$

Solving for $x(0)$ gives

$$x(0) = \phi^{-1}(t_0)x(t_0) - \phi^{-1}(t_0)\int_0^{t_0} \phi(t_0 - \tau)Bu(\tau)d\tau \qquad (6.107)$$

By using the state-transition matrix property $\phi^{-1}(t) = \phi(-t)$, we get

$$x(0) = \phi(-t_0)x(t_0) - \phi(-t_0)\int_0^{t_0} \phi(t_0 - \tau)Bu(\tau)d\tau \qquad (6.108)$$

Substituting equation (6.108) in equation (6.105), we get

$$x(t) = \phi(t)\,\phi(-t_0)\,x(t_0) - \phi(t)\,\phi(-t_0)\int_0^{t_0} \phi(t_0 - \tau)Bu(\tau)d\tau + \int_0^{t} \phi(t - \tau)Bu(\tau)d\tau \quad (6.109)$$

By using state-transition matrix property $\phi(t_2 - t_1)\,\phi(t_1 - t_0) = \phi(t_2 - t_0)$, and combining the integrals, we get

$$x(t) = \phi(t - t_0)x(t_0) + \int_{t_0}^{t} \phi(t - \tau)Bu(\tau)d\tau \quad \text{for } t \geq t_0 \qquad (6.110)$$

The output response $y(t)$ is obtained by substituting $x(t)$ from equation. (6.110) in $y(t) = Cx(t) + Du(t)$

$$y(t) = C\phi(t - t_0)\,x(t_0) + \int_{t_0}^{t} C\,\phi(t - \tau)Bu(\tau)d\tau + Du(t) \ \text{for } t \geq t_0$$

Note: If we put $t_0 = 0$ in equation (6.110), the equation (6.105) is obtained.

6.16 CHARACTERISTIC EQUATION

The characteristic equation is very significant in the study and analysis of the control systems. It determines the character and behaviour of the time response of the control system. The roots of the characteristic equation points towards the stability of the system. Characteristic equation of a control system can be obtained from its differential equation, transfer function and transfer function matrix. The roots of the characteristic equation are referred as eigenvalues. The differential equation is written/converted in s-domain and roots ascertained. The denominator polynomial of the transfer function when equated to zero gives the characteristic equation. Characteristic equation can also be obtained by equating $|sI - A| = 0$.

Example 6.2 Obtain characteristic equation from the following:

(a) $\dfrac{d^3 y}{dt^3} + 5\dfrac{d^2 y}{dt^2} + 2\dfrac{dy}{dt} + 6 = 0$

(b) $\dfrac{Y(s)}{y(s)} = \dfrac{s+4}{s^3 + 6s^2 + 11s + 6}$

(c) $A = \begin{bmatrix} 0 & 1 & 0 \\ 3 & 0 & 2 \\ -12 & -7 & -6 \end{bmatrix}$

Solution

(a) Writing the equation is s-domain gives

$s^3 + 5s^2 + 2s + 6 = 0$ which is the characteristic equation

(b) Putting the denominator of the transfer function to zero gives the characteristic equation i.e.

$s^3 + 5s^2 + 11s + 6 = 0$

(c) The transfer function matrix is given by equation (6.103). Putting the denominator i.e. $|sI - A| = 0$ gives the characteristic equation

$$A = \begin{bmatrix} 0 & 1 & 0 \\ 3 & 0 & 2 \\ -12 & -7 & -6 \end{bmatrix}$$

$$[sI - A] = \begin{bmatrix} s & 0 & 0 \\ 0 & s & 0 \\ 0 & 0 & s \end{bmatrix} - \begin{bmatrix} 0 & 1 & 0 \\ 3 & 0 & 2 \\ -12 & -7 & -6 \end{bmatrix} = \begin{bmatrix} s & -1 & 0 \\ -3 & s & -2 \\ 12 & 7 & s+6 \end{bmatrix}$$

Characteristic equation is

$|sI - A| = s\{s(s+6) - (-2)7\} - (-1)\{-3(s+6) - (-2)(12)\} + 0 = 0$

or $(s+1)(s+2)(s+3) = 0$

or $s^3 + 6s^2 + 11s + 6 = 0$

6.17 MATLAB

MATLAB in-built functions can be used to compute desired parameters. Use of some of the in-built functions has been illustrated is solved Examples and problems

- *eig(A)*   Computes eigenvalues of matrix A

- *[V, D] = eig(A)* Computes eigenvalues eigenvectors and stores as diagonal elements in D and Column vectors in V, respectively.

- *eye(n)*   Construct Identity matrix of order $n \times n$.

- *roots(P)*   Computes roots of polynomial P.

- *det(A)*   Computes determinant of matrix A.

- *sym2poly(A)*   Returns a row vector containing the numeric coefficents of symbolic polynomial.

- *ss(A, B, C, D)*   Creates the continuous-time state-space model

$$\dot{x} = Ax + Bu$$
$$y = Cx + Du$$

- *tf* It is used to create real-or complex-valued transfer function model or to convert state-space or zero-pole-gain models to transfer function form.

- *zpk(sys)*   Converts LTI model *sys* to zero-pole-gain form.

- *ilapace*   Computes inverse Laplace of the element positioned as $(1, 1)$ in matrix A i.e., element common to the first row and first column of matrix A.

- *sym or syms*   Constructs symbolic numbers, variables and objects.

- *subs(Phi, t, t − τ)*   Replaces t with $t - \tau$ in symbolic expression *Phi*.

- *int(X*B*D, tau, 0, t)*   Integrates term $X*B*D$ with respect to *tau* with limits

0 to t i.e., $\int\limits_{0}^{t} XBDd\tau$.

Example 6.3   A control system is described by the following matrices

$$A = \begin{bmatrix} -1 & 0 \\ 0 & -2 \end{bmatrix}; \quad B = [1 \ 1]^T; \quad C = [4 \ -2]; \quad D = 0.$$

Obtain a state-space representation. Also find the transfer function in polynomial and factored form.

```
'Example 6.3'                          %display enclosed text.
'Input Matrices A,B,C and D'           %display enclosed text.
A=[-1 0;0 -2]                          %input matrix A and
                                       %suppress output.
B=[1,1];                               %input matrix B and
                                       %suppress output.
C=[4 -2];                              %input matrix C and
                                       %suppress output.
D=0;                                   %input matrix C and
                                       %supress output.
'State Space Representation;           %display enclosed text
TFss=ss(A,B,C,D)                       %create state-space.
                                       %object.
'Transfer Function Polynomial Form'%display enclosed text.
TFpoly=tf(TFss)                        %convert to polynomial
                                       %form.
'Transfer Function Factored Form'      %display enclosed text.
TFfac=tf(TFss)                         %convert to factored
                                       %form.


Output
Example 6.3
Input Matrices A,B,C and D
State Space Representation
a =
        x1  x2
   x1   -1   0
   x2    0  -2
b =
        u1
   x1    1
   x2    1
C =
        x1  x2
   y1    4  -2

d =
        u1
   y1    0
Continuous-time model.
Transfer Function Polynomial Form
Transfer function:
   2 s + 6
  -------------
s^2 + 3 s + 2
Transfer Function Factored Form
Zero/pole/gain:
   2 (s+3)
  -------------
(s+2)(s+1)
```

SUMMARY

- Mathematically, the dynamics of a control system is described by input, output and the state variables.

- The different variables may be represented by input vector ($u(t)$), output vector ($y(t)$) and state vector ($x(t)$).

- Differential equations governing the system dynamics give the state variable representation.

- The state and output equations constitute the state model of the system. Information about the system state and output is obtained by solving the state-equations.

- Solution of the state-equations consists of response due to initial conditions (free response) and input (forced response).

$$x(t) = e^{A(t-t_0)} x(t_0) + \int_{t_0}^{t} e^{A(t-\tau)} Bu(\tau) d\tau.$$

- Output response is given as

$$y(t) = Ce^{At}x(0) + \int_{0}^{t} Ce^{A(t-\tau)} Bu(\tau) d\tau.$$

- State-equation consists of homogenous and non-homogenous parts
$$\dot{x}(t) = Ax(t) + Bu(t).$$
$Ax(t)$ is termed as the homogenous and $Bu(t)$ is called as the non-homogenous part.

- State transition matrix is ($n \times n$) order matrix which relates the initial state of the system to a state after time t considering that the input is zero and represents the free response. Various methods are available to compute state transition matrix
  - Laplace transform approach.
  - By use of Λ matrix.
  - By Cayley-Hamilton method.

- Laplace transform solution of state and output equations is represented as
$$x(t) = L^{-1}[(sI - A)^{-1}]x(0) + L^{-1}[(sI - A)^{-1}BU(s)]$$
$$Y(s) = C(sI - A)^{-1}x(0) + [(sI - A)^{-1}B + D]U(s)$$

- Transfer function matrix relates the output vector $Y(s)$ to the input vector $U(s)$
$$\frac{Y(s)}{U(s)} = G(s) = \frac{C \, \text{adjoint}\,[sI - A]}{\det\,[sI - A]} B + D.$$

- In terms of state transition matrix ($\phi(t)$)

$$x(t) = \phi(t - t_0)x(t_0) + \int_{t_0}^{t} \phi(t - \tau)Bu(\tau)d\tau \quad \text{for} \quad t \geq t_0$$

$$y(t) = C\phi(t - t_0)x(t_0) + \int_{t_0}^{t} C\phi(t - \tau)Bu(\tau)d\tau + Du(t) \quad \text{for} \quad t \geq t_0.$$

◆ Characteristic equation determines the character and behaviour of a control system. It can be obtained from
  • Differential equation.
  • Transfer function.
  • Transfer function matrix.
◆ Roots of characteristic equation are referred as eigenvalues.

PROBLEMS AND SOLUTIONS

Problem 6.1

The state-equation of a linear time-invariant system are represented by $\dfrac{dx(t)}{dt} = Ax(t) + Bu(t)$. Find the state transition matrix $\phi(t)$, the characteristic equation and the eigenvalues of A for the following cases.

(a) $A = \begin{bmatrix} 0 & 1 \\ -2 & -1 \end{bmatrix}$ $B = \begin{bmatrix} 0 & 1 \\ 1 & 0 \end{bmatrix}$ (b) $A = \begin{bmatrix} 0 & 1 \\ -4 & -5 \end{bmatrix}$ $B = \begin{bmatrix} 1 \\ 1 \end{bmatrix}$

(c) $A = \begin{bmatrix} -3 & 0 \\ 0 & -3 \end{bmatrix}$ $B = \begin{bmatrix} 0 \\ 1 \end{bmatrix}$ (d) $A = \begin{bmatrix} -1 & 0 & 0 \\ 0 & -2 & 1 \\ 0 & 0 & -2 \end{bmatrix}$ $B = \begin{bmatrix} 0 \\ 1 \\ 0 \end{bmatrix}$

Solution

Characteristic equation $= |sI - A| = 0$

$$(sI - A) = \begin{bmatrix} s & 0 \\ 0 & s \end{bmatrix} \begin{bmatrix} 0 & 1 \\ -2 & -1 \end{bmatrix} = \begin{bmatrix} s & -1 \\ 2 & s+1 \end{bmatrix}$$

$$|sI - A| = s(s+1) - (-1) \times 2 = 0 \quad \text{or} \quad s^2 + s + 2 = 0$$

Eigenvalues The roots of characteristic equation are the eigenvalues.

i.e. $\quad\quad\quad\quad s = -0.5 + j1.323 \quad$ and $s = -0.5 - j1.323 \quad$ are the eigenvalues

State transition matrix $(\phi(t))$

$$\phi(t) = e^{At}$$

$$= L^{-1}[(sI - A)^{-1}] = L^{-1}\left[\frac{\text{adjoint}[sI - A]}{|sI - A|}\right]$$

$$= L^{-1} \left[\frac{1}{s^2+s+2} \begin{bmatrix} s+1 & 1 \\ -2 & s \end{bmatrix} \right]$$

$$= L^{-1} \begin{bmatrix} \dfrac{s+1}{s^2+s+2} & \dfrac{1}{s^2+s+2} \\ \dfrac{-2}{s^2+s+2} & \dfrac{s}{s^2+s+2} \end{bmatrix} \qquad (1)$$

Let us find Inverse Laplace

$$L^{-1}\left[\frac{1}{s^2+s+2}\right] = \frac{A}{s+0.5+j1.323} + \frac{B}{s+0.5-j1.323}$$

$$A = \frac{1}{s+0.5-j1.323}\bigg|_{s=-0.5-j1.323} = \frac{1}{-j2.646} = j\,0.378$$

$$B = \frac{1}{s+0.5+j1.323}\bigg|_{s=-0.5+j\,1.323} = \frac{1}{j2.646} = -\,j\,0.378$$

Therefore,

$$L^{-1}\left[\frac{1}{s^2+s+2}\right] = \frac{j0.378}{s+0.5+j1.323} + \frac{-j0.378}{s+0.5-j1.323}$$

$$= j0.378\,(e^{-0.5t-j\,1,323t} - e^{-0.5t+j\,1,323t})$$

$$= j0.378 \times e^{-0.5t}(e^{-j1,323t} - e^{j1,323t})$$

$$= 0.756e^{-0.5t}\left(\frac{e^{j1,323t} - e^{-j1.323t}}{2j}\right)$$

$$= 0.756e^{-0.5t} \sin 1.323t \qquad (2)$$

$$L^{-1}\left[\frac{-2}{s^2+s+2}\right] = -2\times0.756\,e^{-0.5t}\sin1.323t$$

$$= -1.512\,e^{-0.5t}\sin1.323t \qquad (3)$$

$$L^{-1} = \left[\frac{s}{s^2+s+2}\right] = \left[\frac{A}{s+0.5+j1.323} + \frac{B}{s+0.5-j1.323}\right]$$

$$A = \frac{s}{s+0.5-j1.323}\bigg|_{s=-0.5-j1.323}$$

$$= j0.378(-0.5-j1.323)$$

$$= 0.5-j0.189 \qquad (4)$$

Therefore,
$$B = \frac{s}{s+0.5+j1.323}\bigg|_{s=-0.5+j1.323}$$

$$= 0.5+j0.189 \qquad (5)$$

substituting the values of A & B in equation (3) above we get

$$= \frac{0.5 - j0.189}{s + 0.5 + j1.323} + \frac{0.5 + j0.189}{s + 0.5 - j1.323}$$

$$= \frac{0.5}{s + 0.5 + j1.323} + \frac{0.5}{s + 0.5 - j1.323} + \frac{j0.189}{s + 0.5 - j1.323} - \frac{j0.189}{s + 0.5 + j1.323}$$

Inverse Laplace transform gives

$$= 0.5(e^{-0.5t - j1.323t} + e^{-0.5t + j1.323t}) + j0.189(e^{-0.5t + j1.323t} - e^{-0.5t - j1.323t})$$

$$= 2 \times 0.5 e^{-0.5t} \left(\frac{e^{-j1.323t} + e^{j1.323t}}{2} \right) + 2 \times 0.189 \times j^2 \left(\frac{e^{j1.323t} - e^{-j1.323t}}{2j} \right) e^{-0.5t}$$

$$= (\cos 1.323t - 0.378 \sin 1.323t)e^{-0.5t} \tag{6}$$

Now, we will find out inverse Laplace of

$$L^{-1}\left[\frac{s+1}{s^2+s+2} \right] = L^{-1}\left[\frac{s}{s^2+s+2} \right] + L^{-1}\left[\frac{1}{s^2+s+2} \right]$$

Substituting values from equation (6) and (2), we get

$$= (\cos 1.323t - 0.378 \sin 1.323t)e^{-0.5t} + 0.756 e^{-0.5t} \sin 1.323t$$

$$= (\cos 1.323t + 0.378 \sin 1.323t)e^{-0.5t} \tag{7}$$

Substituting the values from equations (7), (6), (3) and (2) in equation (1), the state transition matrix is

$$\phi(t) = \begin{bmatrix} \cos 1.323t + 0.378 \sin 1.323t & 0.756 \sin 1.323t \\ -1.512 \sin 1.323t & \cos 1.323t - 0.378 \sin 1.323t \end{bmatrix} e^{-0.5t}$$

(b) Characteristic equation $= | sI - A | = 0$

$$\left| \begin{bmatrix} s & 0 \\ 0 & s \end{bmatrix} - \begin{bmatrix} 0 & 1 \\ -4 & -5 \end{bmatrix} \right| = 0$$

$$\begin{vmatrix} s & -1 \\ 4 & s+5 \end{vmatrix} = 0$$

$$s(s+5) - (-1) \times 4 = 0$$

$$s^2 + 5s + 4 = 0$$

Eigenvalues   The roots of characteristic equation *i.e.* the eigenvalues are

$(s + 4) (s + 1) = 0$ *i.e.* $s = -4, -1$.

State transition matrix $\phi(t)$

$$\phi(t) = e^{At} = L^{-1}[(sI - A)^{-1}]$$

$$= L^{-1}\left[\frac{adj[sI - A]}{|sI - A|} \right] = L^{-1}[\phi(s)]$$

$$adj[sI - A] = \begin{bmatrix} s+5 & 1 \\ -4 & s \end{bmatrix}$$

$$|sI - A| = \frac{1}{s^2 + 5s + 4}$$

$$\phi(s) = \frac{1}{s^2 + 5s + 4} \begin{bmatrix} s+5 & 1 \\ -4 & s \end{bmatrix}$$

$$= \begin{bmatrix} \dfrac{s+5}{s^2+5s+4} & \dfrac{1}{s^2+5s+4} \\ \dfrac{-4}{s^2+5s+4} & \dfrac{s}{s^2+5s+4} \end{bmatrix}$$

$$= \begin{bmatrix} \dfrac{s+5}{(s+1)(s+4)} & \dfrac{s}{(s+1)(s+4)} \\ \dfrac{-4}{(s+1)(s+4)} & \dfrac{s}{(s+1)(s+4)} \end{bmatrix}$$

$$= \begin{bmatrix} \dfrac{4/3}{(s+1)} + \dfrac{-1/3}{(s+4)} & \dfrac{1/3}{(s+1)} + \dfrac{-1/3}{(s+4)} \\ \dfrac{-4/3}{(s+1)} + \dfrac{4/3}{(s+4)} & \dfrac{-1/3}{(s+1)} + \dfrac{4/3}{(s+4)} \end{bmatrix}$$

$$\phi(t) = L^{-1}[\phi(s)]$$

$$= \begin{bmatrix} \dfrac{4}{3}e^{-t} - \dfrac{1}{3}e^{-4t} & \dfrac{1}{3}e^{-t} - \dfrac{1}{3}e^{-4t} \\ -\dfrac{4}{3}e^{-t} + \dfrac{4}{3}e^{-4t} & -\dfrac{1}{3}e^{-t} + \dfrac{4}{3}e^{-4t} \end{bmatrix}$$

$$= \begin{bmatrix} 1.33e^{-t} - 0.33e^{-4t} & 0.33e^{-t} - 0.33e^{-4t} \\ -1.33e^{-t} + 1.33e^{-4t} & -0.33e^{-t} + 1.33e^{-4t} \end{bmatrix}$$

(c) Characteristic equation

$$|sI - A| = 0$$

$$\begin{bmatrix} s & 0 \\ 0 & s \end{bmatrix} - \begin{bmatrix} -3 & 0 \\ 0 & -3 \end{bmatrix} = 0$$

$$\begin{vmatrix} s+3 & 0 \\ 0 & s+3 \end{vmatrix} = 0$$

$$(s+3)^2 = 0$$

Eigenvalues   The roots of characteristic equation i.e. eigenvalues are s = –3, –3.

State transition matrix $(\phi(t))$

$$\phi(t) = L^{-1}[(\phi(s)]$$

$$\phi(s) = \frac{\text{adj}(sI - A)}{|sI - A|}$$

$$= \frac{1}{(s+3)^2}\begin{bmatrix} s+3 & 0 \\ 0 & s+3 \end{bmatrix} = \begin{bmatrix} \dfrac{s+3}{(s+3)^2} & 0 \\ 0 & \dfrac{s+3}{(s+3)^2} \end{bmatrix} = \begin{bmatrix} \dfrac{1}{s+3} & 0 \\ 0 & \dfrac{1}{s+3} \end{bmatrix}$$

$$\phi(t) = L[\phi(s)]$$

$$\phi(s) = \begin{bmatrix} e^{-3t} & 0 \\ 0 & e^{-3t} \end{bmatrix}$$

(d) Characteristic equation

$$\left| \begin{bmatrix} s & 0 & 0 \\ 0 & s & 0 \\ 0 & 0 & s \end{bmatrix} - \begin{bmatrix} -1 & 0 & 0 \\ 0 & -2 & 1 \\ 0 & 0 & -2 \end{bmatrix} \right| = 0$$

$$\begin{vmatrix} s+1 & 0 & 0 \\ 0 & s+2 & -1 \\ 0 & 0 & s+2 \end{vmatrix} = 0$$

or
$$(s+1)(s+2)(s+2) = 0$$

or
$$s^3 + 5s^2 + 8s + 4 = 0$$

Eigenvalues   The roots of the characteristic equation i.e eigenvalues are

$$(s + 1)(s + 2)(s + 2) = 0$$

or
$$s = -1, -2, -2.$$

State transition matrix

$$\phi(t) = L^{-1}[\phi(s)]$$

$$\phi(s) = \frac{adj[sI - A]}{|sI - A|}$$

$$adj[sI - A] = \begin{bmatrix} (s+2)(s+2) & 0 & 0 \\ 0 & (s+1)(s+2) & (s+1) \\ 0 & 0 & (s+1)(s+2) \end{bmatrix}$$

$$|sI - A| = (s+1)(s+2)(s+2)$$

Therefore,

$$\phi(s) = \frac{1}{(s+1)(s+2)(s+2)} \begin{bmatrix} (s+2)(s+2) & 0 & 0 \\ 0 & (s+1)(s+2) & (s+1) \\ 0 & 0 & (s+1)(s+2) \end{bmatrix}$$

$$= \begin{bmatrix} \dfrac{1}{s+1} & 0 & 0 \\ 0 & \dfrac{1}{s+2} & \dfrac{1}{(s+2)^2} \\ 0 & 0 & \dfrac{1}{s+2} \end{bmatrix}$$

$$\phi(t) = L^{-1}[\phi(s)]$$

or

$$\phi(t) = \begin{bmatrix} e^{-t} & 0 & 0 \\ 0 & e^{-2t} & te^{-2t} \\ 0 & 0 & e^{-2t} \end{bmatrix}$$

Problem 6.2

Check if the following matrices can be state transition matrices or not

(a) $\begin{bmatrix} e^{-t} & 0 \\ 0 & 1-e^{-t} \end{bmatrix}$ (b) $\begin{bmatrix} 1 & 0 \\ 1-e^{-t} & e^{-t} \end{bmatrix}$ (c) $\begin{bmatrix} e^{-2t} & 0 & 0 \\ te^{-2t} & c^{-2t} & t^2 e^{-2t} \\ t^3 e^{-2t} & \dfrac{t^2}{2} e^{-3t} & e^{-2t} \end{bmatrix}$

Solution State transition matrix has a property stated as $\phi(t) = I$ when $t = 0$ where $I =$ Identity matrix. We shall apply and check this property for the given matrices

(a) $\phi(t) = \begin{bmatrix} e^{-t} & 0 \\ 0 & 1-e^{-t} \end{bmatrix}$

$\phi(0) = \begin{bmatrix} 1 & 0 \\ 0 & 0 \end{bmatrix} \neq I$

Hence, the given matrix cannot be a state transition matrix.

(b) $\phi(t) = \begin{bmatrix} 1 & 0 \\ 1-e^{-t} & e^{-t} \end{bmatrix}$

$\phi(0) = \begin{bmatrix} 1 & 0 \\ 0 & 1 \end{bmatrix} = I$

Hence, it can be a state transition matrix.

(c) $\phi(t) = \begin{bmatrix} e^{-2t} & 0 & 0 \\ te^{-2t} & e^{-2t} & t^2 e^{-2t} \\ t^3 e^{-2t} & \dfrac{t^2}{2} e^{-2t} & e^{-2t} \end{bmatrix}$

$\phi(0) = \begin{bmatrix} 1 & 0 & 0 \\ 0 & 1 & 0 \\ 0 & 0 & 1 \end{bmatrix} = I$

Hence, the given matrix can be a state transition matrix.

Problem 6.3

Given that state transition matrix

$$\phi(t) = L^{-1}[(sI - A)^{-1}], \quad \text{show that} \, \phi(t) = I + At + \frac{1}{2!} A^2 t^2 + \frac{1}{3!} A^3 t^3 + \dots.$$

Solution

$\phi(t) = L^{-1}[(sI - A)^{-1}]$ where

$\phi(s) = (sI - A)^{-1}$

$\qquad = \dfrac{1}{s}\left(I - \dfrac{A}{s}\right)^{-1} = \dfrac{I}{s} + \dfrac{A}{s^2} + \dfrac{A^2}{s^3} + \cdots$

Taking inverse Laplace transform gives $\phi(t) = I + At + \dfrac{A^2 t^2}{2!} + \dfrac{A^3 t^3}{3!} + \cdots$

Problem 6.4

The state-equation of a linear time-invariant system is represented by

$$\dot{x}(t) = Ax(t) + Bu(t) \text{ where } A = \begin{bmatrix} 0 & 1 \\ -2 & 3 \end{bmatrix}$$

Find state transition matrix and verify the result by other methods.

Solution

The characteristic equation is $|sI - A| = 0$

$$\left| \begin{bmatrix} s & 0 \\ 0 & s \end{bmatrix} \begin{bmatrix} 1 & 0 \\ 0 & 1 \end{bmatrix} - \begin{bmatrix} 0 & 1 \\ -2 & 3 \end{bmatrix} \right| = 0$$

$$\left| \begin{bmatrix} s & -1 \\ 2 & s-3 \end{bmatrix} \right| = 0$$

$$s(s-3) - (-1)(2) = 0 \text{ or } s^2 - 3s + 2 = 0$$

The eigenvalues are $s^2 - 3s + 2 = (s-1)(s-2)$ or $s = 1, 2$.

$$\phi(s) = [sI - A]^{-1}$$

$$= \frac{adj[sI - A]}{|sI - A|}$$

$$= \frac{1}{s^2 - 3s + 2} \begin{bmatrix} s-3 & 1 \\ -2 & s \end{bmatrix}$$

$$= \frac{1}{s^2 - 3s + 2} \begin{bmatrix} s-3 & 1 \\ -2 & s \end{bmatrix}$$

$$= \begin{bmatrix} \dfrac{s-3}{(s-1)(s-2)} & \dfrac{1}{(s-1)(s-2)} \\ \dfrac{-2}{(s-1)(s-2)} & \dfrac{s}{(s-1)(s-2)} \end{bmatrix}$$

$$\phi(t) = L^{-1}(\phi(s))$$

$$= L^{-1} \begin{bmatrix} \dfrac{2}{s-1} + \dfrac{-1}{s-2} & \dfrac{-1}{s-1} + \dfrac{1}{s-2} \\ \dfrac{2}{s-1} + \dfrac{-2}{s-2} & \dfrac{-1}{s-1} + \dfrac{2}{s-2} \end{bmatrix}$$

$$= \begin{bmatrix} 2e^t - e^{2t} & -e^t + e^{2t} \\ 2e^t - 2e^{2t} & -e^t + 2e^{2t} \end{bmatrix}$$

Verification

$$\phi(t) = e^{At} = Pe^{\Lambda t}P^{-1}$$

$$\Lambda = \begin{bmatrix} \lambda_1 & 0 \\ 0 & \lambda_2 \end{bmatrix}$$

where $\lambda_1 = 1$ & $\lambda_2 = 2$ (eigenvalues). Therefore,

$$\Lambda = \begin{bmatrix} 1 & 0 \\ 0 & 2 \end{bmatrix}$$

Hence

$$e^{\Lambda t} = \begin{bmatrix} e^t & 0 \\ 0 & e^{2t} \end{bmatrix}$$

Eigenvectors

(a) Corresponding to $\lambda = 1$, the eigenvector is

$$\begin{bmatrix} s & -1 \\ 2 & s-3 \end{bmatrix}\begin{bmatrix} x_1 \\ x_2 \end{bmatrix} = 0$$

putting $s = \lambda_1 = 1$, we get

$$\begin{bmatrix} 1 & -1 \\ 2 & -2 \end{bmatrix}\begin{bmatrix} x_1 \\ x_2 \end{bmatrix} = 0$$

which gives

$$x_1 = x_2 = 0 \text{ or } x_2 = x_1$$
$$2x_1 = 2x_2 = 0 \text{ or } x_2 = x_1$$

Therefore if $x_1 = 1$ then $x_2 = x_1$ hence, $[1, 1]^T$ is an eigenvector

(b) Corresponding to $\lambda = 2$, the eigenvector

$$\begin{bmatrix} s & -1 \\ 2 & s-3 \end{bmatrix}\begin{bmatrix} x_1 \\ x_2 \end{bmatrix} = 0$$

putting $s = \lambda_2 = 2$, we get

$$\begin{bmatrix} 2 & -1 \\ 2 & -1 \end{bmatrix}\begin{bmatrix} x_1 \\ x_2 \end{bmatrix} = 0$$

which gives

$$2x_1 - x_2 = 0 \text{ or } x_2 = 2x_1$$

and $2x_1 - x_2 = 0$ or $x_2 = 2x_1$ hence, if $x_1 = 1$, then $x_2 = 2$
Therefore, $[1, 2]^T$ is another eigenvector

Matrix P   It is formed with eigenvectors $P = \begin{bmatrix} 1 & 1 \\ 1 & 2 \end{bmatrix}$

Therefore, state transition matrix

$$\phi(t) = P e^{\Lambda t} P^{-1}$$

$$= \begin{bmatrix} 1 & 1 \\ 1 & 2 \end{bmatrix} \begin{bmatrix} e^{t} & 0 \\ 0 & e^{2t} \end{bmatrix} \begin{bmatrix} 1 & 1 \\ 1 & 2 \end{bmatrix}^{-1}$$

$$= \begin{bmatrix} e^{t} & e^{2t} \\ e^{t} & 2e^{2t} \end{bmatrix} \begin{bmatrix} 1 & 1 \\ 1 & 2 \end{bmatrix}^{-1}$$

$$= \begin{bmatrix} e^{t} & e^{2t} \\ e^{t} & 2e^{2t} \end{bmatrix} \begin{bmatrix} 2 & -1 \\ -1 & 1 \end{bmatrix}$$

$$= \begin{bmatrix} 2e^{t} - e^{2t} & -e^{t} + e^{2t} \\ 2e^{t} - 2e^{2t} & -e^{t} + 2e^{2t} \end{bmatrix}$$

Another Method By using Sylvesters interpolation formula, state transition matrix can be obtained by solving the determinant.

$$\begin{vmatrix} 1 & \lambda_1 & \lambda_1^2 & \cdots\cdots & \lambda_1^{n-1} & e^{\lambda_1 t} \\ 1 & \lambda_2 & \lambda_2^2 & \cdots\cdots & \lambda_2^{n-1} & e^{\lambda_2 t} \\ \cdots & \cdots & \cdots & & \cdots & \cdots \\ 1 & \lambda_n & \lambda_n^2 & \cdots\cdots & \lambda_n^{n-1} & e^{\lambda_n t} \\ I & A & A^2 & \cdots\cdots & A^{n-1} & e^{At} \end{vmatrix} = 0$$

Using the above determinant, we get

$$\begin{vmatrix} 1 & \lambda_1 & e^{\lambda_1 t} \\ 1 & \lambda_2 & e^{\lambda_2 t} \\ I & A & e^{At} \end{vmatrix} = 0$$

Substituting the values of λ_1 and λ_2, we get

$$\begin{vmatrix} 1 & 1 & e^{t} \\ 1 & 2 & e^{2t} \\ I & A & e^{At} \end{vmatrix} = 0$$

which gives

$$(2e^{At} - Ae^{2t}) - (e^{At} - Ie^{2t}) + e^{t}(A - 2I) = 0$$

$$2e^{At} - Ae^{2t} - e^{At} + Ie^{2t} + Ae^{t} - 2Ie^{t} = 0$$

or
$$e^{At} + (I - A)e^{2t} + e^t(A - 2I) = 0$$

or
$$e^{At} = (A - I)e^{2t} + (2I - A)e^t$$

$$= \left\{ \begin{bmatrix} 0 & 1 \\ -2 & 3 \end{bmatrix} - \begin{bmatrix} 1 & 0 \\ 0 & 1 \end{bmatrix} \right\} e^{2t} + \left\{ \begin{bmatrix} 2 & 0 \\ 0 & 2 \end{bmatrix} - \begin{bmatrix} 0 & 1 \\ -2 & 3 \end{bmatrix} \right\} e^t$$

$$= \begin{bmatrix} -1 & 1 \\ -2 & 2 \end{bmatrix} e^{2t} + \begin{bmatrix} 2 & -1 \\ 2 & -1 \end{bmatrix} e^t$$

$$= \begin{bmatrix} -e^{2t} & e^{2t} \\ -2e^{2t} & 2e^{2t} \end{bmatrix} + \begin{bmatrix} 2e^t & -e^t \\ 2e^t & -e^t \end{bmatrix}$$

$$= \begin{bmatrix} 2e^t - 2e^{2t} & -e^t + 2e^t \\ 2e^t - 2e^{2t} & -e^t + 2e^{2t} \end{bmatrix}$$

Problem 6.5

A system is described by the dynamic equation

$$\frac{dx(t)}{dt} = Ax(t) + Bu(t)$$

$$y(t) = Cx(t)$$

where
$$A = \begin{bmatrix} 0 & 1 & 0 \\ 0 & 0 & 1 \\ -1 & -2 & -3 \end{bmatrix}; \quad B = \begin{bmatrix} 0 \\ 0 \\ 1 \end{bmatrix}; \quad C = \begin{bmatrix} 1 & 0 & 0 \end{bmatrix}$$

Find

(a) the transfer function relation between $X(s)$ and $U(s)$. (b) the transfer function.

Solution

(a) Transfer Function Relation

The transfer function relation is given by $X(s) = (sI - A)^{-1} BU(s)$

$$[sI - A]^{-1} = \frac{adj[sI - A]}{|sI - A|}$$

$$[sI - A] = \begin{bmatrix} s & -1 & 0 \\ 0 & s & -1 \\ 1 & 2 & s+3 \end{bmatrix}$$

$$adj\,[sI-A] = \begin{bmatrix} s(s+3)-(2s-1) & -1 & -3 \\ s+3 & s(s+3) & -2s-1 \\ 1 & s & s^2 \end{bmatrix}$$

$$= \begin{bmatrix} s^2+3s+2 & s+3 & 1 \\ -1 & s(s+3) & s \\ -s & -2s-1 & s^2 \end{bmatrix}$$

$$|sI-A| = s\{s(s+3)-(2\times1)\}-(-1)\{0-(-1\times1)\}$$
$$= s(s^2+3s+2)+1$$
$$= s^3+3s^2+2s+1$$

Substituting we get

$$X(s) = \frac{1}{s^3+3s^2+2s+1} \begin{bmatrix} s^2+3s+2 & s+3 & 1 \\ -1 & s(s+3) & s \\ -s & -2s-1 & s^2 \end{bmatrix} \begin{bmatrix} 0 \\ 0 \\ 1 \end{bmatrix} U(s)$$

or
$$X(s) = \frac{1}{s^3+3s^2+2s+1} \begin{bmatrix} 1 \\ s \\ s^2 \end{bmatrix} U(s)$$

Transfer Function

$$T(s) = \frac{Y(s)}{U(s)} = C(sI-A)^{-1}B$$

$$= \begin{bmatrix} 1 & 0 & 0 \end{bmatrix} \frac{1}{s^3+3s^2+2s+1} \begin{bmatrix} 1 \\ s \\ s^2 \end{bmatrix} = \frac{1}{s^3+3s^2+2s+1}$$

Problem 6.6

A control system has system matrix $A = \begin{bmatrix} 0 & 1 \\ -4 & -5 \end{bmatrix}$. Obtain eigenvalues, eigenvectors characteristic equation and poles/roots.

Solution MATLAB program is given below:

```
'Problem 6.6'                                %display enclosed text.
'Input Matrices A'                           %display enclosed text.
A=[0 1;-4 -5]                                 %input matrix A
'Eigenvalues'                                 %display enclosed text.
[V,D]=eig(A)                                  %Compute eigenvalues and
                                             %store in D as diagonal
                                             %elements.
'Alternatively Eigenvalues'                  %display enclosed text.
eig([0 1;-4 -5])                             %Compute eigenvalues and
                                             %display as column vector.
'Characteristic Equation'                    %display enclosed text.
syms s                                        %construct symbolic
                                             %object s.
I=eye(2)                                      %create Identity matrix of
                                             %order 2X2.
DET=det(s*I-A);                               %compute determinant which
                                             %is also the charateristic
                                             %equation of matrix A and
                                             %store in DET.
DET=simplify(DET);                            %simplify DET.
pretty(DET)                                   %pretty print DET.
'Alternatively Eigenvalues/Poles'            %display enclosed text.
P=sym2poly(DET)                               %convert symbolic
                                             %polynomial DET to a vector
                                             %and store in P.
R=roots(P)                                    %compute roots of
                                             %polynomial and store in R.
                                             %Note eigenvalues and poles
                                             %are the same.


Output
Problem 6.6
Input Matrix A
A =
      0      1
     -4     -5
Eigenvalues
V =
     0.7071    -0.2425
    -0.7071     0.9701
D =
     -1      0
      0     -4
Alternatively Eigenvalues
     -1
     -4
Charateristic Equation
I =
      1      0
      0      1
DET =
s^2+5*s+4
Note: It is the charateristic
equation also.

  2
 s  + 5 s + 4
 Note: The output in this form is because
 of function pretty
Alternatively Eigenvalues/Poles
P =
     1     5     4
Note: This represents equations^2+5*s+4
which is also the charateristic equation
R =
     -4
     -1
```

Problem 6.7

Obtain state-transition matrix for the system matrix $A = \begin{bmatrix} 0 & 1 \\ -4 & -5 \end{bmatrix}$.

Solution MATLAB program is given below:

```
'Problem 6.7'              %display enclosed text.
syms s                     %create symbolic
                           %object s.
A=[0 1;-4 -5];             %input matrix A &
                           %suppress output.
I=[1 0;0 1];               %create Identity matrix
                           %of order 2x2 & suppress
                           %output.
LT=((s*I-A)^-1;            %obtain Laplace transform
                           %of (s*I-A)^-1 and store
                           %in LT & suppress output.
LT11=ilaplace(LT(1,1));    %take inverse Laplace
LT12=ilaplace(LT(1,2));    %transform of each
LT21=ilaplace(LT(2,1));    %elements of LT and store
LT22=ilaplace(LT(2,2));    %in LT11,LT12,LT21 and
                           %LT22 respectively &
                           %suppress output.
STM=[LT11 LT12;LT21 LT22]; %construct state-transition
                           %matrix with LT11,Lt12,LT21
                           %and LT22 & suppress output.
simplify(STM)              %simplify STM.


Output
Problem 6.7
[ -1/3*exp(-4*t)+4/3*exp(-t),   1/3*exp(-t)-1/3*exp(-4*t)]
[ -4/3*exp(-t)+4/3*exp(-4*t),   4/3*exp(-4*t)-1/3*exp(-t)]
```

Problem 6.8

A control system is described by

$$\dot{x}(t) = \begin{bmatrix} 0 & 1 \\ -4 & -5 \end{bmatrix} x(t) + \begin{bmatrix} 1 \\ 0 \end{bmatrix} u(t)$$

Ascertain the state equation solution.

Solution MATLAB program is given below:

```
'Problem 6.8'                          %display enclosed text.
syms s                                 %create symbolic object s,t & tau.
A=[0 1;-4 -5];                         %input matrix A & suppress output.
B=[0;1];                               %input matrix B & suppress output.
X0=[1;0];                              %input initial condition vector.
U=1;                                   %unit step input.
I=eye(2);                              %create Identity matrix
                                       %of order 2x2 & suppress output.
LT=((s*I-A)^-1;                        %obtain Laplace transform
                                       %of (s*I-A)^-1 and store
                                       %in LT & suppress output.
LT11=ilaplace(LT(1,1));               %take inverse Laplace
LT12=ilaplace(LT(1,2));               %transform of each
LT21=ilaplace(LT(2,1));               %elements of LT and store
LT22=ilaplace(LT(2,2));               %in LT11,LT12,LT21 and
                                       %LT22 respectively &
                                       %suppress output.
Phi=[LT11 LT12;LT21 LT22];            %construct state-transition
                                       %matrix with LT11,LT12,LT21
                                       %and LT22 & suppress output.
Phitminustau=subs(Phi,t,t-tau);       %substitute t with t-tau.
x1=Phi*X0;                             %compute first term of
                                       %solution of state equation.
x2=int(Phitminustau*B*U,tau,0,t);%compute second term of
                                       %solution of state equation by
                                       %integrating Phitminustau*B*U
                                       %with respect to tau
                                       %with limits 0 to t.
x=x1+x2;                               %add.
x=simplify(x)                          %simplify x and display output.
pretty(x)                              $pretty print x and display output.


Output
Problem 6.8
[ -1/4*exp(-4*t)+*exp(-t)+1/4]
[         -exp(-t)+exp(-4*t)]

                    [- 1/4 exp(-4 t) + exp(-t) + 1/4]
                    [        -exp(-t) + exp(-4 t)    ]
```

1. Dynamic equation of a control system comprises of what types of matrices.

2. What are the order of various matrices of dynamic equations a single-input signal-output and multi variable control system.

3. Define state, state variable and state vector.

4. What is state-space. What is required to represent a system in state-space.

5. A fifth-order system needs how many first order equations to be represented in vector matrix form.

6. Define linear and nonlinear system. What are the common type of nonlinearities.

7. Write the full form of Taylor's series.

8. Consider that a dynamics of a control system are described by

$$\frac{d^n y}{dt^n} + a_{n-1}\frac{d^{n-1}}{dt^{n-1}} + \ldots + a_1\frac{dy}{dt} + a_0 y = b_0 u$$

   Formulate state equations in the following form

   $$\dot{x} = Ax + Bu$$
   $$y = Cx + Du.$$

9. Define homogenous and non-homogenous parts of state equation. General form of state equation is given as $\dot{x} = Ax + Bu$. Separate it into homogeneus and non-homogeneous parts.

10. Define state transition matrix. List the properties of state transition matrix.

11. What are the methods to compute state transition matrix.

12. Define transfer function matrix. Prove that

$$\frac{Y(s)}{U(s)} = C(sI - A)^{-1}B + D.$$

13. What is state transition equation. Dereve the expression.

$$x(t) = \phi(t - t_0)x(t_0) + \int_{t_0}^{t} \phi(t - \tau)Bu(\tau)d\tau$$

   and

$$y(t) = C\phi(t - t_0)x(t_0) + \int_{t_0}^{t} C\phi(t - \tau)Bu(\tau)d\tau$$

   for $t \geq t_0$

14. Define characteristic equation. What is its significance.

15. Find state transition matrix for this following cases

(a) $A = \begin{bmatrix} 0 & 1 \\ -2 & -3 \end{bmatrix}$ (b) $A = \begin{bmatrix} \lambda & 1 & 0 \\ 0 & \lambda & 1 \\ 0 & 0 & \lambda \end{bmatrix}$ (c) $A = \begin{bmatrix} 0 & 1 \\ 0 & -2 \end{bmatrix}$ (d) $A = \begin{bmatrix} 2 & 1 & 4 \\ 0 & 2 & 0 \\ 0 & 3 & 1 \end{bmatrix}$

16. A control system is described by the following vector matrix model

$$\begin{bmatrix} \dot{x}_1 \\ \dot{x}_2 \end{bmatrix} = \begin{bmatrix} 0 & 1 \\ -2 & -3 \end{bmatrix} \begin{bmatrix} x_1 \\ x_2 \end{bmatrix} + \begin{bmatrix} 0 \\ 1 \end{bmatrix} u(t).$$

where $u(t) = 1$. Determine the state vector $x(t)$ for $t \geq 0$.

17. Determine the response of a control system represented in vector matrix form as

$$\begin{bmatrix} \dot{x}_1 \\ \dot{x}_2 \end{bmatrix} = \begin{bmatrix} 1 & 0 \\ 1 & 1 \end{bmatrix} \begin{bmatrix} x_1 \\ x_2 \end{bmatrix}$$

It is given that $x_1(0) = 1$ and $x_2(0) = 1$.

18. A control system is represented as

$$\begin{bmatrix} \dot{x}_1 \\ \dot{x}_2 \end{bmatrix} = \begin{bmatrix} 1 & 0 \\ 1 & 1 \end{bmatrix} \begin{bmatrix} x_1 \\ x_2 \end{bmatrix} + \begin{bmatrix} 1 \\ 1 \end{bmatrix} u(t).$$

where $u(t) = \text{unit step input} = \begin{matrix} 0 \\ 1 \end{matrix} \Big\} \text{ for } \begin{matrix} t < 0 \\ t \geq 0 \end{matrix}$.

If initial conditions are $x_1(0) = x_2(0) = 0$. Find response of the system.

7

STATE-SPACE MODELLING

7.1 INTRODUCTION

In Chapter 6 we introduced the state-space concepts and terminology associated with state-space analysis of linear time-invariant control systems. In this chapter, we will utilise these concepts in representing the control systems in state-space. The state-space representation is not unique. The process of representation involves selecting a minimum number of linearly independent state variables that describe completely and sufficiently, the state of the control systems. No clear well defined approach or method can be assigned to selection of states variables, although the following guidelines may help:

◆ The number of state variables may equal the number of energy storing elements in a control system.

◆ Derivatives of the energy storage elements offer the best choice. In a R-L-C circuit, the energy storing elements are inductor and capacitor. However, the ideal choice of state variables would be the voltage across the capacitor and the current through the inductor.

◆ In a mechanical system, physical variables such as velocity, position and displacement would be the ideal choice for selecting state variables.

◆ If a transfer function depicting the input-output relationship is known or can be formulated, the order of denominator after pole-zero cancellation can be considered to be the order of differential equation for selecting the minimum number of state variables required to obtain state-space representation.

In this chapter we will discuss formulation of state-space models of a linear time-invariant control system.

LEARNING OBJECTIVES

◆ Introduce various forms of state models: physical variable form, phase variable form and canonical variable form.

◆ Learn to draw state-variable diagrams with the help of block diagram and signal flow diagram techniques.

◆ Learn to perform similarity transformation-controllability canonical form, observability canonical form, diagonal canonical form and Jordan canonical form.

◆ Construction of state-variable diagram from the transfer function by the method of decomposition of transfer functions.

7.2　STATE-VARIABLE DIAGRAM

The dynamics of a control system described by a set of first-order differential equations and the differential equations can be depicted diagrammatically using integrator, summing points and constant multipliers in the form of a block diagram or signal flow graph. Concepts of block diagram and signal flow graph have been explained in chapter 5. In this chapter we shall extend the approach to include state variables and relate it with the transfer function depiction.

Integrators can be drawn in s-domain depicting the time dependence of variables. The initial conditions can be added either at the start, with appropriate integrator or in the end, after completing the simulation diagram, at appropriate points. The integrator can be depicted in a rectangular box or in a triangle as shown in Fig. 7.1, either by inserting the integration sign $\int$ or denoting by $1/s$.

Fig. 7.1　*Depiction of integrator*

For example, eqn (7.1) given below can be depicted by a state diagram as shown in Fig. 7.2.

$$X_1(s) = \frac{1}{s}X_2(s) - \frac{1}{s}X_1(0) \tag{7.1}$$

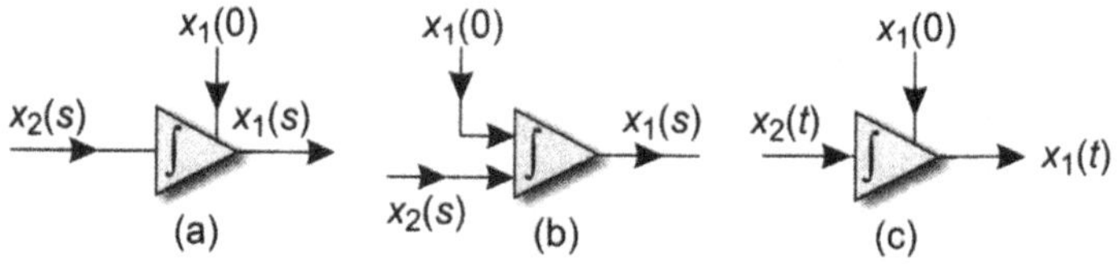

Fig. 7.2　*Simulation of differential equation by three different ways*

Let us consider a linear time-invariant control system defined by a first-order differential equation.

$$\dot{y}(t) = bx(t) - ay(t) \tag{7.2}$$

Laplace transform of the eqn (7.2) yields

$$sY(s) = bX(s) - aY(s) \tag{7.3}$$

The transfer function relating output with input is

$$\frac{Y(s)}{X(s)} = \frac{b}{s+a} \tag{7.4}$$

The first-order differential equation requires only one integrator having output $Y(s)$ and input $sY(s)$, as shown in Fig. 7.3.

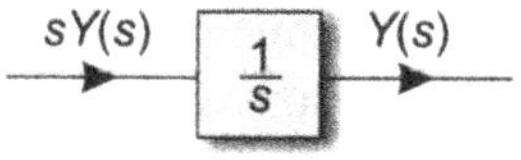

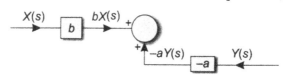

Fig. 7.3   *Use of an Integrator*

The input $sY(s)$ to the integrator is obtained by adding a summing point whose inputs are $bX(s)$ and $aY(s)$. Constants $-a$ and b are inserted in a constant multiplier box, as shown in Fig. 7.4.

Fig. 7.4   *Use of Summing Point*

Fig. 7.5 and 7.6 are then combined to obtain the ultimate state diagram shown in Fig. 7.5

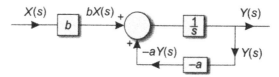

Fig. 7.5   *State variable diagram for* $\dfrac{Y(s)}{X(s)} = \dfrac{b}{s+a}$

Let us now consider a control system having transfer function

$$\frac{Y(s)}{X(s)} = \frac{1}{s^n + a_1 s^{n-1} + a_2 s^{n-2} + \ldots + a_{n-1}s + a_n} \tag{7.5}$$

Equation 7.5 can be rearranged as

$$s^n Y(s) = X(s) - a_1 s^{n-1} Y(s) - a_2 s^{n-2} Y(s) \ldots - a_{n-1} s Y(s) - a_n\, Y(s) \tag{7.6}$$

or

$$\frac{d^n y(t)}{dt^n} + a_1 \frac{d^{n-1} y(t)}{dt^{n-1}} + \ldots + a_n y(t) = x(t) \tag{7.7}$$

From the eqn (7.6) it can be assumed that the term $sY(s)$ can be obtained by subtracting the $Y(s)$ terms associated with constants $a_1 \ldots a_n$ from the input term $X(s)$. Also terms s^n to s can be obtained by sequentially placing n integrators as shown in Fig. 7.6

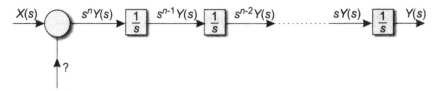

Fig. 7.6   *Sequential Arrangement of Integrators*

The output terms located on the RHS of the eqn (7.6) associated with constants $a_1,....a_n$ are added to the summing point by taking feedback signals from the appropriate forward paths as shown in Fig. 7.7.

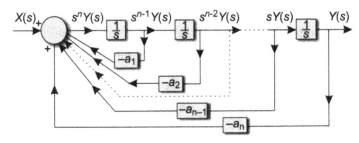

Fig. 7.7   *Addition of feedback paths by taking offshoots from forward paths and their associated constant terms*

Putting

$$
\begin{aligned}
y(t) &= x_1(t) \\
\frac{dy(t)}{dt} &= \frac{dx_1(t)}{dt} = x_2(t) \\
\frac{d^2y(t)}{dt^2} &= \frac{dx_2(t)}{dt} = x_3(t) \\
\frac{d^{n-1}y(t)}{dt^{n-1}} &= \frac{dx_{n-1}(t)}{dt} = x_n(t)
\end{aligned}
\qquad (7.8)
$$

Substituting eqn (7.8) in eqn (7.7) gives

$$
a_n x_1 +......+ a_1 x_n + \frac{d}{dt} x_n(t) = x(t) \qquad (7.9)
$$

Equations (7.8) and (7.9) yield a set of first-order differential equations which can be written as

$$
\frac{dx_1}{dt} = \dot{x}_1 = x_2 = sx_1
$$

$$
\frac{dx_2}{dt} = \dot{x}_2 = x_3 = sx_2
$$

$$
\frac{dx_{n-1}}{dt} = \dot{x}_{n-1} = x_n = sx_{n-1}
$$

and

$$\frac{dx_n}{dt} = -a_n x_1 - a_{n-1} x_2 - \ldots - a_1 x_n + x(t) = sx_n \tag{7.10}$$

Introducing the state variables and adding initial conditions at appropriate places, we get the state diagram shown in Fig. 7.8.

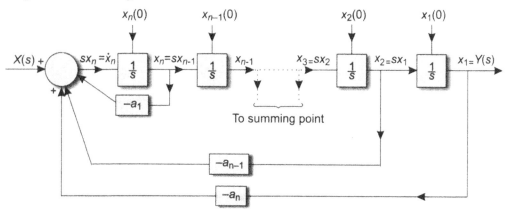

Fig. 7.8 *Ultimate state diagram depicting state variables and initial conditions*

7.3   STATE MODELS: SIGNAL FLOW GRAPHS

A state-model can easily be drawn with the help of Masons signal flow gain formula, explained in Chapter 5. To illustrate the drawing of a signal flow graph state model, let us consider a control system having transfer function

$$\frac{Y(s)}{U(s)} = \frac{b_0 s^n + b_1 s^{n-1} + b_2 s^{n-2} + \ldots + b_{n-1} s + b_n}{s^n + a_1 s^{n-1} + a_2 s^{n-2} + \ldots + a_{n-1} s + a_n}$$

$$= \frac{b_0 s^n + b_1 s^{n-1} + b_2 s^{n-2} + \ldots + b_{n-1} s + b_n}{s^n (1 + a_1 s^{-1} + a_2 s^{-2} + \ldots + a_{n-1} s^{-n+1} + a_n s^{-n})}$$

$$= \frac{b_0 + b_1 s^{-1} + b_2 s^{-2} + \ldots + b_{n-1} s^{-n+1} + b_n s^{-n}}{1 + a_1 s^{-1} + a_2 s^{-2} + \ldots + a_{n-1} s^{-n+1} + a_n s^{-n}}$$

$$= \frac{b_0 + b_1 s^{-1} + b_2 s^{-2} + \ldots + b_{n-1} s^{-n+1} + b_n s^{-n}}{1 - (-a_1 s^{-1} - a_2 s^{-2} - \ldots - a_{n-1} s^{-n+1} - a_n s^{-n})} \tag{7.11}$$

In the above expression, the denominator has been expressed as one minus the algebraic sum of the loop gains, and the terms of the numerator depict the forward path factors as defined in the Mason's gain formula. The system is a nth-order system and hence we identify n state variable $(x_1, x_2, \ldots, x_n)$ and so on. The signal flow graph showing the state variables is depicted in Fig. 7.9. This is achieved by n feedback touching loops involving the a_n coefficient.

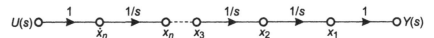

Fig. 7.9   *Signal flow graph showing state variables*

Next, we will add the denominator terms of the transfer function to Fig. 7.9

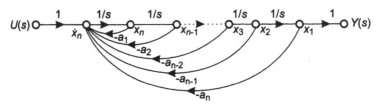

Fig. 7.10   *Signal flow graph with feedback paths multiplication factors*

The n forward path multiplication factors involving the b_n coefficients b_0, b_1/s, b_2/s^2 ....b_{n-1}/s^{n-1}, b_n/s^n are now added at appropriate places on the signal flow diagram shown in Fig. 7.10. The forward paths will touch loops.

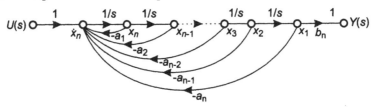

Fig. 7.11(a)   *Addition of forward path factor b_n*

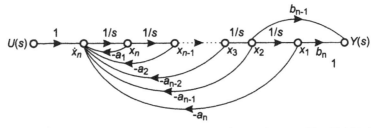

Fig. 7.11(b)   *Addition of forward path factor b_{n-1} to Fig. 7.11(a)*

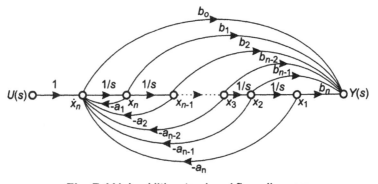

Fig. 7.11(c)   *Ultimate signal flow diagram*

7.4   STATE MODELS - VARIOUS FORMS

State model of a linear time-invariant system is represented in a general from by eqn (7.11). Specifically, state model of a control system can be represented in

- Physical variable form.
- Phase variable form.
- Canonical variable form.

The general procedure to formulate the state variable models of a control system is given below:

- Describe the control system by formulating the equations that describe the dynamics of the control system.
- Convert the equations into a set of first-order differential equations.
- Rearrange the converted equations to obtain matrices A, B, C and D.

7.4.1   Physical Variable Form

In this form, the state model is obtained by selecting physical quantities which can be measured as state variables. Let us consider an R–L–C electric circuit shown in Fig. 7.12.

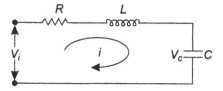

Fig. 7.12   *RLC Circuit*

Application of Kirchoff's Law to the RLC circuit gives

$$L\frac{di}{dt} + Ri + v_c = v_i \tag{7.12}$$

and
$$C\frac{dv_c}{dt} = i \tag{7.13}$$

Choosing i and v_c as state variables, the eqns (7.12) and (7.13) can be written as

$$\frac{di}{dt} = \dot{i} = \frac{-R}{L}i - \frac{v_c}{L} + \frac{v_i}{L} \tag{7.14}$$

and
$$\frac{dv_c}{dt} = \dot{v}_c = \frac{i}{C} \tag{7.15}$$

State model in matrix form can be expressed as

$$\begin{bmatrix} \dot{i} \\ \dot{v}_c \end{bmatrix} = \begin{bmatrix} \dfrac{-R}{L} & \dfrac{1}{L} \\ \dfrac{1}{C} & 0 \end{bmatrix} \begin{bmatrix} i \\ v_c \end{bmatrix} + \begin{bmatrix} \dfrac{1}{L} \\ 0 \end{bmatrix} v_i \tag{7.16}$$

The state variables chosen above i.e. i and v_c can be easily measured . The solution of state equations gives time variance of the measurable state variables having direct relevance to the physical system. Phase variables which are obtained from one of the system variables and its derivatives; are chosen as state variables. Generally, system output is taken as the state variable and other state variables considered are then the derivatives of the output. The disadvantage in choosing phase variables as state variables is that phase variable unlike physical variables are immeasurable parameters and have no similarity with the physical quantities of the system rendering it unsuitable for system analysis and study. However, the phase variables is potent tool for creating state variable representation as it institutes a connection between transfer function approach and time domain approach while designing a control system. Thus the design implementation of the control system becomes easy and straight forward.

7.4.2 Phase Variable Form

This type of state model can be easily obtained if the system model is readily available in the differential equation or transfer function form. Let us consider the general equation for a linear time-invariant system

$$\frac{d^n}{dt^n} y(t) + a_1 \frac{d^{n-1}}{dt^{n-1}} y(t) + \dots + a_n y(t) = b_0 \frac{d^n}{dt^n} u(t) + b_1 \frac{d^{n-1}}{dt^{n-1}} u(t) + \dots + b_n u(t) \qquad (7.17)$$

The transfer function relating output with input in *s-domain* is

$$\frac{Y(s)}{U(s)} = \frac{b_0 s^n + b_1 s^{n-1} + \dots + b_n}{s^n + a_1 s^{n-1} + \dots + a_n} \qquad (7.18)$$

7.4.2.1 First Canonical Form

We will choose the state variables in such a way that the derivatives of the input $u(t)$ are eliminated. Let us introduce a new variable $z(t)$ and express it in terms of $u(t)$ and $y(t)$. The numerator and denominator have been separated into two transfer functions as shown in Fig. 7.13.

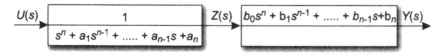

Fig. 7.13 *Separation into transfer functions with Z(s)*

Let

$$x_1(t) = z(t)$$
$$x_2(t) = \dot{x}_1(t) = \dot{z}(t)$$
$$x_3(t) = \dot{x}_2(t) = \ddot{z}(t)$$
$$\dots$$
$$x_n(t) = \dot{x}_{n-1}(t) = z^{(n-1)}(t)$$

$$(7.19)$$

Now $Z(s)$ is

$$Z(s) = \frac{U(s)}{s^n + a_1 s^{n-1} + \dots + a_{n-1}s + a_n}$$

or

$$Z(s)(s^n + a_1 s^{n-1} + \dots + a_{n-1}s + a_n) = U(s)$$
$$z^n(t) + a_1 z^{n-1}(t) + \dots + a_{n-1}\dot{z}(t) + a_n z(t) = u(t)$$

or

$$z^n(t) = -a_n z(t) - a_{n-1}\dot{z}(t) - \dots - a_1 z^{n-1}(t) + u(t)$$

which gives

$$z^n(t) = \dot{x}_n(t) = -a_n x_1(t) - a_{n-1}x_2(t) - \dots - a_1 x_n(t) + u(t) \qquad (7.20)$$

The output $Y(s)$ is

$$Y(s) = (b_0 s^n + b_1 s^{n-1} + \dots + b_{n-1}s + b_n)Z(s)$$

which gives

$$y(t) = b_0 z^n(t) + b_1 z^{n-1}(t) + \dots + b_{n-1}\dot{z}(t) + b_n z(t)$$

or

$$y(t) = b_0 z^n(t) + b_1 x_n(t) + \dots + b_{n-1}x_2(t) + b_n x_1(t) \qquad (7.21)$$

Substituting $z^n(t)$ from eqn (7.20) in eqn (7.21) gives

$$y(t) = b_0[-a_n x_1(t) - a_{n-1}x_2(t) - \dots - a_1 x_n(t)] + b_0 u(t) + b_1 x_n(t) + \dots + b_{n-1}x_2(t) + b_n x_1(t) \qquad (7.22)$$

or

$$y(t) = [b_n - b_0 a_n]x_1(t) + [b_{n-1} - b_0 a_{n-1}]x_2(t) + \dots + [b_1 - b_0 a_1]x_n(t) + b_0 u(t) \qquad (7.23)$$

Expressing eqns (7.20) and (7.23) in matrix form gives

$$\begin{bmatrix} \dot{x}_1(t) \\ \dot{x}_2(t) \\ \vdots \\ \dot{x}_{n-1}(t) \\ \dot{x}_n(t) \end{bmatrix} = \begin{bmatrix} 0 & 1 & 0 & \dots & 0 \\ 0 & 0 & 1 & \dots & 0 \\ \vdots & \vdots & \vdots & \vdots & \vdots \\ 0 & 0 & 0 & \dots & 1 \\ -a_n & -a_{n-1} & -a_{n-2} & \dots & -a_1 \end{bmatrix} \begin{bmatrix} x_1(t) \\ x_2(t) \\ \vdots \\ x_{n-1}(t) \\ x_n(t) \end{bmatrix} + \begin{bmatrix} 0 \\ 0 \\ \vdots \\ 0 \\ 1 \end{bmatrix} u(t) \qquad (7.24)$$

and

$$y(t) = [b_n - b_0 a_n \quad b_{n-1} - b_0 a_{n-1} \dots b_1 - b_0 a_1] \begin{bmatrix} x_1(t) \\ x_2(t) \\ \vdots \\ x_n(t) \end{bmatrix} + b_0 u(t) \qquad (7.25)$$

or

$$\dot{x}(t) = \begin{bmatrix} 0 & 1 & 0 & \dots & 0 \\ 0 & 0 & 1 & \dots & 0 \\ \vdots & \vdots & \vdots & \vdots & \vdots \\ 0 & 0 & 0 & \dots & 1 \\ -a_n & -a_{n-1} & -a_{n-2} & \dots & -a_1 \end{bmatrix} x(t) + \begin{bmatrix} 0 \\ 0 \\ \vdots \\ 0 \\ 1 \end{bmatrix} u(t) \qquad (7.26)$$

and

$$y(t) = [b_n - b_0 a_n \quad b_{n-1} - b_0 a_{n-1} \cdots b_1 - b_0 a_1] x(t) + b_0 u(t) \qquad (7.27)$$

or

In short

$$\begin{aligned} \dot{x}(t) &= Ax(t) + Bu(t) \\ \text{and} \\ y(t) &= Cx(t) + Du(t) \end{aligned} \qquad (7.28)$$

Phase variable model of a system described by eqns (7.24) and (7.25) when $n = 4$ is

$$\begin{bmatrix} \dot{x}_1(t) \\ \dot{x}_2(t) \\ \dot{x}_3(t) \\ \dot{x}_4(t) \end{bmatrix} = \begin{bmatrix} 0 & 1 & 0 & 0 \\ 0 & 0 & 1 & 0 \\ 0 & 0 & 0 & 1 \\ -a_4 & -a_3 & -a_2 & -a_1 \end{bmatrix} \begin{bmatrix} x_1(t) \\ x_2(t) \\ x_3(t) \\ x_4(t) \end{bmatrix} + \begin{bmatrix} 0 \\ 0 \\ 0 \\ 1 \end{bmatrix} u(t)$$

and

$$y(t) = \begin{bmatrix} b_4 - a_4 b_0 & b_3 - a_3 b_0 & b_2 - a_2 b_0 & b_1 - a_1 b_0 \end{bmatrix} x(t) + b_0 u(t)$$

Generalised state model is shown in Fig. 7.14 below

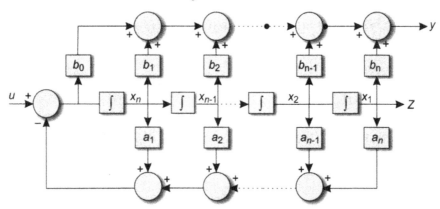

Fig. 7.14 *Phase variable state model*

7.4.2.2 Second Canonical Form

In the first canonical form, the last row of matrix A gets constituted from the coefficients of the denominator of the transfer function. However, in the second canonical form, the coefficients show as a column vector in the matrix A.

Let us consider eqn (7.18). Cross multiplications gives

$$(s^n + a_1 s^{n-1} + \ldots\ldots + a_n)Y(s) = (b_0 s^n + b_1 s^{n-1} + \ldots..+ b_n)U(s) \qquad (7.29)$$

or $\qquad s^n(Y(s) - b_0 U(s)) + s^{n-1}(a_1 Y(s) - b_1 U(s)) + \ldots\ldots + (a_n Y(s) - b_n U(s)) = 0 \qquad (7.30)$

Dividing eqn (7.30) by s^n gives

$$(Y(s) - b_0 U(s)) + \frac{1}{s}(a_1 Y(s) - b_1 U(s)) + \ldots\ldots + \frac{1}{s^n}(a_n Y(s) - b_n U(s)) = 0 \qquad (7.31)$$

Rearranging we get

$$Y(s) = b_0 U(s) + \frac{1}{s}(b_1 U(s) - a_1 Y(s)) + \frac{1}{s^2}(b_2 U(s) - a_2 Y(s)) + \ldots\ldots +$$

$$+ \frac{1}{s^{n-1}}(b_{n-1} U(s) - a_{n-1} Y(s)) + \ldots\ldots + \frac{1}{s^n}(b_n U(s) - a_n Y(s)) \tag{7.32}$$

Equation (7.32) can be represented by a block-diagram representation as shown in Fig. 7.15. below.

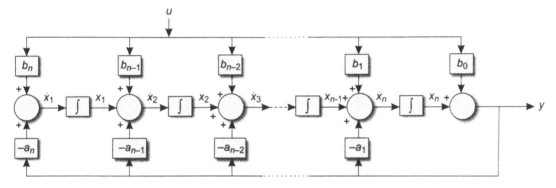

Fig 7.15 *Simulation diagram for eqn (7.32)*

From the simulation diagram, we get

Output equation $y = x_n + b_0 u$; and (7.33)

State equations $\dot{x}_n = x_{n-1} - a_1 y + b_1 u$

or $\dot{x}_n = x_{n-1} - a_1(x_n + b_0 u) + b_1 u$ (7.34)

or $\dot{x}_n = x_{n-1} - a_1 x_n + (b_1 - a_1 b_0)u$ (7.35)

Also $\dot{x}_{n-1} = x_{n-2} - a_2(x_n + b_0 u) + b_2 u$ (7.36)

or $\dot{x}_{n-1} = x_{n-2} - a_2 x_n + (b_2 - a_2 b_0)u$ (7.37)

Similarly $\dot{x}_2 = x_1 - a_{n-1}(x_n + b_o u) + b_{n-1}u$

or $\dot{x}_2 = x_1 - a_{n-1}x_n + (b_{n-1} - a_{n-1}b_0)u$ (7.38)

and $\dot{x}_1 = -a_n(x_n + b_0 u) + b_n u$ (7.39)

or $\dot{x}_1 = -a_n x_n + (b_n - a_n b_0)u$ (7.40)

Conversion of eqns (7.40), (7.38), (7.37) and (7.35) into vector matrix form gives the following state model

$$\begin{bmatrix} \dot{x}_1 \\ \dot{x}_2 \\ \vdots \\ \dot{x}_{n-1} \\ \dot{x}_n \end{bmatrix} = \begin{bmatrix} 0 & 0 & \ldots & 0 & -a_n \\ 1 & 0 & \ldots & 0 & -a_{n-1} \\ \vdots & \vdots & \vdots & \vdots & \vdots \\ 0 & 0 & \ldots & 0 & -a_2 \\ 0 & 0 & \ldots & 1 & -a_1 \end{bmatrix} \begin{bmatrix} x_1 \\ x_2 \\ \vdots \\ x_{n-1} \\ x_n \end{bmatrix} + \begin{bmatrix} b_n - a_n b_0 \\ b_{n-1} - a_{n-1} b_0 \\ \vdots \\ b_2 - a_2 b_0 \\ b_1 - a_1 b_0 \end{bmatrix} u \tag{7.41}$$

Equation (7.33) yields the following output equation in vector matrix form

$$y = \begin{bmatrix} 0 & 0 & \dots & 0 & 1 \end{bmatrix} \begin{bmatrix} x_1 \\ x_2 \\ \vdots \\ x_{n-1} \\ x_n \end{bmatrix} + b_0 u \tag{7.42}$$

In short

$$\dot{x}(t) = Ax(t) + Bu(t) \; ; \text{ and} \tag{7.43(a)}$$
$$y = Cx(t) + Du(t) \tag{7.43(b)}$$

The state variable equations for a control system having input-output relation described by eqn (7.43) with $n = 4$ are

$$y = x_4 + b_0 u$$
$$\dot{x}_4 = x_3 - a_1 x_4 + (b_1 - a_1 b_0)u$$
$$\dot{x}_3 = x_2 - a_2 x_4 + (b_2 - a_2 b_0)u$$
$$\dot{x}_2 = x_1 - a_3 x_4 + (b_3 - a_3 b_0)u$$
$$\dot{x}_1 = -a_4 x_4 + (b_4 - a_4 b_o)u$$

The state diagram is shown in Fig. 7.16

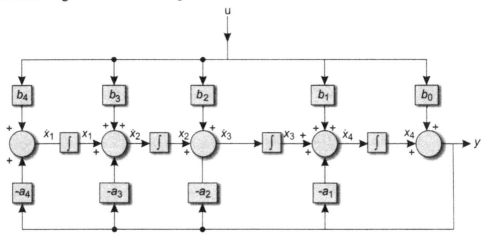

Fig. 7.16   *State diagram in second canonical form with n = 4*

The state model in vector matrix form is

$$\begin{bmatrix} \dot{x}_1(t) \\ \dot{x}_2(t) \\ \dot{x}_3(t) \\ \dot{x}_4(t) \end{bmatrix} = \begin{bmatrix} 0 & 0 & 0 & -a_4 \\ 1 & 0 & 0 & -a_3 \\ 0 & 1 & 0 & -a_2 \\ 0 & 0 & 1 & -a_1 \end{bmatrix} \begin{bmatrix} x_1(t) \\ x_2(t) \\ x_3(t) \\ x_4(t) \end{bmatrix} + \begin{bmatrix} b_4 - a_4 b_0 \\ b_3 - a_3 b_0 \\ b_2 - a_2 b_0 \\ b_1 - a_1 b_0 \end{bmatrix} u(t)$$

$$y = \begin{bmatrix} 0 & 0 & 0 & 1 \end{bmatrix} \begin{bmatrix} x_1(t) \\ x_2(t) \\ x_3(t) \\ x_4(t) \end{bmatrix} + b_0 u(t)$$

COMPARISON OF PHASE VARIABLE FORMS

There are certain similarities as well as differences in the phase variable state models represented by the first and second canonical forms.

◆ In the first canonical form, the a's appear in the last row of matrix A, while in the second canonical form they appear in the last column of matrix A.

◆ The derivatives of the input which appear as multiplication terms in the form of column vector (matrix B) in the second canonical form, shift to output equation in the form of row vector (matrix C) in the first canonical form.

◆ In the first canonical form matrix B represented by $[0 \ 0 \ .... \ 0 \ 1]^T$ in the state equation shifts as $[0 \ 0 \ .... \ 0 \ 1]$ as matrix C in the output equation.

◆ There is no change in matrix D.

◆ In the first canonical form, the first column of matrix A does not have the number 1 (one), while in the second canonical form, the first column has the number 1 (one) in the second row.

◆ Matrix A in the first canonical form becomes its transpose in the second canonical form.

◆ Matrix C in the first canonical form becomes matrix B in the second canonical form as transpose of matrix C.

Note: Noting the pattern of 1's in both the forms and with the similarities mentioned above, one can easily formulate one of the canonical forms, if the other is known.

7.4.3 Jordan Canonical Form

Let us once again consider the transfer function of a control system described as

$$G(s) = \frac{Y(s)}{U(s)} = \frac{b_0 s^n + b_1 s^{n-1} + ... + b_n}{s^n + a_1 s^{n-1} + ... + a_n} \tag{7.44}$$

Let us assume that the partial fraction of eqn (7.44) yields

$$G(s) = \frac{Y(s)}{U(s)} = A_0 + \left[\frac{A_1}{s - \lambda_1} + \frac{A_2}{s - \lambda_2} + \frac{A_3}{s - \lambda_3} + ... + \frac{A_n}{s - \lambda_n} \right] \tag{7.45}$$

where $A_1, ...A_n$ are the residues of the transfer function at the corresponding poles $\lambda_1 .... \lambda_n$ respectively. The transfer function in the form of eqn (7.45), indicates n first-order transfer functions in parallel and a forward path with gain of A_0. Eqn (7.45) has a unique simulation diagram as shown in Fig. 7.17. The state variables are decoupled i.e. each $x_i(t)$ depends only on the *input* $u(t)$. This feature helps the control engineer in system analysis. The state model in the vector matrix form becomes

$$\dot{x}(t) = \begin{bmatrix} \lambda_1 & 0 & 0 & .... & 0 \\ 0 & \lambda_2 & 0 & .... & 0 \\ 0 & 0 & \lambda_3 & .... & 0 \\ \vdots & \vdots & \vdots & \vdots & \vdots \\ 0 & 0 & 0 & .... & \lambda_n \end{bmatrix} \begin{bmatrix} x_1(t) \\ x_2(t) \\ x_3(t) \\ .... \\ x_n(t) \end{bmatrix} + \begin{bmatrix} 1 \\ 1 \\ 1 \\ ... \\ 1 \end{bmatrix} u(t)$$

and

$$y(t) = [A_1 \ A_2 \ A_3 \ .... \ A_n] \begin{bmatrix} x_1(t) \\ x_2(t) \\ x_3(t) \\ .... \\ x_n(t) \end{bmatrix} + A_0 u(t) \tag{7.46}$$

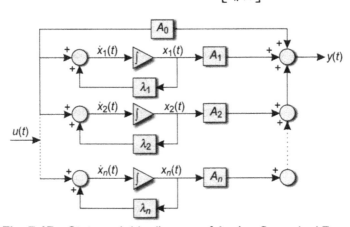

Fig. 7.17 *State variable diagram of Jordan Canonical Form*

The state equations are

$$y(t) = A_1 x_1(t) + A_2 x_2(t) .....A_n x(t) + A_0 u(t)$$
$$\dot{x}_1(t) = \lambda_1 x_1(t) + u(t)$$
$$\dot{x}_2(t) = \lambda_2 x_2(t) + u(t)$$
$$\vdots \qquad \vdots \qquad \vdots \qquad \vdots$$
$$\dot{x}_n(t) = \lambda_n \dot{x}_n(t) + u(t)$$

The state model realised above is true for a system transfer function having distinct eigenvalues (roots). Let us new consider a general case of multiple roots wherein the transfer function has roots λ_1 (with multiplicity p), λ_2 (with multiplicity r), λ_3, ..., λ_n. The transfer function will be of the form

$$\frac{Y(s)}{U(s)} = A_0 + \left[\frac{A_{p_1}}{(s-\lambda_1)^p} + \frac{A_{p_2}}{(s-\lambda_1)^{p-1}} + \ldots\ldots + \frac{A_p}{(s-\lambda_1)} \right]$$

$$+ \left[\frac{A_{r_1}}{(s-\lambda_2)^r} + \frac{A_{r_2}}{(s-\lambda_2)^{p-1}} + \ldots\ldots + \frac{A_r}{(s-\lambda_2)} \right]$$

$$+ \frac{A_1}{s-\lambda_3} + \ldots\ldots + \frac{A_n}{s-\lambda_n} \tag{7.47}$$

State model realisation is shown in Fig. 7.17(a).

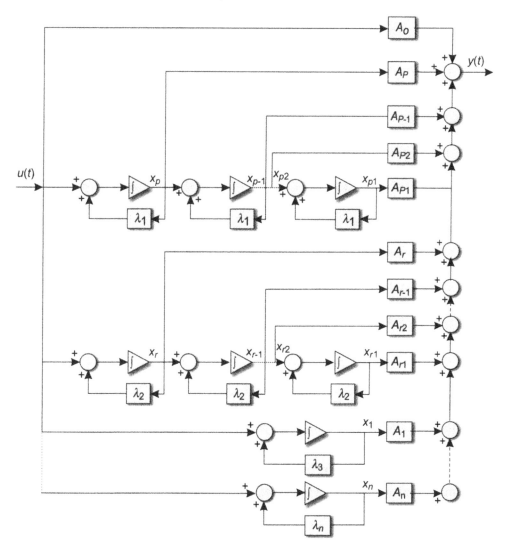

Fig. 7.17(a) *State model (multiple eigenvalues) in Jordan Form*

The state equations are

$$
\left.
\begin{aligned}
\dot{x}_{p_1}(t) &= \lambda_1 x_{p_1}(t) + x_{p_2}(t) \\
\dot{x}_{p_2}(t) &= \lambda_1 x_{p_2}(t) + x_{p_3}(t) \\
&\vdots \\
\dot{x}_{(p-1)}(t) &= \lambda_1 x_{(p-1)}(t) + x_p(t) \\
\dot{x}_p(t) &= \lambda_1 x_p(t) + u(t)
\end{aligned}
\right\}
\tag{7.48}
$$

$$
\left.
\begin{aligned}
\dot{x}_{r_1}(t) &= \lambda_2 x_{r_1}(t) + x_{r_2}(t) \\
\dot{x}_{r_2}(t) &= \lambda_2 x_{r_2}(t) + x_{r_3}(t) \\
&\vdots \\
\dot{x}_r(t) &= \lambda_2 x_r(t) + u(t)
\end{aligned}
\right\}
\tag{7.49}
$$

$$
\left.
\begin{aligned}
\dot{x}_1(t) &= \lambda_3 x_1(t) + u(t) \\
&\vdots \\
\dot{x}_n(t) &= \lambda_n x_n(t) + u(t)
\end{aligned}
\right.
\tag{7.50}
$$

and output equation is

$$
\begin{aligned}
y(t) = {} & A_{p_1} x_{p_1}(t) + A_{p_2} x_{p_2}(t) + \ldots\ldots + A_p x_p(t) + A_{r_1} x_{r_1}(t) + A_{r_2} x_{r_2}(t) \\
& + \ldots + A_r x_r(t) + \lambda_3 x_1(t) + \ldots + \lambda_n x_n(t) + A_0 u(t)
\end{aligned}
\tag{7.51}
$$

The state-space model in matrix form is

$$
\dot{x}(t) =
\begin{bmatrix}
\underline{J_1} & & & & 0 \\
& \underline{J_2} & & & \\
& & \lambda_3 & & \\
& & & \ddots & \\
0 & & & & \lambda_n
\end{bmatrix}
x(t) +
\begin{bmatrix}
\underline{b_1} \\
\underline{b_2} \\
1 \\
\vdots \\
1
\end{bmatrix}
u(t)
\tag{7.52}
$$

where

$\underline{J_1} = p \times p$ matrix with $\lambda_1's$ on the diagonal, ones above the diagonal and zeros at other places.

$\underline{J_2} = r \times r$ matrix with $\lambda_2's$ on the diagonal, ones above the diagonal and zeros at other places.

$\underline{b_1} = $ a column vector of p elements with last element as one and other elements zeros.

$\underline{b_2} = $ a column vector of r elements with last element as one and other elements zeros.

and

$$
y(t) = \left[A_{p_1} \ A_{p_2} \ldots A_p \ \vdots \ A_{r_1} \ A_{r_2} \ldots A_r \ \vdots \ldots \vdots \ A_1 \ldots A_n \right] x(t) + A_0 u(t)
\tag{7.53}
$$

Let us take a case of a control system whose transfer function is represented as

$$G(s) = \frac{Y(s)}{U(s)} = \frac{A_1}{(s-\lambda_1)^2} + \frac{A_2}{s-\lambda_1} + \frac{A_3}{(s-\lambda_2)^2} + \frac{A_4}{s-\lambda_2} + \frac{A_5}{s-\lambda_3} + \frac{-A_6}{s-\lambda_4} \qquad (7.54)$$

The state equations will then be

$$\dot{x}_1(t) = \lambda_1 x_1(t) + x_2(t)$$
$$\dot{x}_2(t) = \lambda_1 x_2(t) + u(t)$$
$$\dot{x}_3(t) = \lambda_2 x_3(t) + x_4(t)$$
$$\dot{x}_4(t) = \lambda_2 x_4(t) + u(t)$$
$$\dot{x}_5(t) = \lambda_3 x_5(t) + u(t)$$
$$\dot{x}_6(t) = \lambda_4 x_6(t) + u(t)$$

and output equation

$$y(t) = A_1 x_1(t) + A_2 x_2(t) + A_3 x_3(t) + A_4 x_4(t) + A_5 x_5(t) - A_6 x_6(t) \qquad (7.56)$$

The state-space model in matrix form can be written as

$$\dot{x}(t) = \begin{bmatrix} \dot{x}_1(t) \\ \dot{x}_2(t) \\ \dot{x}_3(t) \\ \dot{x}_4(t) \\ \dot{x}_5(t) \\ \dot{x}_6(t) \end{bmatrix} \begin{bmatrix} \lambda_1 & 1 & 0 & 0 & 0 & 0 \\ 0 & \lambda_1 & 0 & 0 & 0 & 0 \\ 0 & 0 & \lambda_2 & 1 & 0 & 0 \\ 0 & 0 & 0 & \lambda_2 & 0 & 0 \\ 0 & 0 & 0 & 0 & \lambda_3 & 0 \\ 0 & 0 & 0 & 0 & 0 & \lambda_4 \end{bmatrix} \begin{bmatrix} x_1(t) \\ x_2(t) \\ x_3(t) \\ x_4(t) \\ x_5(t) \\ x_6(t) \end{bmatrix} + \begin{bmatrix} 0 \\ 1 \\ 0 \\ 1 \\ 1 \\ 1 \end{bmatrix} u(t) \qquad (7.57)$$

and

$$y(t) = [A_1 \; A_2 \; A_3 \; A_4 \; A_5 \; -A_6]x(t) \qquad (7.58)$$

where

$$x(t) = [x_1(t) \; x_2(t) \; x_3(t) \; x_4(t) \; x_5(t) \; x_6(t)]' \qquad (7.59)$$

We can also realise the state model by drawing a signal flow diagram

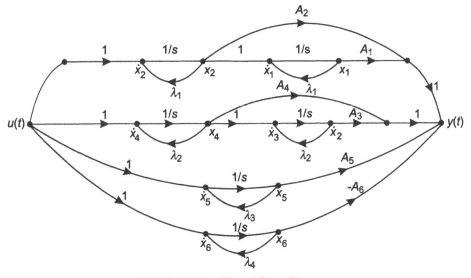

Fig 7.17(b) *Signal flow diagram*

KEY POINTS LEARNT

- ◆ System dynamics can be represented by a set of first-order differential equations and the differential equations can be depicted pictorially with the aid of
  - Block diagrams.
  - Signal flow graphs.
- ◆ State model representation is not unique. It can be represented in
  - Physical variable form.
  - Phase variable form-first canonical form and second canonical form.
  - Canonical variable form (Jordan canonical form).

Example 7.1   Obtain the state-space form for the control system whose dynamics are described by the following differential equation

$$\ddot{y} + 10\,\dot{y} + 9y = \dot{u} + 5u$$

Solution   A simulation diagram for the given differential equation can be derived form

$$\ddot{y} = -10\,\dot{y} - 9y + \dot{u} + 5u$$

or

$$y = -\frac{10}{s}\,y - \frac{9}{s^2}\,y + \frac{u}{s} + \frac{5}{s^2}\,u$$

or

$$y = -10\!\int y - 9\!\int\!\int y + \int u + 5\!\int\!\int u$$

or

$$y = \underbrace{-9\!\int\!\int y + 5\!\int\!\int u}_{\substack{\text{terms containing}\\ \text{two intergrators}}} \;\; \underbrace{-10\!\int y + \int u}_{\substack{\text{terms containing}\\ \text{one intergrator}}} \tag{1}$$

First considering two integrator terms; the simulation diagram is drawn with the following

(a) two integrator in series,

(b) output is y and input u,

(c) gain of -9 originates from output y; hence it is a feedback gain element, and

(d) gain of 5 associated with input u; hence it is located in the forward path.

The diagram simulating the above mentioned conditions is shown in Fig. 7.18

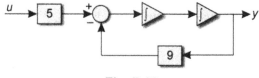

Fig. 7.18

Now considering the terms associated with one integrator in eqn (1) i.e. $-10\!\int y + \int u$

(a) gain of -10 is a gain of feedback element originating from output y and terminating at a point on the arrow between the two integrators; and

(b) gain of 1(one) originates from input u terminates at a point on the arrow connecting the two integrators

(c) Since the point of termination of both the terms is same, a summing point is positioned between the two integrators where the two terms terminate.

The simulation diagram satisfying these conditions is shown in Fig. 7.18(a)

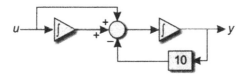

Fig. 7.18(a)

Combining both the diagrams, the simulation diagram satisfying eqn (1) is shown in Fig. 7.18(b)

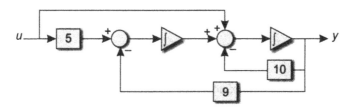

Fig. 7.18(b)

(a) **First Canonical Form**

Eqn (1) can be written as

$$\frac{Y(s)}{U(s)} = \frac{s+5}{s^2 + 10s + 9} \tag{2}$$

The transfer functions is separated into two separate transfer functions as shown in Fig 7.19 which gives

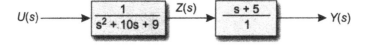

Fig. 7.19

$$\frac{Z(s)}{U(s)} = \frac{1}{s^2 + 10s + 9} \tag{3}$$

and

$$\frac{Y(s)}{Z(s)} = s + 5 \tag{4}$$

From eqn (3), we get
$$Z(s)\,(s^2 + 10s + 9) = U(s)$$

or $\qquad \ddot{z}\,(t) + 10\dot{z}\,(t) + 9z(t) = u(t)$ (5)

Let $\qquad\qquad z(t) = x_1(t)$ (6)

then $\qquad \dot{x}_1(t) = \dot{z}\,(t) = x_2(t)$ (7)

and $\qquad \dot{x}_2(t) = \ddot{z}\,(t)$ (8)

From eqn (5)

$$\ddot{z}(t) = -9z(t) - 10\dot{z}(t) + u(t)$$ (9)

Substitution from eqns (6), (7) and (8) in eqn (9) gives

$$\dot{x}_2(t) = -9x_1(t) - 10x_2(t) + u(t)$$ (10)

Similarly, from eqn (4), we get

$$Y(s) = (s + 5)\,Z(s)$$ (11)

or $\qquad y(t) = \dot{z}\,(t) + 5z(t)$ (12)

Substitution from eqns (6) & (7) in eqn (12) gives

$$y(t) = x_2(t) + 5x_1(t)$$

or $\qquad y(t) = 5x_1(t) + x_2(t)$ (13)

The state model in vector matrix notation is derived from eqns (7), (10) and (13) and is of the form

$$\begin{bmatrix} \dot{x}_1(t) \\ \dot{x}_2(t) \end{bmatrix} = \begin{bmatrix} 0 & 1 \\ -9 & -10 \end{bmatrix} \begin{bmatrix} x_1(t) \\ x_2(t) \end{bmatrix} + \begin{bmatrix} 0 \\ 1 \end{bmatrix} u(t)$$ (14)

and

$$y(t) = \begin{bmatrix} 5 & 1 \end{bmatrix} \begin{bmatrix} x_1(t) \\ x_2(t) \end{bmatrix}$$ (15)

The state diagram is shown in Fig. 7.20

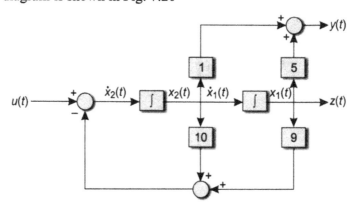

Fig. 7.20

(b) Second Canonical Form

We know that

$$\frac{Y(s)}{U(s)} = \frac{s+5}{s^2 + 10s + 9}$$

Rearranging

$$Y(s)(s^2 + 10s + 9) = (s + 5)U(s)$$

$$s^2 Y(s) + 10s Y(s) + 9Y(s) = sU(s) + 5U(s)$$

or $$s^2 Y(s) + (10Y(s) - U(s))s + 9Y(s) - 5U(s) = 0$$

or $$Y(s) + \frac{1}{s}(10Y(s) - U(s)) + \frac{1}{s^2}(9Y(s) - 5U(s)) = 0$$

or $$Y(s) = \frac{1}{s^2}(5U(s) - 9Y(s)) + \frac{1}{s}(U(s) - 10Y(s)) \qquad (16)$$

Eqn (16) can be represented by a block-diagram representation as shown in Fig 7.21

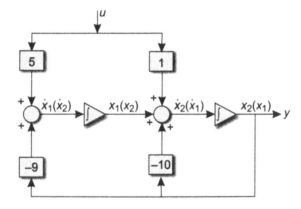

Fig. 7.21 *Block-diagram representation of second canonical form. The state variables enclosed in bracket are to be considered while formulating alternate state model of second canonical form which is illustrated on the next page.*

From the block-diagram, shown in Fig. 7.21, first considering the state variables not enclosed in brackets, we get

Output Equation

$$y = x_2 \qquad (17)$$

State Equations

$$\dot{x}_2 = x_1 - 10y + u \qquad (18)$$

or $$\dot{x}_2 = x_1 - 10x_2 + u \qquad (19)$$

and

$$\dot{x}_1 = -9y + 5u \qquad (20)$$

or $$\dot{x}_1 = -9x_2 + 5u \qquad (21)$$

State model in vector matrix form is derived from eqns (21), (19) and (17); and is of the form

$$\begin{bmatrix} \dot{x}_1 \\ \dot{x}_2 \end{bmatrix} = \begin{bmatrix} 0 & -9 \\ 1 & -10 \end{bmatrix} \begin{bmatrix} x_1 \\ x_2 \end{bmatrix} + \begin{bmatrix} 5 \\ 1 \end{bmatrix} u \tag{22}$$

$$y = \begin{bmatrix} 0 & 1 \end{bmatrix} \begin{bmatrix} x_1 \\ x_2 \end{bmatrix} \tag{23}$$

Alternatively (Ref. Note against Fig. 7.20). Considering the state variables enclosed in brackets, we get

Output Equation

$$y = x_1 \tag{24}$$

State Equations

$$\dot{x}_1 = x_2 - 10y + u$$

or $\qquad \dot{x}_1 = x_2 - 10x_1 + u$

or $\qquad \dot{x}_1 = -10x_1 + x_2 + u \tag{25}$

and

$$\dot{x}_2 = -9y + 5u$$

or $\qquad \dot{x}_2 = -9x_1 + 5u \tag{26}$

State model in vector matrix form is derived from eqns (26), (25) and (24); and is of the form

$$\begin{bmatrix} \dot{x}_1 \\ \dot{x}_2 \end{bmatrix} = \begin{bmatrix} -10 & 1 \\ -9 & 0 \end{bmatrix} \begin{bmatrix} x_1 \\ x_2 \end{bmatrix} + \begin{bmatrix} 1 \\ 5 \end{bmatrix} u \tag{27}$$

$$y = \begin{bmatrix} 1 & 0 \end{bmatrix} \begin{bmatrix} x_1 \\ x_2 \end{bmatrix} \tag{28}$$

(c) Jordan Canonical Form

$$\frac{Y(s)}{U(s)} = \frac{s+5}{s^2 + 10s + 9}$$

Partial fraction yields

$$\frac{Y(s)}{U(s)} = \frac{A}{s+9} + \frac{B}{s+1} = \frac{1/2}{s+9} + \frac{1/2}{s+1} \tag{29}$$

State diagram for the Jordan canonical form is shown in Fig. 7.22

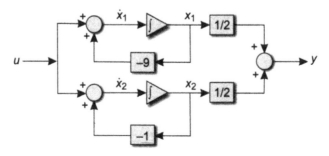

Fig. 7.22 *Block diagram of Jordan Canonical Form*

Assigning the outputs of the integrators as the state variables, results in

$$\dot{x}_1 = -9x_1 + u \tag{30}$$

and
$$\dot{x}_2 = -x_2 + u \tag{31}$$

and
$$y = \frac{1}{2}x_1 + \frac{1}{2}x_2 \tag{32}$$

or
$$\begin{bmatrix} \dot{x}_1 \\ \dot{x}_2 \end{bmatrix} = \begin{bmatrix} -9 & 0 \\ 0 & -1 \end{bmatrix} \begin{bmatrix} x_1 \\ x_2 \end{bmatrix} + \begin{bmatrix} 1 \\ 1 \end{bmatrix} u \tag{33}$$

$$y = \begin{bmatrix} \frac{1}{2} & \frac{1}{2} \end{bmatrix} \begin{bmatrix} x_1 \\ x_2 \end{bmatrix} \tag{34}$$

(d) Another Representation

$$\frac{Y(s)}{U(s)} = \frac{s+5}{s^2 + 10s + 9}$$

or
$$\ddot{y} + 10\dot{y} + 9y = \dot{u} + 5u \tag{35}$$

Let

$$x_1 = y \tag{36}$$

$$\dot{x}_1 = \dot{y} = x_2 + u \tag{37}$$

$$\dot{x}_2 + \dot{u} = \ddot{y} \tag{38}$$

or
$$\dot{x}_2 = \ddot{y} - \dot{u}$$

Substituting the value of $\ddot{y}$ from eqn (35), we get

$$\dot{x}_2 = -10\dot{y} - 9y + \dot{u} + 5u - \dot{u} \tag{39}$$

or
$$\dot{x}_2 = -10\dot{y} - 9y + 5u$$
$$= -10(x_2 + u) - 9x_1 + 5u$$
$$= -10x_2 - 10u - 9x_1 + 5u$$
$$= -9x_1 - 10x_2 - 5u \tag{40}$$

Representation in vector matrix form is derived from eqns (37), (40) and (36); and is of the form

$$\begin{bmatrix} \dot{x}_1 \\ \dot{x}_2 \end{bmatrix} = \begin{bmatrix} 0 & 1 \\ -9 & -10 \end{bmatrix} \begin{bmatrix} x_1 \\ x_2 \end{bmatrix} + \begin{bmatrix} 1 \\ -5 \end{bmatrix} u \tag{41}$$

$$y = \begin{bmatrix} 1 & 0 \end{bmatrix} \begin{bmatrix} x_1 \\ x_2 \end{bmatrix} \tag{42}$$

State diagram representation is shown in Fig. 7.23

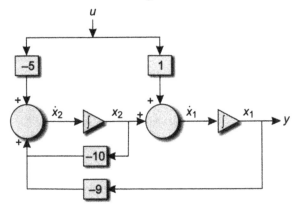

Fig. 7.23

Example 7.2   A MIMO Control system having two inputs and two outputs is described by

$$\ddot{y}_1 + 5\dot{y}_1 + y_2 = 3\dot{u}_1 + 2u_2 + \dot{u}_2$$

$$\ddot{y}_2 + 4\dot{y}_2 + y_1 = \dot{u}_2 + 6u_1$$

Obtain state-space representation by selecting a suitable set of state variables.

Solution   Integrating each equation twice, we get

$$y_1 + 5\int y_1 + \int\int y_2 = 3\int u_1 + 2\int\int u_2 + \int u_2$$

and $\qquad y_2 + 4\int y_2 + \int\int y_1 = \int u_2 + 6\int\int u_1$

or $\qquad\qquad y_1 = -5\int y_1 - \int\int y_2 + 3\int u_1 + 2\int\int u_2 + \int u_2$ $\qquad\qquad$ (1)

and $\qquad\qquad y_2 = -4\int y_2 - \int\int y_1 + \int u_2 + 6\int\int u_1$ $\qquad\qquad$ (2)

State diagram derived from eqn (1) is shown in Fig. 7.24(a)

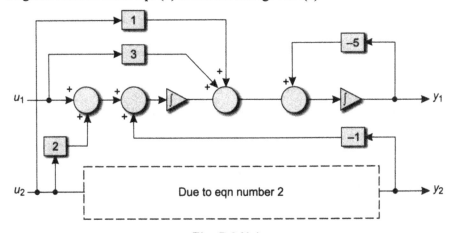

Fig. 7.24(a)

State diagram derived from eqn (2) is shown in Fig. 7.24(b)

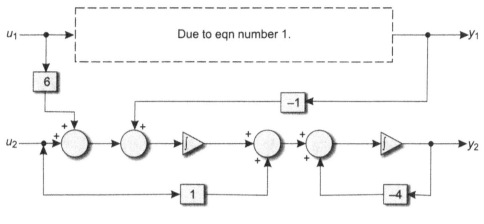

Fig. 7.24(b)

Combining Figs 7.24(a) and 7.24(b); the complete state diagram is shown in Fig. 7.24(c)

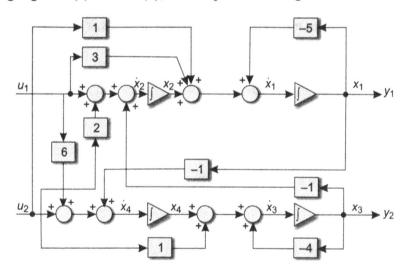

Fig. 7.24(c) *Complete state-diagram*

State equations derived from Fig. 7.24(c) are

$$\dot{x}_1 = -5x_1 + x_2 + 3u_1 + u_2$$
$$\dot{x}_2 = -x_3 + u_1 + 2u_2$$
$$\dot{x}_3 = -4x_3 + x_4 + u_2$$
$$\dot{x}_4 = -x_1 + 6u_1 + u_2$$
$$y_1 = x_1$$
$$y_2 = x_3$$

in matrix form

$$
\begin{bmatrix} \dot{x}_1 \\ \dot{x}_2 \\ \dot{x}_3 \\ \dot{x}_4 \end{bmatrix} = \begin{bmatrix} -5 & 1 & 0 & 0 \\ 0 & 0 & -1 & 0 \\ 0 & 0 & -4 & 1 \\ -1 & 0 & 0 & 0 \end{bmatrix} \begin{bmatrix} x_1 \\ x_2 \\ x_3 \\ x_4 \end{bmatrix} + \begin{bmatrix} 3 & 1 \\ 1 & 2 \\ 0 & 1 \\ 6 & 1 \end{bmatrix} \begin{bmatrix} u_1 \\ u_2 \end{bmatrix}
$$

$$
\begin{bmatrix} y_1 \\ y_2 \end{bmatrix} = \begin{bmatrix} 1 & 0 & 0 & 0 \\ 0 & 0 & 1 & 0 \end{bmatrix} \begin{bmatrix} x_1 \\ x_2 \\ x_3 \\ x_4 \end{bmatrix}
$$

7.5 SIMILARITY TRANSFORMATIONS

Sometimes during the analysis and design of a control system, the state model formulated may not be suitable and convenient in a particular form. Therefore, the need may arise to transform the state model into other forms without having to derive the state model again. Let us assume that the state model represented in matrix form by the matrices A, B, C and D are required to be transformed to $\bar{A}, \bar{B}, \bar{C}$ and $\bar{D}$ respectively. This change is called *similarity transformation* and the change of the state variables into another set is represented as

$$x(t) = P\bar{x}(t) \tag{7.60}$$

where

$\quad\quad x(t)$ = original state vector.

$\quad\quad \bar{x}(t)$ = newly transformed state vectors

$\quad\quad P$ = nonsingular $n \times n$ constant matrix.

Equation (7.60) can also be written as

$$\bar{x}(t) = P^{-1}x(t) \tag{7.61}$$

The transformed state and output equations can be written as

$$\frac{d\bar{x}(t)}{dt} = \dot{\bar{x}}(t) = \bar{A}\bar{x}(t) + \bar{B}u(t) \tag{7.62}$$

$$y(t) = \bar{C}\bar{x}(t) + \bar{D}u(t) \tag{7.63}$$

Differentiation of eqn (7.61) gives

$$\frac{d\bar{x}(t)}{dt} = P^{-1}\frac{dx(t)}{dt} \tag{7.64}$$

or $\quad\quad\quad\quad\quad\quad \dfrac{dx(t)}{dt} = P\dfrac{d\bar{x}(t)}{dt} \tag{7.65}$

The original state equation and output equation describing the dynamics of a linear time-invariant control system is written as

$$\dot{x}(t) = \frac{dx(t)}{dt} = Ax(t) + Bu(t) \tag{7.66}$$

$$y(t) = Cx(t) + Du(t) \tag{7.67}$$

Substituting for $x(t)$ from eqn (7.60) in eqns (7.66) and (7.67) gives

$$\dot{x}(t) = AP\,\bar{x}(t) + Bu(t) \tag{7.68}$$

and
$$y(t) = CP\bar{x}(t) + Du(t) \tag{7.69}$$

Substituting $\dfrac{dx(t)}{dt}$ from eqn (7.65) in eqn (7.68) gives

$$\frac{Pd\,\bar{x}(t)}{dt} = AP\bar{x}(t) + Bu(t)$$

or
$$P\dot{\bar{x}}(t) = AP\bar{x}(t) + Bu(t) \tag{7.70}$$

or
$$\dot{\bar{x}}(t) = P^{-1}AP\bar{x}(t) + P^{-1}Bu(t) \tag{7.71}$$

or
$$\dot{\bar{x}}(t) = \bar{A}\,\bar{x}(t) + \bar{B}u(t) \tag{7.72}$$

and, from eqn (7.69),

$$y(t) = CP\bar{x}(t) + Du(t)$$

or
$$y(t) = \bar{C}\,\bar{x}(t) + \bar{D}u(t)$$

where
$$\bar{A} = P^{-1}AP, \quad \bar{B} = P^{-1}B, \quad \bar{C} = CP \text{ and } \bar{D} = D \tag{7.73}$$

Similarity transformation does not alter the properties associated with the transfer function, characteristic equation, eigenvalues and eigenvectors and is invariant under the transformations.

7.5.1 Characteristic Equation

The characteristic equation is defined as $|sI - A| = 0$. Under transformation it is

$$|sI - \bar{A}| = |sI - P^{-1}AP|$$

$$= |sP^{-1}P - P^{-1}AP|$$

$$= |P^{-1}||(sI - A)||P|$$

$$= |sI - A| \tag{7.74}$$

This shows that the characteristic equation is preserved and not altered, which also implies that the eigenvalues and eigenvectors remain the same.

7.5.2 Transfer Function Matrix (G(s))

The transfer function matrix is given by

$$G(s) = C(sI - A)^{-1}B + D \tag{7.75}$$

Therefore, the transformed transfer function is

$$\bar{G}(s) = \bar{C}(sI - \bar{A})^{-1}\bar{B} + \bar{D} \tag{7.76}$$

Substituting the values of $\bar{A}, \bar{B}, \bar{C}$ and $\bar{D}$ from eqn (7.73) in eqn (7.76), gives

$$
\begin{aligned}
\bar{G}(s) &= CP(sI - P^{-1}AP)^{-1}P^{-1}B + D \\
&= CP(sP^{-1}P - P^{-1}AP)^{-1}P^{-1}B + D \\
&= CP[P^{-1}(sI - A)P]^{-1}P^{-1}B + D \\
&= CPP^{-1}(sI - A)^{-1}PP^{-1}B + D \\
&= C(sI - A)^{-1}B + D \\
&= G(s)
\end{aligned}
$$

Since $\bar{G}(s) = G(s)$, the transfer function matrix remains the same.

7.6   OTHER USEFUL TRANSFORMATION

Some Other useful transformations which are useful in the study, analysis and design of control systems are discussed in the following paragraphs.

7.6.1   Controllability Canonical Form (CCF)

Let us assume that the characteristic equation of system matrix A is

$$|sI - A| = s^n + a_1 s^{n-1} + a_2 s^{n-2} + \ldots\ldots + a_{n-2} s^2 + a_{n-1} s + a_n = 0 \qquad (7.77)$$

The dynamic equations are defined as

$$\dot{x}(t) = Ax(t) + Bu(t) \qquad (7.78)$$
$$y(t) = Cx(t) + Du(t) \qquad (7.79)$$

and are transformed into CCF equations defined as

$$
\begin{aligned}
\dot{\bar{x}}(t) &= \bar{A}\,\bar{x}(t) + \bar{B}\,u(t) \\
y(t) &= \bar{C}\,\bar{x}(t) + \bar{D}u(t)
\end{aligned}
\qquad (7.80)
$$

by transformation

$$x(t) = P\bar{x}(t) \qquad (7.81)$$

where

$$P = Q_c M \qquad (7.81)$$

Q_c is a nonsingular matrix and is given as

$$Q_c(n \times n) = \begin{bmatrix} B & AB & A^2 B & \ldots & A^{n-1}B \end{bmatrix} \qquad (7.83)$$

and

$$M = \begin{bmatrix} a_{n-1} & a_{n-2} & \ldots & a_1 & 1 \\ a_{n-2} & a_{n-3} & \ldots & 1 & 0 \\ \vdots & \vdots & \ldots & \vdots & \vdots \\ a_1 & 1 & \ldots & 0 & 0 \\ 1 & 0 & \ldots & 0 & 0 \end{bmatrix} \qquad (7.84)$$

Q_c is also called *controllability matrix* and is useful in determining controllability of the control systems. $\bar{A}$ & $\bar{B}$ are of the form

$$\bar{A} = P^{-1}AP = \begin{bmatrix} 0 & 1 & 0 & \cdots & 0 \\ 0 & 0 & 1 & \cdots & 0 \\ \vdots & \vdots & \vdots & \cdots & \vdots \\ 0 & 0 & 0 & \cdots & 1 \\ -a_n & -a_{n-1} & -a_{n-2} & \cdots & -a_1 \end{bmatrix} \tag{7.85}$$

and

$$\bar{B} = P^{-1}B = \begin{bmatrix} 0 \\ 0 \\ \vdots \\ 0 \\ 1 \end{bmatrix} \tag{7.86}$$

Matrices $\bar{C}$ and $\bar{D}$ given in eqn (7.79) do not follow any pattern.

7.6.2 Observability Canonical Form (OCF)

Let us again consider a control system described by eqns (7.78) and (7.79) and is transformed to the Observability canonical form (OCF) by the transformation

$$x(t) = P\bar{x}(t) \tag{7.87}$$

with

$$P = (MQ_o)^{-1} \tag{7.88}$$

The transformed equations are described as

$$\dot{\bar{x}}(t) = \bar{A}\,\bar{x}(t) + \bar{B}u(t)$$

$$y(t) = \bar{C}\bar{x}(t) + \bar{D}u(t) \tag{7.89}$$

where

$$\bar{A} = P^{-1}AP = \begin{bmatrix} 0 & 0 & \cdots & 0 & -a_n \\ 1 & 0 & \cdots & 0 & -a_{n-1} \\ 0 & 1 & \cdots & 0 & -a_{n-2} \\ \vdots & \vdots & \cdots & \vdots & \vdots \\ 0 & 0 & \cdots & 0 & -a_1 \end{bmatrix} \tag{7.90}$$

$$\bar{C} = CP = [0 \quad 0 \quad \cdots \quad 0 \quad 1] \tag{7.91}$$

The matrix $\bar{B}$ and $\bar{D}$ do not follow any pattern. Also

$$Q_0 = \text{Observability Matrix } (n \times n) = \begin{bmatrix} C \\ CA \\ CA^2 \\ \vdots \\ CA^{n-1} \end{bmatrix} \tag{7.92}$$

and matrix M is defined as given in eqn (7.84).

COMPARISON OF CCF AND OCF

- Modal matrix M remains the same in both the forms
- Controllability matrix Q_c and Observability matrix Q_0 are $n \times n$ nonsingular matrices.
- Matrix $\bar{A}$ in OCF is transpose of matrix $\bar{A}$ in CCF
- Matrix $\bar{C}$ in OCF is transpose of matrix $\bar{B}$ in CCF.
- Matrices $\bar{C}$ and $\bar{D}$ in CCF do not follow any particular pattern whereas matrices $\bar{B}$ and $\bar{D}$ in OCF do not follow any particular pattern.
- The CCF transformation is possible if Q_c^{-1} exists and OCF transformation is possible if Q_0^{-1} exists.
- Both transformations require that P^{-1} exists.

7.6.3 Diagonal Canonical Form (DCF)

Considering the state-space equations which describe the dynamics of a linear time-invariant system

$$\dot{x}(t) = Ax(t) + Bu(t) \tag{7.93}$$

$$y(t) = Cx(t) + Du(t) \tag{7.94}$$

where A is the system matrix having distinct eigenvalues. The transformation defined by

$$x(t) = Q_d \bar{x}(t)) \tag{7.95}$$

transforms the eqns (7.94) into

$$\dot{x}(t) = \bar{A}\bar{x}(t) + \bar{B}u(t)$$

$$y(t) = \bar{C}\bar{x}(t) + \bar{D}u(t) \tag{7.96}$$

where $\qquad \bar{A} = Q_d^{-1} A Q_d$ (7.97)

$$\bar{B} = Q_d^{-1} B$$

$$\bar{C} = C Q_d \text{ and}$$

$$\bar{D} = D$$

The matrix Q_d is formed by the use of eigenvectors of matrix A,

$$Q_d = \begin{bmatrix} P_1 & P_2 & P_3 & \dots & P_n \end{bmatrix} \text{ where}$$

$P_i = 1, 2, 3....n$ denotes the eigenvectors associated with λ_i.

The matrix $\bar{A}$ is a diagonal matrix of order $n \times n$

$$A = \begin{bmatrix} \lambda_1 & 0 & 0 & \dots & 0 \\ 0 & \lambda_2 & 0 & \dots & 0 \\ 0 & 0 & \lambda_3 & \dots & 0 \\ \vdots & \vdots & \vdots & \ddots & \vdots \\ 0 & 0 & 0 & \dots & \lambda_n \end{bmatrix}$$ (7.98)

where

$\lambda_i = 1, 2, 3.....n$ are the n distinct eigenvalues of matrix A. The transformed matrices $\bar{B}$, $\bar{C}$ and $\bar{D}$ do not follow any rigid pattern.

Transformation from CCF to DCF

If the state-space equations described in eqn (7.94) have matrix A in the CCF form and has distinct eigenvalues

$$A = \begin{bmatrix} 0 & 1 & 0 & \dots & 0 \\ 0 & 0 & 1 & \dots & 0 \\ \vdots & \vdots & \vdots & & \vdots \\ 0 & 0 & 0 & \dots & 0 \\ -a_n & -a_{n-1} & -a_{n-2} & \dots & -a_1 \end{bmatrix}$$ (7.99)

then the transformation matrix Q_d is of the form called *Vandermonde matrix*, and its pattern is

$$Q_d = \begin{bmatrix} 1 & 1 & 1 & \dots & 1 \\ \lambda_1 & \lambda_2 & \lambda_3 & \dots & \lambda_n \\ \lambda_1^2 & \lambda_2^2 & \lambda_3^2 & \dots & \lambda_n^2 \\ \vdots & \vdots & \vdots & & \vdots \\ \lambda_1^{n-1} & \lambda_2^{n-1} & \lambda_3^{n-1} & \dots & \lambda_n^{n-1} \end{bmatrix}$$ (7.100)

Where $\lambda_i = 1, 2, 3 .... \lambda_n$ are the eigenvalues of A.

7.6.4 Jordan Canonical Form (JCF)

If the system matrix A has multiple order eigenvalues, or if the transfer function is formed in such a way that it has repeated real roots, it cannot be transformed into a diagonal matrix unless the matrix is symmetric with real elements. However, one can achieve an almost diagonal similarity transformation. Such a diagonal form of matrix $\bar{A}$ is called *Jordan canonical form*.

A typical form of matrix $\bar{A}$ of the order four in JCF having third-order eigenvalues λ_1 and a distinct eigenvalue λ_2 is shown below:

$$\bar{A} = \begin{bmatrix} \lambda_1 & 1 & 0 & 0 \\ 0 & \lambda_1 & 1 & 0 \\ 0 & 0 & \lambda_1 & 0 \\ \hline 0 & 0 & 0 & \lambda_2 \end{bmatrix} \quad \begin{array}{l} \text{— Super diagonal} \\ \\ \text{— Main diagonal} \end{array} \tag{7.101}$$

The properties associated with the Jordan form are:

- Matrix $\bar{A}$ ($n \times n$) constitutes Jordan blocks.
- Each Jordan block has same eigenvalues for diagonal elements and *ones* in the elements one position above the diagonal called the *super diagonal*. In eqn (7.101), the Jordan blocks are enclosed by dashed lines.
- The number of Jordan blocks is equal to the number of independent eigenvectors. There is only one linearly independent eigenvector associated with each Jordan block. Let us assume that there are r linearly independent eigenvectors.
- The main diagonal is formed by eigenvalues.
- All the elements below the main diagonal are zero.
- Some of the elements of the super diagonal are 1's (ones)
- Elements above the super diagonal are zero.
- The number of 1's (ones) in the super diagonal = $n - r$

where n = order of matrix A, and r = number of linearly independent eigenvectors.

The transformation matrix Q_d is formed by eigenvectors as given in sec 7.6.3.

Example 7.3 The state equations of a linear time-invariant system are given by

$$\dot{x}(t) = \begin{bmatrix} 0 & 1 & 0 \\ 3 & 0 & 2 \\ -12 & -7 & -6 \end{bmatrix} x(t) + \begin{bmatrix} 1 \\ 0 \\ 2 \end{bmatrix} u(t)$$

$$y(t) = \begin{bmatrix} 1 & 0 & 0 \end{bmatrix} x(t)$$

Transform the state equation into CCF.

Solution

$$A = \begin{bmatrix} 0 & 1 & 0 \\ 3 & 0 & 2 \\ -12 & -7 & -6 \end{bmatrix}$$

The characteristic equation of A is

$$[sI - A] = \begin{bmatrix} s & -1 & 0 \\ -3 & s & -2 \\ 12 & 7 & s+6 \end{bmatrix} = s^3 + 6s^2 + 11s + 6 = 0$$

The coefficients of the characteristic equation are identified by comparing with the characteristic equation.

Now

$$M = \begin{bmatrix} a_{n-1} & a_{n-2} & 1 \\ a_{n-2} & 1 & 0 \\ 1 & 0 & 0 \end{bmatrix}$$

a_{n-1} is the coefficient associated with the s term and a_{n-2} is the coefficient associated with the s^2 term

Therefore

$$a_n = 6, a_{n-1} = 11 \text{ and } a_{n-2} = 6$$

Hence

$$M = \begin{bmatrix} 11 & 6 & 1 \\ 6 & 1 & 0 \\ 1 & 0 & 0 \end{bmatrix}$$

The controllability matrix $Q_c = \begin{bmatrix} B & AB & A^2B \end{bmatrix}$

$$B = \begin{bmatrix} 1 \\ 0 \\ 2 \end{bmatrix};$$

$$AB = \begin{bmatrix} 0 & 1 & 0 \\ 3 & 0 & 2 \\ -12 & -7 & -6 \end{bmatrix} \begin{bmatrix} 1 \\ 0 \\ 2 \end{bmatrix} = \begin{bmatrix} 0 \\ 7 \\ -24 \end{bmatrix}$$

$$A^2B = \begin{bmatrix} 0 & 1 & 0 \\ 3 & 0 & 2 \\ -12 & -7 & -6 \end{bmatrix} \begin{bmatrix} 0 \\ 7 \\ -24 \end{bmatrix} = \begin{bmatrix} 7 \\ -48 \\ 95 \end{bmatrix}$$

Therefore

$$Q_c = \begin{bmatrix} 1 & 0 & 7 \\ 0 & 7 & -48 \\ 2 & -24 & 95 \end{bmatrix}$$

$$P = Q_c M$$

$$= \begin{bmatrix} 1 & 0 & 7 \\ 0 & 7 & -48 \\ 2 & -24 & 95 \end{bmatrix} \begin{bmatrix} 11 & 6 & 1 \\ 6 & 1 & 0 \\ 1 & 0 & 0 \end{bmatrix} = \begin{bmatrix} 18 & 6 & 1 \\ -6 & 7 & 0 \\ -27 & -12 & 2 \end{bmatrix}$$

$$\bar{A} = P^{-1}AP$$

$$= \begin{bmatrix} 18 & 6 & 1 \\ -6 & 7 & 0 \\ -27 & -12 & 2 \end{bmatrix}^{-1} \begin{bmatrix} 0 & 1 & 0 \\ 3 & 0 & 2 \\ -12 & -7 & -6 \end{bmatrix} \begin{bmatrix} 18 & 6 & 1 \\ -6 & 7 & 0 \\ -27 & -12 & 2 \end{bmatrix}$$

$$= \begin{bmatrix} 0 & 1 & 0 \\ 1 & 0 & 1 \\ -6 & -11 & -6 \end{bmatrix} = \begin{bmatrix} 0 & 1 & 0 \\ 1 & 0 & 1 \\ -a_n & -a_{n-1} & -a_{n-2} \end{bmatrix}$$

7.7 DECOMPOSITION OF TRANSFER FUNCTION

In the preceding sections, we formed state-model equations and learnt how to draw state diagrams from the differential equations. The state diagrams and the state equations can also be obtained from the transfer functions of the control systems. *The process by which state diagrams are drawn from the transfer functions is termed as decomposition.* A flow chart showing various methods of decomposition is given in Fig. 7.25.

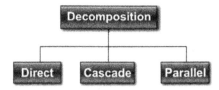

Fig. 7.25 *Various methods of decomposition of transfer functions*

7.7.1 Direct Decomposition

When the transfer function is expressed in polynomial form, then the *direct decomposition* method is utilized to obtain the state diagram leading to the Controllable Canonical Form (CCF) and Observable Canonical Form (OCF). Let us assume that the transfer function is expressed as given below:

$$\frac{Y(s)}{U(s)} = \frac{b_0 s^n + b_1 s^{n-1} + b_2 s^{n-2} + \ldots\ldots + b_{n-1} s + b_n}{s^n + a_1 s^{n-1} + a_2 s^{n-2} + \ldots.. + a_{n-1} s + a_n} \tag{7.102}$$

State diagram leading to CCF

Multiplying and dividing the RHS of eqn (7.102) by s^{-n} we get

$$\frac{Y(s)}{U(s)} = \frac{b_0 + b_1 s^{-1} + b_2 s^{-2} + b_3 s^{-3} + \ldots\ldots + b_{n-1} s^{-n+1} + b_n s^{-n}}{1 + a_1 s^{-1} + a_2 s^{-2} + a_3 s^{-3} + \ldots\ldots + a_{n-1} s^{-n+1} + a_n s^{-n}}$$

Creating a function with unity numerator by putting

$$X(s) = \frac{Y(s)}{b_0 + b_1 s^{-1} + b_2 s^{-2} + b_3 s^{-3} + \ldots.. + b_{n-1} s^{-n+1} + b_n s^{-n}} \tag{7.103}$$

gives

$$\frac{X(s)}{Y(s)} = \frac{1}{b_0 + b_1 s^{-1} + b_2 s^{-2} + b_3 s^{-3} + \ldots.. + b_{n-1} s^{-n+1} + b_n s^{-n}} \tag{7.104}$$

or

$$Y(s) = (b_0 + b_1 s^{-1} + b_2 s^{-2} + b_3 s^{-3} + \ldots.. + b_{n-1} s^{-n+1} + b_n s^{-n}) X(s) \tag{7.105}$$

and

$$X(s)(1 + a_1 s^{-1} + a_2 s^{-2} + a_3 s^{-3} + \ldots\ldots + a_{n-1} s^{-n+1} + a_n s^{-n}) = U(s) \tag{7.106}$$

Rearranging eqn (7.106), we get

$$X(s) = U(s) - (a_1 s^{-1} + a_2 s^{-2} + a_3 s^{-3} + \ldots.. + a_{n-1} s^{-n+1} + a_n s^{-n}) X(s) \tag{7.107}$$

Since the system is nth order, we identify n state variables $(x_1, x_2, x_3 \ldots x_{n-1}, x_n, \dot{x}_n)$ and arrange them in the ascending order from right to left on the state diagram as shown in Fig. 7.26(a). Each node is an output of the integrator of the previous mode. Next, we locate the output and input, $Y(s)$ and $U(s)$ respectively, by addition of two nodes with unit transmittance as shown in Fig. 7.28(b).

Equation (7.105) defines output $Y(s)$ as the sum of the forward path terms on the RHS. Let us take the terms on the RHS one by one.

1) b_0 **term** It does not have s associated with it. This forward path can be drawn by connecting the $\dot{x}_n$ node with the $Y(s)$ node directly, having b_0 transmittance.

2) $b_1 s^{-1}$ **term** The term $b_1 s^{-1}$ or b_1/s depicting a forward path with transmittances b_1 can be drawn by connecting x_n node with node $Y(s)$. This gives a forward path gain from input $(U(s))$ to output $(Y(s))$

$$= 1 \times \frac{1}{s} \times b_1 \times 1 = \frac{b_1}{s} \tag{7.108}$$

The above mentioned forward paths are depicted on the state diagram shown in Fig. 7.26(c)

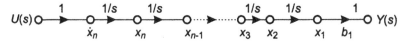

Fig. 7.26(a) *n state variables arranged from right to left*

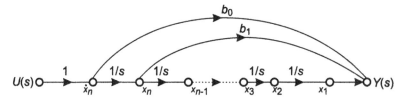

Fig. 7.26(b) *Addition of Y(s) and U(s) with unit transmittance*

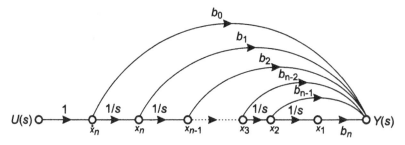

Fig. 7.26(c) *Addition of forward paths b_0 and $b_1 s^{-1}$*

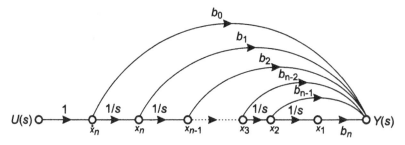

Fig. 7.26 (d) *Simplified state diagram with forward paths satisfying eqn 7.105*

3) $b_2 s^{-2}$**term**   This is obtained by adding a forward path having transmittance b_2, which connects node x_{n-1} with node $Y(s)$ giving forward path gain from $U(s)$ to $Y(s)$

$$= 1 \times \frac{1}{s} \times \frac{1}{s} \times b_2 \times 1 = \frac{b_2}{s^2} = b_2 s^{-2} \qquad (7.109)$$

4) $(b_{n-1} s^{-n+1})$ **term**   This is achieved by adding a forward path having trasmittance b_{n-1} and which connects node x_2 with node $Y(s)$ giving a forward path gain of

$$\frac{b_{n-1}}{s^{n-1}} = b_{n-1} s^{-n+1} \qquad (7.110)$$

5) $b_n s^n$ **term**   This is obtained by adding a forward path having transmittance b_n between nodes x_1 and $Y(s)$

The state diagram depicting the forward paths of eqn (7.105) is shown in Fig. 7.26(d).

Equation (7.107) represents $X(s)$ equal to the input minus the terms associated with the product of $X(s)$ with coefficients $a_1, a_2, \ldots a_n$ times $X(s)$. This implies the existence of feedback paths as explained by Mason's gain formula, starting from different nodes and terminating at a node located after $U(s)$, which in this case is $\dot{x}_n$.

1) Feedback loop (Fig. 7.27(a)) starting from node x_1 and also terminating at node x_1 via node $\dot{x}_n....x_2$ has transmittance of $-a_n$ and gives the loop gain

$$= -a_n \times \frac{1}{s} \times \frac{1}{s} \times ..... \times \frac{1}{s} \left(n \text{ times } \frac{1}{s} \right)$$

$$= \frac{-a_n}{s^n} = -a_n s^{-n} \qquad (7.111)$$

2) Similarly, the feedback loop (7.27(b)) starting from node x_n and terminating at node x_n, *via* node $\dot{x}_n$ has transmittance of $-a_1$. The loop gain is

$$= -a_1 \times \frac{1}{s} = -\frac{a_1}{s} = -a_1 s^{-1} \qquad (7.112)$$

3) Other feedback loops are drawn in a similar fashion. The state diagram depicting all feedback path loops is shown in Fig. 7.27(c).

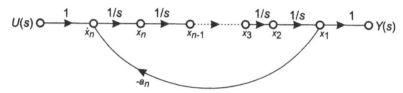

Fig. 7.27(a) *State diagram with feedback path associated with coefficient $-a_n$ and with loop gain of $-a_n s^{-n}$*

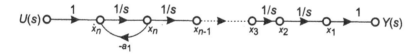

Fig.7.27(b) *State diagram with feedback path associated with coefficient $-a_1$ and with loop gain of $a_1 s^{-1}$*

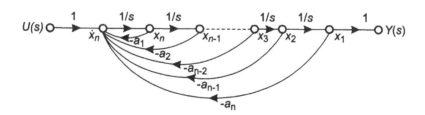

Fig. 7.27(c) *State diagram with all the feedback paths satisfying eqn (7.106)*

The complete state-diagram is obtained by combining Figs. 7.26(d) and 7.27(c) and is shown in Fig. 7.28.

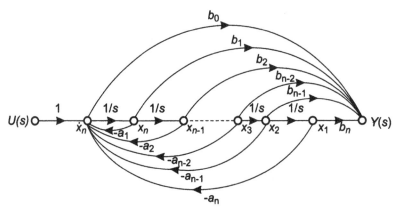

Fig. 7.28 *Complete state diagram*

To obtain the set of first order differential equations which describe the dynamics of the system, a new set of nodes are introduced before each $1/s$ term representing an integrator and shown by the darkened circles (7.21). The paths connecting the additional nodes and the original nodes have unit transmittance.

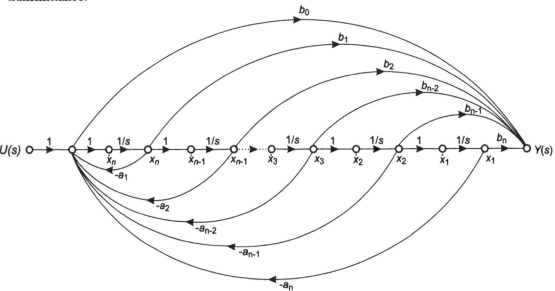

Fig. 7.29 *State diagram with additional nodes (•) introduced preceding each integrator*

Using signal flow graph algebra explained in chapter 5, we can find the relationship between the nodes and obtain a set of first-order differential equations.

Node $\dot{x}_1$ The transmittance is 1 (one) as indicated on the arrow of the branch connecting nodes $\dot{x}_1$ and x_2. Since $\dot{x}_1$ is sink node and x_2 is a

Fig. 7.30

source node, the relation $\dot{x}_1(t) = 1 \times x_2(t)$ holds good. $\hspace{2cm}$ (7.113)

Similarly

$$
\left.
\begin{array}{l}
\dot{x}_2(t) = x_3(t) \\
\dot{x}_3(t) = x_4(t) \\
\text{and} \\
\dot{x}_n = -a_n x_1(t) - a_{n-1} x_2(t) \dots a_2 x_{n-1} - a_1 x_n + u(t)
\end{array}
\right\}
\hspace{1.5cm} (7.114)
$$

where $x_1(t)$, $x_2(t) \dots x_n(t)$ are the state variables.

In matrix form, eqns (7.113) and (7.114) can be written as

$$
\begin{bmatrix}
\dot{x}_1(t) \\
\dot{x}_2(t) \\
\dot{x}_3(t) \\
\vdots \\
\dot{x}_n(t)
\end{bmatrix}
=
\begin{bmatrix}
0 & 1 & 0 & \cdots & 0 \\
0 & 0 & 1 & \cdots & 0 \\
0 & 0 & 0 & \cdots & 1 \\
\vdots & \vdots & \vdots & \vdots & \vdots \\
-a_n & -a_{n-1} & -a_{n-2} & \cdots & -a_1
\end{bmatrix}
\begin{bmatrix}
x_1(t) \\
x_2(t) \\
x_3(t) \\
\vdots \\
x_n(t)
\end{bmatrix}
+
\begin{bmatrix}
0 \\
0 \\
0 \\
\vdots \\
1
\end{bmatrix}
u(t)
\hspace{1cm} (7.115)
$$

or $\hspace{4cm} \dot{x}(t) = Ax(t) + Bu(t) \hspace{3cm}$ (7.116)

Output Equation

At the output node $Y(s)$, the inputs are from

1) node x_1 with transmittance b_n contributing forward path gain $b_n x_1(t)$
2) node x_2 with transmittance b_{n-1} contributing forward path gain $b_{n-1} x_2(t)$
3) node x_3 with transmittance b_{n-2} contributing forward path gain $b_{n-2} x_3(t)$; and so on, and
4) node x_n with transmittance b_1 contributing forward path gain $b_1 x_n(t)$
5) node $U(s)$ with transmittance b_0 contributing forward path gain $b_0 u(t)$.

Therefore

$$
y(t) = b_n x_1(t) + b_{n-1} x_2(t) + b_{n-2} x_3(t) + \dots + b_1 x_n(t) + b_0 u(t)
\hspace{1cm} (7.117)
$$

In matrix form $\hspace{9cm}$ (7.118)

$$
y(t) = \begin{bmatrix} b_n & b_{n-1} & b_{n-2} & \cdots & b_1 \end{bmatrix}
\begin{bmatrix}
x_1(t) \\
x_2(t) \\
x_3(t) \\
\cdots \\
x_n(t)
\end{bmatrix}
+ b_0\, u(t)
\hspace{1cm} (7.119)
$$

It can be seen that eqns (7.115) and (7.119) are in *controllable canonical* form (CCF).

State diagram leading to OCF

Considering the transfer function in polynomial form as given in eqn (7.102), i.e.

$$\frac{Y(s)}{U(s)} = \frac{b_0 s^n + b_1 s^{n-1} + b_2 s^{n-2} + \ldots + b_{n-1} s + b_n}{s^n + a_1 s^{n-1} + a_2 s^{n-2} + \ldots + a_{n-1} s + a_n} \qquad (7.120)$$

or

$$\frac{Y(s)}{U(s)} = \frac{b_0 + b_1 s^{-1} + b_2 s^{-2} + \ldots + b_{n-1} s^{-n+1} + b_n s^{-n}}{1 + a_1 s^{-1} + a_2 s^{-2} + \ldots + a_{n-1} s^{-n+1} + a_n s^{-n}} \qquad (7.121)$$

or $Y(s)(1 + a_1 s^{-1} + a_2 s^{-2} + \cdots + a_{n-1} s^{-n+1} + a_n s^{-n}) = U(s)(b_0 + b_1 s^{-1} + b_2 s^{-2} + \cdots + b_{n-1} s^{-n+1} + b_n s^{-n})$ (7.122)

or $Y(s) = -(a_1 s^{-1} + a_2 s^{-2} + \cdots + a_{n-1} s^{-n+1} + a_n s^{-n})Y(s) + (b_0 + b_1 s^{-1} + b_2 s^{-2} + \cdots + b_{n-1} s^{-n+1} + b_n s^{-n})U(s)$
$$\qquad (7.123)$$

To obtain the state diagram from the transfer function leading to OCF, the state variables are arranged from right to left in descending order as shown in Fig. 7.31(a). The output node $Y(s)$ and input node $U(s)$ are added with unit transmittance.

$$U(s) \circ \xrightarrow{1} \underset{\dot{x}_1}{\circ} \xrightarrow{1/s} \underset{x_1}{\circ} \xrightarrow{1} \underset{\dot{x}_2}{\circ} \xrightarrow{1/s} \underset{x_2}{\circ} \cdots \underset{x_{n-2}}{\circ} \xrightarrow{1} \underset{\dot{x}_{n-1}}{\circ} \xrightarrow{1/s} \underset{x_{n-1}}{\circ} \xrightarrow{1} \underset{\dot{x}_n}{\circ} \xrightarrow{1/s} \underset{x_n}{\circ} \xrightarrow{1} \circ Y(s)$$

Fig. 7.31(a)

The RHS of the eqn (7.123) indicates that the output $Y(s)$ is equal to the difference between the terms associated with the input and the terms associated with the output.

Terms associated with $Y(s)$

All the terms associated with $Y(s)$ are negative, and hence it can be deduced that these terms result in output from node $Y(s)$. All the feedback paths are shown in Fig. 7.31(b).

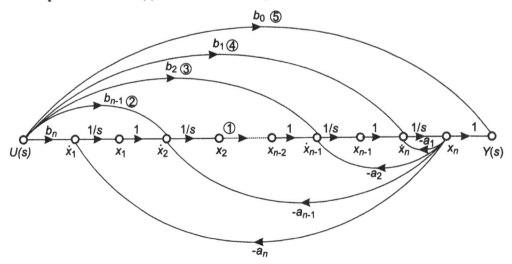

Fig. 7.31(b)

- **Term $-a_1 s^{-1} Y(s)$**

  This term contributes a feedback loop gain of $-a_1 s^{-1}$. On the state diagram it can be depicted by joining node x_n with node $\dot{x}_n$ with branch transmittance of $-a_1$. The loop $(x_n - \dot{x}_n - x_n)$ gives the required loop gain

  $$= -a_1 \times \frac{1}{s} = -a_1 s^{-1} \qquad (7.124)$$

- **Term $-a_2 s^{-2} Y(s)$**

  The loop gain of $-a_2 s^{-2}$ is obtained by connecting node x_n with node $\dot{x}_{n-1}$ with branch transmittance of $-a_2$ and forming the loop $(x_n - \dot{x}_{n-1} - x_{n-1} - \dot{x}_n - x_n)$. The loop gain

  $$= -a_2 \times \frac{1}{s} \times 1 \times \frac{1}{s} = -a_2 s^{-2} \qquad (7.125)$$

- **Term $-a_{n-1} s^{-n+1} Y(s)$**

  Loop $(x_n - \dot{x}_2 - x_2 \ldots \ldots x_{n-2} - \dot{x}_{n-1} - x_{n-1} - \dot{x}_n - x_n)$ having feedback path transmittance of $-a_{n-1}$ gives the required loop gain

  $$= -a_{n-1} \times \frac{1}{s} \times \frac{1}{s} \ldots \ldots \frac{1}{s} \times \frac{1}{s} \left((n-1) \text{ times} \frac{1}{s} \right)$$

  $$= -a_{n-1} s^{-n+1} \qquad (7.126)$$

- **Term $-a_n s^{-n} Y(s)$**

  Loop $(x_n - \dot{x}_1 - x_1 - \dot{x}_2 - x_2 \ldots \ldots x_{n-2} - \dot{x}_{n-1} - x_{n-1} - \dot{x}_n - x_n)$ with feedback path transmittance of $-a_n$ gives the required loop gain

  $$= -a_n \times \frac{1}{s} \times \frac{1}{s} \ldots \ldots \frac{1}{s} \times \frac{1}{s} \left(n \text{ times} \frac{1}{s} \right)$$

  $$= -a_n s^{-n} \qquad (7.127)$$

Terms associated with $U(s)$

Positive terms indicate forward paths emerging out from node $U(s)$ and terminating at different nodes. All the forward paths have been marked by *encircled number* on the state diagram shown in Fig. 7.31(b).

- **Forward path term $b_n s^{-n} Y(s)$**

  The branch transmittance of unity between nodes $U(s)$ and node $\dot{x}_1$ is replaced by b_n. The forward path having gain of $b_n s^{-n}$ has been marked ① on the state diagram shown in Fig. 7.31(b). It passes through the nodes $U(s), \dot{x}_1, x_1, \dot{x}_2, x_2 \ldots x_{n-2}, \dot{x}_{n-1}, x_{n-1}, \dot{x}_n, x_n, Y(s)$.

- **Forward path term $b_{n-1} s^{-n+1} Y(s)$**

  Forward path is created by joining nodes $U(s)$ and $\dot{x}_2$ having transmittance of b_{n-1}. This forward path originates form $U(s)$ and terminates at $Y(s)$ with intermediate node $\dot{x}_2, x_2, \ldots \ldots x_{n-2}, \dot{x}_{n-1}, x_{n-1}, \dot{x}_n, x_n$ and has been marked ② on the state diagram. The forward path gain is

$$= b_{n-1} \times \frac{1}{s} \times \frac{1}{s} \times \ldots \times \frac{1}{s} \times \frac{1}{s} \left((n-1) \text{ times } \frac{1}{s} \right)$$

$$= b_{n-1} s^{-n+1} \tag{7.128}$$

♦ **Forward path term $b_2 s^{-2} Y(s)$**

Another forward path is created by joining nodes $U(s)$ with $\dot{x}_{n-1}$ having transmittance of b_2. Intermediate nodes on the forward path between nodes $U(s)$ and $Y(s)$ are $\dot{x}_{n-1}, x_{n-1}, \dot{x}_n, x_n$. The forward path has been marked ③ on the state diagram shown in Fig. 7.32 and has forward path gain of

$$= b_2 \times \frac{1}{s} \times \frac{1}{s} = b_2 s^{-2} \tag{7.129}$$

Similarly, forward paths marked ④ and ⑤ on the state diagram shown in Fig. 7.32 are created by joining node $U(s)$ with nodes $\dot{x}_n$ and x_n respectively. Forward paths ④ has gain of $b_1 s^{-1}$ and forward path ⑤ has gain of b_0.

State Equation

State equation is obtained with the help of signal flow graph algebra, explained in chapter 5, by writing the input-output relationship at relevant nodes.

node $\dot{x}_1$ $\qquad \dot{x}_1(t) = -a_n x_n(t) + b_n u(t)$ $\hfill (7.130)$

node $\dot{x}_2$ $\qquad \dot{x}_2(t) = x_1(t) - a_{n-1} x_n(t) + b_{n-1} u(t)$ $\hfill (7.131)$

node $\dot{x}_{n-1}$ $\qquad \dot{x}_{n-1} = x_{n-2}(t) - a_2 x_n(t) + b_2 u(t)$ $\hfill (7.132)$

node $\dot{x}_n$ $\qquad \dot{x}_n = x_{n-1}(t) - a_1 x_n(t) + b_1 u(t)$ $\hfill (7.133)$

In the Matrix form, eqns (7.130), (7.131), (7.132) and (7.133) is

$$
\begin{bmatrix} \dot{x}_1(t) \\ \dot{x}_2(t) \\ \dot{x}_3(t) \\ \vdots \\ \dot{x}_{n-1}(t) \\ \dot{x}_n(t) \end{bmatrix} =
\begin{bmatrix}
0 & 0 & \ldots & 0 & -a_n \\
1 & 0 & \ldots & 0 & -a_{n-1} \\
0 & 1 & \ldots & 0 & -a_{n-2} \\
\vdots & \vdots & \ldots & \vdots & \vdots \\
0 & 0 & \ldots & 0 & -a_2 \\
0 & 0 & \ldots & 1 & -a_1
\end{bmatrix}
\begin{bmatrix} x_1(t) \\ x_2(t) \\ x_3(t) \\ \vdots \\ x_{n-1}(t) \\ x_n(t) \end{bmatrix} +
\begin{bmatrix} b_n \\ b_{n-1} \\ b_{n-2} \\ \vdots \\ b_2 \\ b_1 \end{bmatrix} u(t) \tag{7.134}
$$

or $\qquad \dot{x}(t) = Ax(t) + Bu(t)$ $\hfill (7.135)$

Output Equation

At node $Y(s)$

$$y(t) = x_n(t) + b_0 u(t) \tag{7.136}$$

In matrix form

$$y(t) = \begin{bmatrix} 0 & 0 & .... & 0 & 1 \end{bmatrix} \begin{bmatrix} x_1(t) \\ x_2(t) \\ \vdots \\ x_{n-1}(t) \\ x_n(t) \end{bmatrix} + b_0 u(t) \qquad (7.137)$$

or
$$y(t) = Cx(t) + Du(t) \qquad (7.138)$$

Another State-Space Representation

The state diagram is shown in Fig. 7.32. The state-space representation in vector matrix form

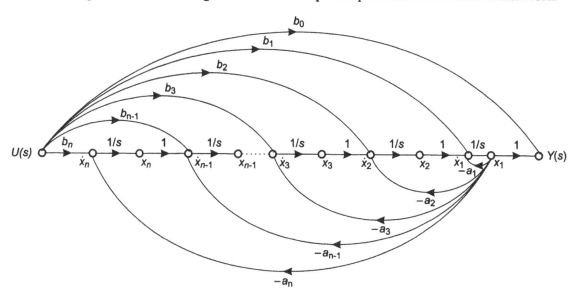

Fig. 7.32

is
$$\dot{x}(t) = \begin{bmatrix} -a_1 & 1 & 0 & \dots & 0 \\ -a_2 & 0 & 1 & \dots & 0 \\ -a_3 & 0 & 0 & \dots & 1 \\ \vdots & \vdots & \vdots & \dots & \vdots \\ -a_n & 0 & 0 & \dots & 0 \end{bmatrix} x(t) + \begin{bmatrix} b_1 \\ b_2 \\ b_3 \\ \vdots \\ b_n \end{bmatrix}$$

$$y(t) = \begin{bmatrix} 1 & 0 & 0 & \cdots & 0 \end{bmatrix} x(t) + b_0 u(t).$$

7.7.2 Cascade Decomposition

When the transfer functions are given in the factored form i.e. both numerator and denominator are expressed as products of first or second order factors, then the state diagram can be easily obtained by cascade decomposition. Let us consider the following transfer function

$$\frac{Y(s)}{U(s)} = \frac{K(s+b_1)(s+b_2)}{(s+a_1)(s+a_2)} \tag{7.139}$$

$$\frac{Y(s)}{Y_1(s)} \times \frac{Y_1(s)}{U(s)} = \frac{K(s+b_1)(s+b_2)}{(s+a_1)(s+a_2)} \tag{7.140}$$

Let

$$\frac{Y(s)}{Y_1(s)} = \frac{(s+b_1)}{(s+a_1)} \quad \text{and} \quad \frac{Y_1(s)}{U(s)} = \frac{K(s+b_2)}{(s+a_2)} \tag{7.141}$$

We will first draw state diagram of $\dfrac{Y_1(s)}{U(s)}$

$$\frac{Y_1(s)}{U(s)} = \frac{K(s+b_2)}{(s+a_2)} = \frac{K(1+b_2 s^{-1})}{1-(-a_2 s^{-1})} \tag{7.142}$$

The state diagram drawn with the help of Masons gain formula is shown in Fig. 7.33. The numerator

$$K(1+b_2 s^{-1}) = K + K b_2 s^{-1} \tag{7.143}$$

indicates that there are two forward paths having forward path gain of K and $K b_2 s^{-1}$. Forward path touching the nodes $u - \dot{x}_2 - y_1$ has path gain of $K \times 1 = K$ and the forward path touching the nodes $u - \dot{x}_2 - x_2 - y_1$ has path gain of $K \times \dfrac{1}{s} \times b_2 = k b_2 s^{-1}$.

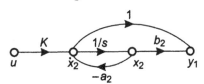

Fig. 7.33 *State function of $Y_1(s)/U(s)$*

The denominator of the transfer function is

$$\frac{Y_1(s)}{U(s)} = \frac{1}{\left[1-(-a_2 s^{-1})\right]} \tag{7.144}$$

Equation (7.144) indicates a feedback path with a loop having loop gain of $-a_2 s^{-1}$. This is accomplished by connecting node x_2 with $\dot{x}_2$ having branch transmittance of $-a_2$ as shown in Fig. 7.33. The loop $x_2 - \dot{x}_2 - x_2$ has

$$\text{loop gain} = -a_2 \times \frac{1}{s} = -a_2 s^{-1} \tag{7.145}$$

Next, we will draw the state diagram for

$$\frac{Y(s)}{Y_1(s)} = \frac{s+b_1}{s+a_1} = \frac{1+b_1 s^{-1}}{1-(-a_1 s^{-1})} \tag{7.146}$$

The state diagram is shown in Fig. 7.34

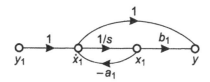

Fig. 7.34 *State diagram of Y(s)/U(s)*

There are two forward paths:

1) forward path $y_1 - \dot{x}_1 - y$ has path gain of 1; and

2) forward path $y_1 - \dot{x}_1 - x_1 - y$ has path gain of $1 \times \dfrac{1}{s} \times b_1 = b_1 s^{-1}$.

There is only one feedback loop $x_1 - \dot{x}_1 - x_1$ and the loop gain is

$$= -a_1 \times \frac{1}{s} = -a_1 s^{-1} \tag{7.147}$$

Next step is to combine Figs. 7.33 and 7.34 to obtain state diagram for the transfer function given in eqn (7.143). The combined state diagram is shown in Fig. 7.35.

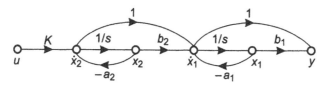

Fig. 7.35

State Equation

State equation is obtained by writing the input-output relationship at nodes $\dot{x}_1$ and $\dot{x}_2$ by using signal flow graph algebra.

Node $\dot{x}_1$

The input to sink node $\dot{x}_1$ is from source nodes x_1, x_2 and $\dot{x}_2$. Hence

$$\dot{x}_1(t) = -a_1 x_1(t) + b_2 x_2(t) + \dot{x}_2(t) \tag{7.148}$$

Node $\dot{x}_2$

The input to the sink node $\dot{x}_2$ is from nodes x_2 and u. Hence

$$\dot{x}_2(t) = -a_2 x_2(t) + Ku(t) \tag{7.149}$$

Substituting value of $\dot{x}_2(t)$ from eqn (7.149) in eqn (7.148), we get the following equations:

$$\dot{x}_1(t) = -a_1 x_1(t) + (b_2 - a_2)x_2(t) + Ku(t) \tag{7.150}$$

$$\dot{x}_2(t) = -a_2 x_2(t) + Ku(t) \tag{7.151}$$

In vector matrix form, the eqns (7.150) and (7.151) can be expressed as

$$\begin{bmatrix} \dot{x}_1(t) \\ \dot{x}_2(t) \end{bmatrix} = \begin{bmatrix} -a_1 & b_2-a_2 \\ 0 & -a_2 \end{bmatrix} \begin{bmatrix} x_1(t) \\ x_2(t) \end{bmatrix} + \begin{bmatrix} K \\ K \end{bmatrix} u(t) \tag{7.152}$$

or

$$\dot{x}(t) = Ax(t) + Bu(t) \tag{7.153}$$

where

$$A = \begin{bmatrix} -a_1 & b_2-a_2 \\ 0 & -a_2 \end{bmatrix} \text{ and } B = \begin{bmatrix} K \\ K \end{bmatrix} \tag{7.154}$$

Output Equation

Output equation is obtained by writing the node equation at sink node y. Input to node y is from two source nodes i.e. x_1 and $\dot{x}_1$. Hence

$$y(t) = b_1 x_1(t) + \dot{x}_1(t) \tag{7.155}$$

Substituting the value of $\dot{x}_1(t)$ from eqn (7.150) in eqn (7.155), we get

$$y(t) = b_1 x_1(t) - a_1 x_1(t) + (b_2 - a_2)x_2(t) + Ku(t) \tag{7.156}$$

or

$$y(t) = (b_1 - a_1)x_1(t) + (b_2 - a_2)x_2(t) + Ku(t) \tag{7.157}$$

In vector matrix form

$$y(t) = \begin{bmatrix} b_1 - a_1 & b_2 - a_2 \end{bmatrix} \begin{bmatrix} x_1(t) \\ x_2(t) \end{bmatrix} + Ku(t) \tag{7.158}$$

or

$$y(t) = Cx(t) + Du(t) \tag{7.159}$$

where

$$C = \begin{bmatrix} b_1 - a_1 & b_2 - a_2 \end{bmatrix} \text{ and } D = K \tag{7.160}$$

7.7.3 Parallel Decomposition

When the transfer function can be expanded by partial fraction expansion, parallel decomposition method is employed to generate the state diagram. The complete state diagram is obtained by combining in parallel the state diagrams for each fraction. Let us consider the transfer function of a system expressed as a ratio of polynomials

$$\frac{Y(s)}{U(s)} = \frac{N(s)}{D(s)} \tag{7.161}$$

Let

$$\frac{N(s)}{D(s)} = \frac{C_1}{s+a_1} + \frac{C_2}{s+a_2} \tag{7.162}$$

where $N(s)$ is a polynomial of order less than two, a_1 and a_2 are constants either real or complex and C_1 and C_2 are residues.

◆ **State diagram** $\dfrac{C_1}{s+a_1}$

The expression can also be expressed as

$$= \frac{C_1 s^{-1}}{[1-(-a_1 s^{-1})]} \tag{7.163}$$

The eqn (7.163) indicates that the state diagram will have

1) a forward path with gain $C_1 s^{-1}$; and
2) a feedback loop with gain $-a_1 s^{-1}$.

The state diagram shown in Fig. 7.36 satisfies the above conditions.

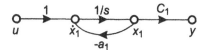

Fig. 7.36

In the same fashion, the state diagram for the next term, $\dfrac{C_2}{s+a_2}$, can be drawn and is shown in Fig. 7.37

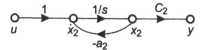

Fig. 7.37

The final state diagram is obtained by combining Figs. 7.36 and 7.37 in parallel as shown in Fig. 7.38.

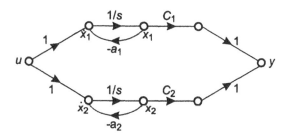

Fig. 7.38

State Equation and Output Equation

State equation is obtained by writing node equations at nodes $\dot{x}_1$ and $\dot{x}_2$:

$$\dot{x}_1(t) = a_1 x_1(t) + u(t)$$
$$\dot{x}_2(t) = a_2 x_2(t) + u(t)$$
$$y(t) = C_1 x_1(t) + C_2 x_2(t) \tag{7.164}$$

In matrix form,

$$\begin{bmatrix} \dot{x}_1(t) \\ \dot{x}_2(t) \end{bmatrix} = \begin{bmatrix} -a_1 & 0 \\ 0 & -a_2 \end{bmatrix} \begin{bmatrix} x_1(t) \\ x_2(t) \end{bmatrix} + \begin{bmatrix} 1 \\ 1 \end{bmatrix} u(t) \tag{7.165}$$

and

$$y(t) = \begin{bmatrix} C_1 & C_2 \end{bmatrix} \begin{bmatrix} x_1(t) \\ x_2(t) \end{bmatrix} \tag{7.166}$$

or

$$\dot{x}(t) = Ax(t) + Bu(t)$$
$$y(t) = Cx(t) \tag{7.167}$$

where

$$A = \begin{bmatrix} -a_1 & 0 \\ 0 & -a_2 \end{bmatrix}; \ B = \begin{bmatrix} 1 \\ 1 \end{bmatrix} \text{ and } C = \begin{bmatrix} C_1 & C_2 \end{bmatrix} \tag{7.168}$$

The state transition equations resemble the diagonal canonical form. In the case of control systems having transfer function with multiple eigenvalues, the state transition equation would resemble the Jordan canonical form.

7.8 MATLAB

MATLAB built-in functions help manipulate state-space representations with Case.

◆ **[A B C D] = tf2ss(num,den)** It converts the system in the transfer function form

$$\frac{Y(s)}{U(s)} = \frac{\text{num}}{\text{den}} = C(sI - A)^{-1}B + D \tag{7.169}$$

into the state-space form represented by
$$\dot{x}(t) = Ax(t) + Bu(t)$$
$$y(t) = Cx(t) + Du(t) \tag{7.170}$$

◆ **[num, den] = ss2tf(A, B, C, D)** It converts the state-space form represented by eqn (7.170) into the transfer function form defined by eqn (7.169).

◆ **[num, den] = ss2tf(A, B, C, D, iu)** If the system has more than one input, then this MATLAB command converts the state-space representation given by eqn (7.170) into the transfer function form

$$\frac{Y(s)}{U_i(s)} = i\text{th element of transfer function represented by eqn (7.169). i.e.}$$

$$= i\text{th element of } [C(sI - A)^{-1}B + D].$$

◆ **[num, den] = tfdata(sys, 'V')** It returns the numerator and denominator as row vectors.

♦ **sysT = *ss2ss*(sys, T)** Given a state-space model *sys* with equation given in eqn (7.170), the command *ss2ss* performs the similarity transformation $\bar{x} = Tx$ on the state vector x and generates the equivalent state-space model *sysT* with equation

$$\dot{x} = TAT^{-1}x + TBu$$
$$y = CT^{-1}x + Du \qquad (7.171)$$

The output generated is the transformed state-space model *sysT*, given *sys* and the state coordinate transformation T

♦ **csyss = *canon*(sys, 'type')** If *sys* represents the continuous linear time-invariant system, the command *canon* computes its canonical state-space. Two types of canonical state-space model supported are:

♦♦ **Companion form** For a system whose dynamics are described by the characteristic polynomial $s^n + a_1 s_{n-1} + \dots + a_{n-1}s + a_n$; the corresponding system matrix (matrix A) returned is of the form

$$A = \begin{bmatrix} 0 & 0 & \dots & 0 & -a_n \\ 1 & 0 & \dots & 0 & -a_{n-1} \\ \vdots & \vdots & \dots & \vdots & \vdots \\ 0 & 0 & \dots & 1 & -a_2 \\ 0 & 0 & \dots & 0 & -a_1 \end{bmatrix} \qquad (7.172)$$

The companion transformation requires that the system be controllable form the first input. Since, this form is poorly conditioned for most state-space computations; its use should be avoided.

♦♦ **Modal form** The modal form is represented by system matrix A (considering eigenvalues of the system are λ_1, λ_2, $\sigma \pm j\omega$)

$$A = \begin{bmatrix} \lambda_1 & 0 & 0 & 0 \\ 0 & \lambda_2 & 0 & 0 \\ 0 & 0 & \sigma & \omega \\ 0 & 0 & -\omega & \sigma \end{bmatrix}$$

The modal form of transformation requires that matrix A is diagonalizable i.e. matrix A has no repeated eigenvalues

♦ **[csys, T] = *canon*(a, b, c, d, 'type)** For state-space models *sys*, the command returns the state coordinate transformation T related by

$$x_c = Tx_c \qquad (7.172)$$

where x is state vector and x_c is the canonical state vector. If the *sys* is not a state-space model, this syntax returns $T = [\ ]$

♦ **ss(sys)** It converts an arbitrary transfer function or zero-pole gain model *sys* to an equivalent state-space model.

SUMMARY

◆ State-space representation is not unique. By selecting minimum number of linearly independent state variables, a control system can be represented in state-space.

◆ Differential equations which help in describing dynamics of a control system can be simulated by state-variable diagrams (block as well signal flow diagrams)

◆ State model of a control system can be represented in

- Physical variable form.
- Phase variable form.
- Canonical variable form.

◆ A state-model can be transformed to another form by similarity transformation.

◆ Controllability and Observability Canonical forms are two useful transformation forms.

◆ The process by which state diagrams are drawn form the transfer functions is called decomposition. Various methods are

- Direct.
- Cascade.
- Parallel.

PROBLEMS AND SOLUTIONS

Problem 7.1

Obtain the state-space representation of the mechanical system shown in Fig 7.39

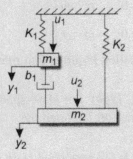

Fig. 7.39

Solution The mechanical system is described by the following equations:

$$m_1 \ddot{y}_1 + b_1 (\dot{y}_1 - \dot{y}_2) + k_1 y_1 = u_1$$

and

$$m_2 \ddot{y}_2 + b_1 (\dot{y}_2 - \dot{y}_1) + k_2 y_2 = u_2$$

State-Space representation

Let

$$\begin{array}{ll} y_1 = x_1 & y_2 = x_3 \\ \dot{y}_1 = x_2 = \dot{x}_1 \quad \text{and} \quad & \dot{y}_2 = x_4 = \dot{x}_3 \\ \ddot{y}_1 = \dot{x}_2 & \ddot{y}_2 = \dot{x}_4 \end{array}$$

Substituting the values in the equations give

$$m_1 \dot{x}_2 + b_1 (x_2 - x_4) + k_1 x_1 = u_1$$
$$m_2 \dot{x}_4 + b_1 (x_4 - x_2) + k_2 x_3 = u_2$$

which gives

$$\dot{x}_2 = \frac{-b_1 (x_2 - x_4)}{m_1} - \frac{k_1 x_1}{m_1} + \frac{1}{m_1} u_1$$

and

$$\dot{x}_4 = \frac{-b_1 (x_4 - x_2)}{m_2} - \frac{k_2 x_3}{m_2} + \frac{1}{m_2} u_2$$

Expressing in vector matrix form by relating $\dot{x}_1$, $\dot{x}_2$, $\dot{x}_3$ $\dot{x}_4$, y_1 and y_2, with x_1, x_2, x_3 x_4, u_1 and u_2,

$$\begin{bmatrix} \dot{x}_1 \\ \dot{x}_2 \\ \dot{x}_3 \\ \dot{x}_4 \end{bmatrix} = \begin{bmatrix} 0 & 1 & 0 & 0 \\ \dfrac{-k_1}{m_1} & \dfrac{-b_1}{m_1} & 0 & \dfrac{b_1}{m_1} \\ 0 & 0 & 0 & 1 \\ 0 & \dfrac{b_1}{m_2} & \dfrac{-k_2}{m_2} & \dfrac{-b_1}{m_2} \end{bmatrix} \begin{bmatrix} x_1 \\ x_2 \\ x_3 \\ x_4 \end{bmatrix} + \begin{bmatrix} 0 & 0 \\ \dfrac{1}{m_1} & 0 \\ 0 & 0 \\ 0 & \dfrac{1}{m_2} \end{bmatrix} \begin{bmatrix} u_1 \\ u_2 \end{bmatrix}$$

and

$$\begin{bmatrix} y_1 \\ y_2 \end{bmatrix} = \begin{bmatrix} 1 & 0 & 0 & 0 \\ 0 & 0 & 1 & 0 \end{bmatrix} \begin{bmatrix} x_1 \\ x_2 \\ x_3 \\ x_4 \end{bmatrix}$$

Problem 7.2

For the circuit shown in Fig. 7.40, identify a set of state variables

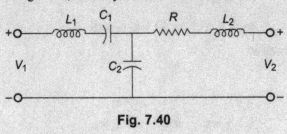

Fig. 7.40

Solution   Assuming the current distribution as shown in Fig. 7.41. The differential equations which describe the electrical circuit based on Kirchoff's laws are given below:

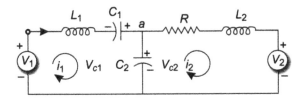

Fig. 7.41

Loop 1
$$\frac{L_1 di_1}{dt} - v_{c1} + v_{c2} - v_1 = 0$$

Loop 2
$$Ri_2 + \frac{L_2 di_2}{dt} + v_2 - v_{c2} = 0$$

Capacitor C_1
$$C_1 \frac{dv_{c1}}{dt} = i_1$$

Capacitor C_2
$$C_2 \frac{dv_{c2}}{dt} = i_1 - i_2 .$$

Generally the inductor currents and capacitor voltages are selected as state variables. Let

$$i_1 = x_1(t)$$
$$i_2 = x_2(t)$$
$$v_{c1} = x_3(t)$$
$$v_{c2} = x_4(t)$$

Substituting the state variables in the differential equations, we get

$$L_1 \dot{x}_1(t) - x_3(t) + x_4(t) - v_1 = 0$$
$$Rx_2(t) + L_2 \dot{x}_2(t) + v_2 - x_4(t) = 0$$
$$C_1 \dot{x}_3(t) = x_1(t)$$
$$C_2 \dot{x}_4(t) = x_1(t) - x_2(t)$$

Rearranging, we get

$$\dot{x}_1(t) = \frac{1}{L_1} x_3(t) - \frac{1}{L_1} x_4(t) + \frac{1}{L_1} v_1$$

$$\dot{x}_2(t) = -\frac{R}{L_2} x_2(t) + \frac{1}{L_2} x_4(t) - \frac{1}{L_2} v_2$$

$$\dot{x}_3(t) = \frac{1}{C_1} x_1(t)$$

$$\dot{x}_4(t) = \frac{1}{C_2} x_1(t) - \frac{1}{C_2} x_2(t)$$

The state-space representation in vector matrix form is

$$\begin{bmatrix} \dot{x}_1(t) \\ \dot{x}_2(t) \\ \dot{x}_3(t) \\ \dot{x}_4(t) \end{bmatrix} = \begin{bmatrix} 0 & 0 & \frac{1}{L_1} & -\frac{1}{L_1} \\ 0 & -\frac{R}{L_2} & 0 & \frac{1}{L_2} \\ \frac{1}{C_1} & 0 & 0 & 0 \\ \frac{1}{C_2} & -\frac{1}{C_2} & 0 & 0 \end{bmatrix} \begin{bmatrix} x_1(t) \\ x_2(t) \\ x_3(t) \\ x_4(t) \end{bmatrix} + \begin{bmatrix} \frac{1}{L_1} & 0 \\ 0 & -\frac{1}{L_2} \end{bmatrix} \begin{bmatrix} v_1 \\ v_2 \end{bmatrix}$$

The output in the network is the voltage across the capacitor C_2

$$y = v_{c2} = x_4(t)$$

$$y = \begin{bmatrix} 0 & 0 & 0 & 1 \end{bmatrix} \begin{bmatrix} x_1(t) \\ x_2(t) \\ x_3(t) \\ x_4(t) \end{bmatrix}$$

Problem 7.3

Obtain a state-space description for the electrical circuit shown in Fig. 7.42.

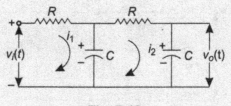

Fig. 7.42

Solution The differential equations for the system shown in Fig. 7.42 are:

$$Ri_1(t)+\frac{1}{C}\int(i_1-i_2(t))dt=v_i(t)$$

$$Ri_2(t)+\frac{1}{C}\int i_2(t)dt+\frac{1}{C}\int(i_2(t)-i_1(t))dt=0$$

The output

$$\frac{1}{C}\int i_2\,dt=v_0(t)$$

Selecting the state variables as,

$$x_1(t)=\frac{1}{C}\int i_2\,dt$$

$$x_2(t)=\frac{1}{C}\int i_1\,dt$$

we get

$$i_1(t)=C\dot{x}_2(t)$$

$$i_2(t)=C\dot{x}_1(t)$$

Substitution of the state variables in the equations, yield

$$RC\dot{x}_2(t)+x_2(t)-x_1(t)=v_i(t)$$

$$RC\dot{x}_1(t)+2x_1(t)-x_2(t)=0$$

$$x_1(t)=v_0(t)=y(t)$$

The state-space representation in matrix notation

$$\begin{bmatrix}\dot{x}_1(t)\\\dot{x}_2(t)\end{bmatrix}=\begin{bmatrix}\dfrac{-2}{RC}&\dfrac{1}{RC}\\[2mm]\dfrac{1}{RC}&\dfrac{-1}{RC}\end{bmatrix}\begin{bmatrix}x_1(t)\\x_2(t)\end{bmatrix}+\begin{bmatrix}0\\\dfrac{1}{RC}\end{bmatrix}v_i(t)$$

and

$$y(t)=v_0(t)=\begin{bmatrix}1&0\end{bmatrix}\begin{bmatrix}x_1(t)\\x_2(t)\end{bmatrix}$$

Problem 7.4

Obtain state-space description for the electrical circuit shown in Fig. 7.43.

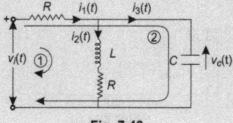

Fig. 7.43

Solution The differential equations for the system shown in Fig 7.43 are

Loop 1 $\qquad\qquad v_i(t) = Ri_1(t) + \dfrac{L di_2(t)}{dt} + Ri_2(t)$

Loop 2 $\qquad\qquad v_i(t) = Ri_1(t) + v_c(t)$

Voltage $v_c(t)$ $\qquad C\dfrac{dv_c(t)}{dt} = i_3(t) = i_1(t) - i_2(t)$

Current $\qquad\qquad i_1(t) - i_2(t) = i_3(t)$

Let us select current through the inductor $i_2(t)$ and voltage across the capacitor $v_c(t)$ as the state variables. Let

$$x_1(t) = i_2(t)$$
$$x_2(t) = v_c(t)$$

Substituting the state-variables in the differential equations,

$$v_i(t) = Ri_1(t) + L\dot{x}_1(t) + Rx_1(t) \tag{1}$$

$$v_i(t) = Ri_1(t) + x_2(t) \tag{2}$$

$$C\dot{x}_2(t) = i_3(t) = i_1(t) - x_1(t) \tag{3}$$

$$i_1(t) - x_1(t) = i_3(t) \tag{4}$$

Substituting eqn (2) in eqn (1), we get

$$Ri_1(t) + x_2(t) = Ri_1(t) + L\dot{x}_1(t) + Rx_1(t)$$

or $\qquad\qquad \dot{x}_1(t) = \dfrac{-R}{L}x_1(t) + \dfrac{1}{L}x_2(t) \tag{5}$

From eqn (2)

$$i_1(t) = \dfrac{1}{R}v_i(t) - \dfrac{1}{R}x_2(t) \tag{6}$$

Substituting $i_1(t)$ from eqn (6) in eqn (3), we get

$$C\dot{x}_2(t) = \dfrac{1}{R}v_i(t) - \dfrac{1}{R}x_2(t) - x_1(t)$$

or $\qquad\qquad \dot{x}_2(t) = -\dfrac{1}{C}x_1(t) - \dfrac{1}{RC}x_2(t) + \dfrac{1}{RC}v_i(t)$

Considering $v_c(t)$ as the output

$$y = v_c(t) = x_2(t) \tag{8}$$

Using eqns (5) and (7) for writing the state equation, and eqn (8) as output equation, the state-space descriptions can be written as

$$\begin{bmatrix} \dot{x}_1(t) \\ \dot{x}_2(t) \end{bmatrix} = \begin{bmatrix} -\dfrac{R}{L} & \dfrac{1}{L} \\ -\dfrac{1}{C} & -\dfrac{1}{RC} \end{bmatrix} \begin{bmatrix} x_1(t) \\ x_2(t) \end{bmatrix} + \begin{bmatrix} 0 \\ \dfrac{1}{RC} \end{bmatrix} v_i(t)$$

$$y(t) = \begin{bmatrix} 0 & 1 \end{bmatrix} \begin{bmatrix} x_1(t) \\ x_2(t) \end{bmatrix}$$

Problem 7.5

An RLC network is shown in Fig. 7.44. Obtain state differential equations for the circuit.

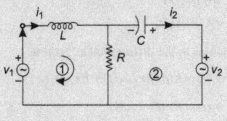

Fig. 7.44

The differential equations for the given electrical circuit are

Loop 1 $$v_1(t) = \frac{L di_1(t)}{dt} + R(i_1(t) - i_2(t)) \tag{1}$$

Loop 2 $$v_c(t) - v_2(t) + R(i_1(t) - i_2(t)) = 0 \tag{2}$$

Combined loop $$\frac{L di_1(t)}{dt} - v_c(t) + v_2(t) = v_1(t) \tag{3}$$

Also $$\frac{1}{C} \int i_2(t) dt = v_c(t)$$

or $$i_2(t) = C \dot{v}_c(t) \tag{4}$$

Let the current through the inductor L and voltage across capacitor C be the state variables.

$$x_1(t) = i_1(t) \tag{5}$$

$$x_2(t) = v_c(t) \tag{6}$$

The inputs are $v_1(t)$ and $v_2(t)$. Equating eqns (1) and (3), we get

$$\frac{L di_1(t)}{dt} + R(i_1(t) - i_2(t)) = \frac{L di_1(t)}{dt} - v_c(t) + v_2(t)$$

or $$v_c(t) - v_2(t) + R(i_1(t) - i_2(t)) = 0$$

or $$v_c(t) - v_2(t) + R(i_1(t) - C \dot{v}_c(t)) = 0$$

or
$$x_2(t) - v_2(t) + Rx_1(t) - RC\dot{x}_2(t) = 0$$

or
$$\dot{x}_2(t) = \frac{1}{C}x_1(t) + \frac{1}{RC}x_2(t) - \frac{1}{RC}v_2(t) = 0 \qquad (7)$$

Substituting state variables in eqn (3), we get

$$L\dot{x}_1(t) - x_2(t) + v_2(t) = v_1(t)$$

or
$$\dot{x}_1(t) = \frac{1}{L}x_2(t) + \frac{1}{L}v_1(t) - \frac{1}{L}v_2(t) \qquad (8)$$

State equation in vector matrix notation is obtained from eqns (7) and (8).

$$\begin{bmatrix} \dot{x}_1(t) \\ \dot{x}_2(t) \end{bmatrix} = \begin{bmatrix} 0 & \frac{1}{L} \\ \frac{1}{C} & \frac{1}{RC} \end{bmatrix} \begin{bmatrix} x_1(t) \\ x_2(t) \end{bmatrix} + \begin{bmatrix} \frac{1}{L} & -\frac{1}{L} \\ 0 & -\frac{1}{RC} \end{bmatrix} \begin{bmatrix} v_1(t) \\ v_2(t) \end{bmatrix}$$

Let the voltage, v_c across the capacitor be the output, then from eqn (6), we get

$$y(t) = \begin{bmatrix} 0 & 1 \end{bmatrix} \begin{bmatrix} x_1(t) \\ x_2(t) \end{bmatrix}$$

where
$$y(t) = v_c(t) = x_2(t)$$

Problem 7.6

Find the state-space represention for the electrical circuit shown in Fig. 7.45 considering the capacitor voltage and the current the inductor as the state variables and $i_1(t)$ as the output variable.

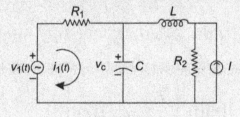

Fig. 7.45

Solution Simplifying the circuit diagram by converting the current source into the voltage source, we get the simplified circuit shown in Fig. 7.46

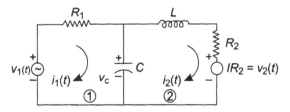

Fig. 7.46

The differential equations for the given circuit diagram are

Loop 1
$$R_1 i_1(t) + v_c(t) - v_1(t) = 0 \tag{1}$$

Loop 2
$$\frac{L di_2(t)}{dt} + R_2 i_2(t) + v_2(t) - v_c(t) = 0 \tag{2}$$

Capacitor voltage
$$\frac{C dv_c(t)}{dt} = i_1(t) - i_2(t) \tag{3}$$

Substituting $i_1(t)$ from eqn (1) in eqn (3), we get

$$C\frac{dv_c(t)}{dt} = -\frac{1}{R_1}v_c(t) + \frac{1}{R_1}v_1(t) - i_2(t)$$

or
$$\frac{dv_c(t)}{dt} = -\frac{1}{CR_1}v_c(t) + \frac{1}{CR_1}v_1(t) - \frac{1}{C}i_2(t) \tag{4}$$

Let the state variables be

$$x_1(t) = i_2(t); \quad x_2(t) = v_c(t); \quad y(t) = i_1(t)$$

Inputs are $v_1(t)$ and $v_2(t) = IR_2$ and output is $y = i_1(t)$.

Substitution of state variables in eqns (2) and (4) gives

$$L\dot{x}_1(t) + R_2 x_1(t) + v_2(t) - x_2(t) = 0$$

or
$$\dot{x}_1(t) = \frac{-R_2}{L}x_1(t) + \frac{1}{L}x_2(t) - \frac{v_2(t)}{L} \tag{5}$$

and
$$\dot{x}_2(t) = -\frac{1}{CR_1}x_2(t) + \frac{1}{CR_1}v_1(t) - \frac{1}{C}x_1(t) \tag{6}$$

or
$$\dot{x}_2(t) = -\frac{1}{C}x_1(t) - \frac{1}{CR_1}x_2(t) + \frac{1}{CR_1}v_1(t) \tag{7}$$

The state equations obtained from eqns (5) and (7) in matrix notation is

$$\begin{bmatrix} \dot{x}_1(t) \\ \dot{x}_2(t) \end{bmatrix} = \begin{bmatrix} -\dfrac{R_2}{L} & \dfrac{1}{L} \\ -\dfrac{1}{C} & -\dfrac{1}{CR_1} \end{bmatrix} \begin{bmatrix} x_1(t) \\ x_2(t) \end{bmatrix} + \begin{bmatrix} 0 & -\dfrac{1}{L} \\ \dfrac{1}{CR_1} & 0 \end{bmatrix} \begin{bmatrix} v_1(t) \\ v_2(t) \end{bmatrix}$$

Substituting the state variables in eqn (1), we get the output equation,

$$R_1 y(t) + x_2(t) - v_1(t) = 0$$

or
$$y(t) = \frac{-x_2(t)}{R_1} + \frac{v_1(t)}{R_1} \tag{8}$$

In vector matrix notation eqn (8) is written as

$$y(t) = \begin{bmatrix} 0 & -\dfrac{1}{R_1} \end{bmatrix} \begin{bmatrix} x_1(t) \\ x_2(t) \end{bmatrix} + \begin{bmatrix} \dfrac{1}{R_1} & 0 \end{bmatrix} \begin{bmatrix} v_1(t) \\ v_2(t) \end{bmatrix}$$

Problem 7.7

Determine the state variable differential matrix for the electrical circuit shown in Fig. 7.47.

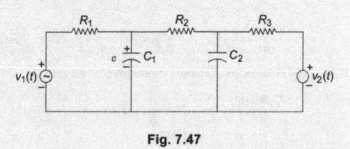

Fig. 7.47

Solution   Assuming $i_1(t)$, $i_2(t)$ and $i_3(t)$ be the loop currents as shown in Fig. 7.48, the differential equations are

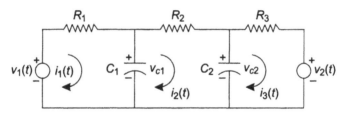

Fig. 7.48

Loop 1

$$R_1 i_1(t) + v_{c1}(t) - v_1(t) = 0 \tag{1}$$

or

$$i_1(t) = -\frac{1}{R_1} v_{c1}(t) + \frac{1}{R_1} v_1(t)$$

Loop 2

$$R_2 i_2(t) + v_{c2}(t) - v_{c1}(t) = 0$$

or

$$i_2(t) = \frac{1}{R_2} v_{c1}(t) - \frac{1}{R_2} v_{c2}(t) \tag{2}$$

Loop 3

$$R_3 i_3(t) + v_2(t) - v_{c2}(t) = 0$$

or

$$i_3(t) = \frac{1}{R_3} v_{c2}(t) - \frac{1}{R_3} v_2(t) \tag{3}$$

Capacitor voltages

$$C_1 \frac{dv_{c1}(t)}{dt} = i_1(t) - i_2(t) \tag{4}$$

and

$$C_2 \frac{dv_{c2}(t)}{dt} = i_2(t) - i_3(t) \tag{5}$$

Substituting $i_1(t)$ and $i_2(t)$ from eqns (1) and (2) in eqn (4), we get

$$\frac{C_1 dv_{c1}(t)}{dt} = \frac{-1}{R_1}v_{c1}(t) + \frac{1}{R_1}v_1(t) - \frac{1}{R_2}v_{c1}(t) + \frac{1}{R_2}v_{c2}(t)$$

or

$$\frac{dv_{c1}(t)}{dt} = \frac{-1}{R_1 C_1}v_{c1}t - \frac{1}{R_2 C_1}v_{c1}(t) + \frac{1}{R_2 C_1}v_{c2}(t) + \frac{1}{R_1 C_1}v_1(t) \tag{6}$$

and substituting $i_2(t)$ and $i_3(t)$ from eqns (2) and (3) in eqn (5), we get

$$\frac{C_2 dv_{c2}(t)}{dt} = \frac{1}{R_2}v_{c1}(t) - \frac{1}{R_2}v_{c2}(t) + \frac{-1}{R_3}v_{c2} + \frac{1}{R_3}v_2(t)$$

or

$$\frac{dv_{c2}(t)}{dt} = \frac{1}{R_2 C_2}v_{c1}(t) - \frac{1}{R_2 C_2}v_{c2}(t) - \frac{1}{R_3 C_2}v_{c2}(t) + \frac{1}{R_3 C_2}v_2(t) \tag{7}$$

The state variables, output and input are as follows:

Let
$$\left.\begin{array}{l} x_1(t) = v_{c1}(t) \\ x_2(t) = v_{c2}(t) \end{array}\right\} \text{ be the state variables}$$

and $v_1(t)$ and $v_2(t)$ be the inputs.

Substituting state variables in eqns (6) and (7), we get

$$\dot{x}_1(t) = -\left(\frac{1}{R_1 C_1} + \frac{1}{R_2 C_1}\right)x_1(t) + \frac{1}{R_2 C_1}x_2(t) + \frac{1}{R_1 C_1}v_1(t) \tag{8}$$

$$\dot{x}_2(t) = \frac{1}{R_2 C_2}x_1(t) - \left(\frac{1}{R_2 C_2} + \frac{1}{R_3 C_2}\right)x_2(t) + \frac{1}{R_3 C_2}v_2(t) \tag{9}$$

State equation in matrix notation as obtained from eqns (8) and (9) is

$$\begin{bmatrix} \dot{x}_1(t) \\ \dot{x}_2(t) \end{bmatrix} = \begin{bmatrix} -\left(\dfrac{1}{R_1 C_1} + \dfrac{1}{R_2 C_1}\right) & \dfrac{1}{R_2 C_1} \\ \dfrac{1}{R_2 C_2} & -\left(\dfrac{1}{R_2 C_2} + \dfrac{1}{R_3 C_2}\right) \end{bmatrix}\begin{bmatrix} x_1(t) \\ x_2(t) \end{bmatrix} + \begin{bmatrix} \dfrac{1}{R_1 C_1} & 0 \\ 0 & \dfrac{1}{R_3 C_2} \end{bmatrix}\begin{bmatrix} v_1(t) \\ v_2(t) \end{bmatrix}$$

Output Equation

Let the capacitor voltage V_{c1} be the output then

$$y(t) = v_{c1}(t) = x_1(t)$$

Then

$$y(t) = \begin{bmatrix} 1 & 0 \end{bmatrix}\begin{bmatrix} x_1(t) \\ x_2(t) \end{bmatrix} \tag{10}$$

If we assume the capacitor voltages $V_{c1}(t)$ and $V_{c2}(t)$ to be the outputs i.e.

$$y_1(t) = v_{c1}(t) = x_1(t)$$
$$y_2(t) = v_{c2}(t) = x_2(t)$$

then the matrix notation for the output equation is

$$y(t) = \begin{bmatrix} y_1(t) \\ y_2(t) \end{bmatrix} = \begin{bmatrix} 1 & 0 \\ 0 & 1 \end{bmatrix} \begin{bmatrix} x_1(t) \\ x_2(t) \end{bmatrix} \qquad (11)$$

If we assume voltage across R_2 i.e., $i_2(t)R_2$ as the output voltage, then

$$y(t) = v_{c1}(t) - v_{c2}(t)$$
$$= x_1(t) - x_2(t)$$

or $$y(t) = \begin{bmatrix} 1 & -1 \end{bmatrix} \begin{bmatrix} x_1(t) \\ x_2(t) \end{bmatrix} \qquad (12)$$

Problem 7.8

Write the state equations for the rotation system shown in Fig. 7.49 in matrix form.

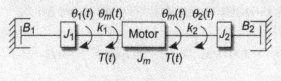

Fig. 7.49

Solution   The network diagram for the given rotational system is shown in Fig. 7.50

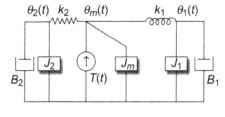

Fig. 7.50

The nodal equations are as follows:

Node $\theta_m(t)$ $T(t) = J_m \dfrac{d^2\theta_m(t)}{dt^2} + k_2(\theta_m(t) - \theta_2(t)) + k_1(\theta_m(t) - \theta_1(t))$

or
$$\frac{d^2\theta_m(t)}{dt^2} = \frac{-k_1}{J_m}(\theta_m(t) - \theta_1(t)) - \frac{k_2}{J_m}(\theta_m(t) - \theta_2(t)) + \frac{T(t)}{J_m}$$
(1)

Node $\theta_1(t)$
$$J_1\frac{d^2\theta_1(t)}{dt^2} + B_1\frac{d\theta_1(t)}{dt} + K_1(\theta_1(t) - \theta_m(t)) = 0$$

or
$$\frac{d^2\theta_1(t)}{dt^2} = \frac{-B_1}{J_1}\frac{d\theta_1(t)}{dt} + \frac{k_1}{J_1}(\theta_m(t) - \theta_1(t))$$
(2)

Node $\theta_2(t)$
$$\frac{d^2\theta_2(t)}{dt^2} = \frac{-B_2}{J_2}\frac{d\theta_2(t)}{dt} + \frac{k_2}{J_2}(\theta_m(t) - \theta_2(t))$$
(3)

Assigning the state variables, let

$$x_1(t) = \theta_m(t) - \theta_1(t); \quad x_2(t) = \theta_m(t) - \theta_2(t)$$
(4)

$$x_3(t) = \frac{d\theta_1(t)}{dt}; \quad x_4(t) = \frac{d\theta_2(t)}{dt}; \quad x_5(t) = \frac{d\theta_m(t)}{dt}$$

Input $\quad u(t) = T(t)$

Substituting the state variables in eqns (1), (2) and (3), we get

$$\frac{dx_5(t)}{dt} = \dot{x}_5(t) = -\frac{k_1}{J_m}x_1(t) - \frac{k_2}{J_m}x_2(t) + \frac{1}{J_m}u(t)$$
(5)

$$\frac{dx_3(t)}{dt} = \dot{x}_3(t) = -\frac{B_1}{J_1}x_3(t) + \frac{k_1}{J_1}x_1(t)$$
(6)

$$\frac{dx_4(t)}{dt} = \dot{x}_4(t) = -\frac{B_2}{J_2}x_4(t) + \frac{k_2}{J_2}x_2(t)$$
(7)

Differentiation of eqns (4) and (5) gives

$$\frac{dx_1(t)}{dt} = \frac{d\theta_m(t)}{dt} - \frac{d\theta_1(t)}{dt}$$

or
$$\dot{x}_1(t) = x_5(t) - x_3(t)$$
(8)

and
$$\frac{dx_2(t)}{dt} = \frac{d\theta_m(t)}{dt} - \frac{d\theta_2(t)}{dt}$$

or
$$\dot{x}_2(t) = x_5(t) - x_4(t)$$
(9)

In vector matrix notation, eqns (5), (6), (7), (8) and (9) can be written as

$$\begin{bmatrix} \dot{x}_1(t) \\ \dot{x}_2(t) \\ \dot{x}_3(t) \\ \dot{x}_4(t) \\ \dot{x}_5(t) \end{bmatrix} = \begin{bmatrix} 0 & 0 & -1 & 0 & 1 \\ 0 & 0 & 0 & -1 & 1 \\ \dfrac{k_1}{J_1} & 0 & \dfrac{-B_1}{J_1} & 0 & 0 \\ 0 & \dfrac{k_2}{J_2} & 0 & -\dfrac{B_2}{J_2} & 0 \\ \dfrac{-k_1}{J_m} & \dfrac{-k_2}{J_m} & 0 & 0 & 0 \end{bmatrix} \begin{bmatrix} x_1(t) \\ x_2(t) \\ x_3(t) \\ x_4(t) \\ x_5(t) \end{bmatrix} + \begin{bmatrix} 0 \\ 0 \\ 0 \\ 0 \\ \dfrac{1}{J_m} \end{bmatrix} u(t)$$

Problem 7.9

Write the state equation and output equation for the electrical circuit shown in Fig. 7.51.

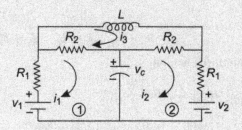

Fig. 7.51

Solution The differential equations for the given circuit are

Loop 1 $v_1(t) = i_1(t)R_1 + R_2(i_1(t) - i_3(t)) + v_c(t)$ (1)

Loop 2 $v_c(t) = R_2(i_2(t) - i_3(t)) + i_2(t)R_1 + v_2(t)$ (2)

Loop 3 $\dfrac{L di_3(t)}{dt} + R_2(i_3(t) - i_2(t)) + R_2(i_3(t) - i_1(t)) = 0$ (3)

or $\dfrac{L di_3(t)}{dt} + 2R_2 i_3(t) - R_2(i_1(t) + i_2(t)) = 0$

or $\dfrac{di_3(t)}{dt} - \dfrac{R_2}{L}(i_1(t) + i_2(t)) + \dfrac{2R_2}{L} i_3(t) = 0$ (4)

Capacitor voltage $\dfrac{C dv_c(t)}{dt} = i_1(t) - i_2(t)$ (5)

or $\dfrac{dv_c(t)}{dt} = \dfrac{1}{C}(i_1(t) - i_2(t))$ (6)

From eqn (4)

$$\frac{di_3(t)}{dt} = \frac{R_2}{L}(i_1(t) + i_2(t)) - \frac{2R_2}{L}i_3(t) \tag{7}$$

From eqn (1)

$$v_1(t) = i_1(t)(R_1 + R_2) - i_3(t)R_2 + v_c(t)$$

or

$$i_1(t) = \frac{v_1(t)}{R_1 + R_2} + \frac{i_3(t)R_2}{R_1 + R_2} - \frac{v_c(t)}{R_1 + R_2} \tag{8}$$

From eqn (2)

$$i_2(t) = \frac{-v_2(t)}{R_1 + R_2} + \frac{v_c(t)}{R_1 + R_2} + \frac{i_3(t)R_2}{R_1 + R_2} \tag{9}$$

From eqns (8) and (9)

$$i_1(t) + i_2(t) = \frac{v_1(t)}{R_1 + R_2} - \frac{v_2(t)}{R_1 + R_2} + \frac{2i_3(t)R_2}{R_1 + R_2} \tag{10}$$

and

$$i_1(t) - i_2(t) = \frac{v_1(t)}{R_1 + R_2} + \frac{v_2(t)}{R_1 + R_2} - \frac{2v_c(t)}{R_1 + R_2} \tag{11}$$

Substituting eqn (10) in eqn (4), we get

$$\frac{di_3(t)}{dt} = \frac{R_2}{L}(i_1(t) + i_2(t)) - \frac{2R_2}{L}i_3(t)$$

$$= \frac{R_2}{L}\left[\frac{v_1(t)}{R_1 + R_2} - \frac{v_2(t)}{R_1 + R_2} + \frac{2R_2i_3(t)}{R_1 + R_2}\right] - \frac{2R_2i_3(t)}{L}$$

$$= \frac{R_2v_1(t)}{L(R_1 + R_2)} - \frac{R_2v_2(t)}{L(R_1 + R_2)} + \left(\frac{2R_2^2}{L(R_1 + R_2)} - \frac{2R_2}{L}\right)i_3(t)$$

$$= \frac{R_2v_1(t)}{L(R_1 + R_2)} - \frac{R_2v_2(t)}{L(R_1 + R_2)} + \left(\frac{2R_2^2 - 2R_1R_2 - 2R_2^2}{L(R_1 + R_2)}\right)i_3(t)$$

or

$$\frac{di_3(t)}{dt} = \frac{R_2v_1(t)}{L(R_1 + R_2)} - \frac{R_2v_2(t)}{L(R_1 + R_2)} - \frac{2R_1R_2i_3(t)}{L(R_1 + R_2)} \tag{12}$$

Substituting eqn (11) in eqn (6), we get

$$\frac{dv_c(t)}{dt} = \frac{1}{C}(i_1(t) - i_2(t))$$

$$= \frac{1}{C}\left(\frac{v_1(t)}{R_1 + R_2} + \frac{v_2(t)}{R_1 + R_2} - \frac{2v_c(t)}{R_1 + R_2}\right) \tag{13}$$

Let the state variables

$$x_1(t) = i_3(t)$$
$$x_2(t) = v_c(t)$$

Substituting state variables in eqns (12) and (13), we get

$$\dot{x}_1(t) = \frac{R_2}{L(R_1 + R_2)} v_1(t) - \frac{R_2}{L(R_1 + R_2)} v_2(t) - \frac{2R_1 R_2}{L(R_1 + R_2)} x_1(t) \qquad (14)$$

$$\dot{x}_2(t) = \frac{1}{C(R_1 + R_2)} v_1(t) + \frac{1}{C(R_1 + R_2)} v_2(t) - \frac{2}{C(R_1 + R_2)} x_2(t) \qquad (15)$$

The vector matrix notation obtained from eqns (14) and (15) are

$$\begin{bmatrix} \dot{x}_1(t) \\ \dot{x}_2(t) \end{bmatrix} = \begin{bmatrix} -\dfrac{2R_1 R_2}{L(R_1 + R_2)} & 0 \\ 0 & \dfrac{-2}{C(R_1 + R_2)} \end{bmatrix} \begin{bmatrix} x_1(t) \\ x_2(t) \end{bmatrix} + \begin{bmatrix} \dfrac{R_2}{L(R_1 + R_2)} & -\dfrac{R_2}{L(R_1 + R_2)} \\ \dfrac{1}{C(R_1 + R_2)} & \dfrac{1}{C(R_1 + R_2)} \end{bmatrix} \begin{bmatrix} v_1(t) \\ v_2(t) \end{bmatrix}$$

Let the voltage across the capacitor C be the output.

$$y = v_c(t) = x_2(t)$$

Therefore

$$y = \begin{bmatrix} 0 & 1 \end{bmatrix} \begin{bmatrix} x_1(t) \\ x_2(t) \end{bmatrix}$$

Problem 7.10

Obtain the state variable matrix in phase variable form for a system described with a differential equation

$$\frac{2d^3 y}{dt^3} + \frac{4d^2 y}{dt^2} + \frac{6dy}{dt} + 8y = 10u(t)$$

Solution Simplification of the given differential equation gives

$$\frac{d^3 y}{dt^3} + \frac{2d^2 y}{dt^2} + \frac{3dy}{dt} + 4y = 5u(t)$$

Let

$$y = x_1$$

$$\dot{x}_1 = x_2 = \frac{dy}{dt} \qquad (1)$$

$$\dot{x}_2 = x_3 = \frac{d^2 y}{dt^2} \qquad (2)$$

$$\dot{x}_3 = \frac{d^3 y}{dt^3} = -4y - \frac{3dy}{dt} - \frac{2d^2 y}{dt^2} + 5u(t)$$

or

$$\dot{x}_3 = -4x_1 - 3x_2 - 2x_3 + 5u(t) \qquad (3)$$

State model in vector matrix notation as obtained from eqns (1), (2) and (3) is

$$
\begin{bmatrix} \dot{x}_1 \\ \dot{x}_2 \\ \dot{x}_3 \end{bmatrix} = \begin{bmatrix} 0 & 1 & 0 \\ 0 & 0 & 1 \\ -4 & -3 & -2 \end{bmatrix} \begin{bmatrix} x_1 \\ x_2 \\ x_3 \end{bmatrix} + \begin{bmatrix} 0 \\ 0 \\ 5 \end{bmatrix} u(t)
$$

and

$$
y = \begin{bmatrix} 1 & 0 & 0 \end{bmatrix} \begin{bmatrix} x_1 \\ x_2 \\ x_3 \end{bmatrix}
$$

Problem 7.11

Determine the state variable description in the phase variable format for the system described by the transfer function

$$
\frac{Y(s)}{U(s)} = \frac{4(s+1)(s+3)}{(s+2)(s+4)(s+5)}
$$

Solution Writing the transfer function as a ratio of polynomials

$$
\frac{Y(s)}{U(s)} = \frac{4(s^2+4s+3)}{s^3+11s^2+38s+40}
$$

$$
= \frac{4s^2+16s+12}{s^3+11s^2+38s+40}
$$

$$
= \frac{4s^{-1}+16s^{-2}+12s^{-3}}{1+11s^{-1}+38s^{-2}+40s^{-3}} \tag{1}
$$

$$
\frac{Y(s)}{X(s)} \times \frac{X(s)}{U(s)} = \frac{4s^{-1}+16s^{-2}+12s^{-3}}{1+11s^{-1}+38s^{-2}+40s^{-3}}
$$

$$
\frac{Y(s)}{X(s)} = 4s^{-1}+16s^{-2}+12s^{-3}
$$

and

$$
\frac{X(s)}{U(s)} = \frac{1}{1+11s^{-1}+38s^{-2}+40s^{-3}}
$$

which gives

$$
Y(s) = (4s^{-1}+16s^{-2}+12s^{-3})X(s) \tag{2}
$$

and

$$
U(s) = (1+11s^{-1}+38s^{-2}+40s^{-3})X(s)
$$

or

$$
U(s) = X(s)+(11s^{-1}+38s^{-2}+40s^{-3})X(s)
$$

or

$$
X(s) = U(s)-(11s^{-1}+38s^{-2}+40s^{-3})X(s) \tag{3}
$$

The state equation is directly obtained from eqn (3) as

$$\begin{bmatrix} \dot{x}_1(t) \\ \dot{x}_2(t) \\ \dot{x}_3(t) \end{bmatrix} = \begin{bmatrix} 0 & 1 & 0 \\ 0 & 0 & 1 \\ -40 & -38 & -11 \end{bmatrix} \begin{bmatrix} x_1(t) \\ x_2(t) \\ x_3(t) \end{bmatrix} + \begin{bmatrix} 0 \\ 0 \\ 1 \end{bmatrix} u(t)$$

and the output equation from eqn (2) as

$$y(t) = \begin{bmatrix} 12 & 16 & 4 \end{bmatrix} \begin{bmatrix} x_1(t) \\ x_2(t) \\ x_3(t) \end{bmatrix}$$

Considering y as the output and u as input, let us write the state equation from the state model formed above.

$$\dot{x}_1(t) = x_1(t)$$
$$\dot{x}_2(t) = x_3(t)$$
$$\dot{x}_3(t) = -40x_1(t) - 38x_2(t) - 11x_3(t) + u(t) \tag{4}$$

and
$$y(t) = 12x_1(t) + 16x_2(t) + 4x_3(t) \tag{5}$$

Let

$$y(t) = x_1(t)$$
$$x_2(t) = \dot{x}_1(t) = \dot{y}(t)$$
$$x_3(t) = \dot{x}_2(t) = \ddot{y}(t)$$
$$\dot{x}_3(t) = \ddot{x}_2(t) = \dddot{y}(t)$$

Let us draw the nodes $y(t)$, $x_1(t)$, $x_2(t)$, $x_3(t)$, $\dot{x}_3(t)$ from right to left as shown in the Fig. 7.52.

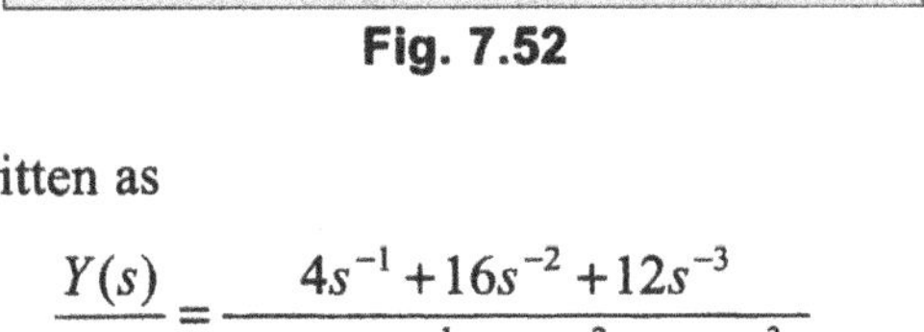

Fig. 7.52

Equation (1) can be written as

$$\frac{Y(s)}{U(s)} = \frac{4s^{-1} + 16s^{-2} + 12s^{-3}}{1 - (-11s^{-1} - 38s^2 - 40s^{-3})} \tag{6}$$

The numerator specifies the forward paths, and the denominator, the feedback loop gains.

Addition of forward path can be done as follows:

The output equation can be written as

$$y(t) = 12x_1(t) + 16x_2(t) + 4x_3(t)$$

This signifies that three paths each originating from nodes $u(t)$ of gain $12s^{-3}$, $16s^{-2}$ and $4s^{-1}$ respectively and terminating at node $y(t)$. The forward paths are shown in Fig. 7.53.

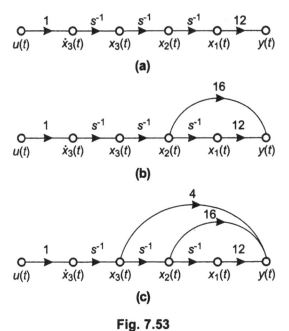

Fig. 7.53

Feedback paths are shown in Fig.7.54 and are obtained from denominator of transfer function described by eqn (4). Three feedback paths originate from nodes $x_1(t)$, $x_2(t)$ and $x_3(t)$ and form the following loops

(a) loop $x_1(t) - \dot{x}_3(t) - x_3(t) - x_2(t) - x_1(t)$ with loop gain of $-40s^{-3}$ as shown in Fig. 7.54 (a).

(b) loop $x_2(t) - \dot{x}_3(t) - x_3(t) - x_2(t)$ with loop gain of $-38s^{-2}$ as shown in Fig. P7.54(b)

(c) loop $x_3(t) - \dot{x}_3(t) - x_3(t)$ with loop gain $-11s^{-1}$ as shown in Fig. 7. 54(c)

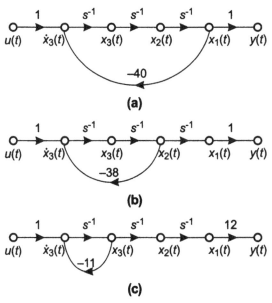

Fig. 7.54

The complete state-diagram is obtained by combining Figs 7.53 and 7.54 and is shown in Fig. 7.55.

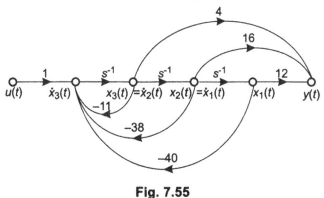

Fig. 7.55

Problem 7.12

A linear time-invariant system in represented by the differential equation

$$\frac{d^3 y(t)}{dt^3} + \frac{10 d^2 y(t)}{dt^2} + \frac{2 dy(t)}{dt} + y(t) + \int_0^t y(\tau) d\tau = u(t)$$

Write the state equation and output equation is vector matrix notation.

Solution

Let

$$x_1(t) = \int_0^t y(\tau) d\tau \tag{1}$$

$$x_2(t) = \frac{dx_1(t)}{dt} = \dot{x}_1(t) = y(t) \tag{2}$$

$$x_3(t) = \frac{dy(t)}{dt} = \dot{x}_2(t) \tag{3}$$

$$x_4(t) = \frac{d^2 y(t)}{dt^2} = \dot{x}_3(t) \tag{4}$$

$$\dot{x}_4(t) = \frac{d^3 y(t)}{dt^3} = -\frac{10 d^2 y}{dt^2} - \frac{2 dy(t)}{dt} - y(t) - \int_0^t y(\tau) d\tau + u(t)$$

$$= -10 x_4(t) - 2 x_3(t) - x_2(t) - x_1(t) + u(t) \tag{5}$$

The state equation is

$$\begin{bmatrix} \dot{x}_1(t) \\ \dot{x}_2(t) \\ \dot{x}_3(t) \\ \dot{x}_4(t) \end{bmatrix} = \begin{bmatrix} 0 & 1 & 0 & 0 \\ 0 & 0 & 1 & 0 \\ 0 & 0 & 0 & 1 \\ -1 & -1 & -2 & -10 \end{bmatrix} \begin{bmatrix} x_1(t) \\ x_2(t) \\ x_3(t) \\ x_4(t) \end{bmatrix} + \begin{bmatrix} 0 \\ 0 \\ 0 \\ 1 \end{bmatrix} u(t)$$

The output equation is obtained from eqn (2)

$$y(t) = \begin{bmatrix} 0 & 1 & 0 & 0 \end{bmatrix} \begin{bmatrix} x_1(t) \\ x_2(t) \\ x_3(t) \\ x_4(t) \end{bmatrix}$$

Problem 7.13

Represent the system consisting of DC servo motor with the load shown in Fig. 7.56, in state space.

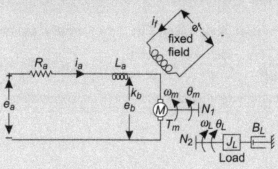

Fig. 7.56

Solution   The torque developed by the motor is given by

$$T_m = \frac{Jd\omega_m}{dt} + B\omega_m \tag{1}$$

where J and B are the equivalent of load converted to motor side.

The torque developed by the motor T_m is given by

$$T_m = k_m i_a \tag{2}$$

Equating (1) and (2), we get

$$\frac{Jd\omega_m}{dt} + B\omega_m = k_m i_a \tag{3}$$

But

$$\omega_m = \frac{N_2}{N_1}\omega_L \tag{4}$$

Therefore

$$J\frac{N_2}{N_1}\frac{d\omega_L}{dt} + \frac{N_2}{N_1}B\omega_L = k_m i_a$$

or
$$\frac{d\omega_L}{dt} = -\frac{B\omega_L}{J} + \frac{k_m}{J}\frac{N_1}{N_2}i_a \qquad (5)$$

Also
$$\frac{d\theta_L}{dt} = \omega_L \qquad (6)$$

and the voltage e_a, after applying Kirchoff's law to the armature circuit, is given by

$$e_a = L_a\frac{di_a}{dt} + R_a i_a + e_b \qquad (7)$$

or
$$e_a = L_a\frac{di_a}{dt} + R_a i_a + k_b\omega_m \qquad (8)$$

or
$$e_a = L_a\frac{di_a}{dt} + R_a i_a + k_b\frac{N_2}{N_1}\omega_L \qquad (9)$$

or
$$\frac{di_a}{dt} = -\frac{k_b}{L_a}\frac{N_2}{N_1}\omega_L - \frac{R_a}{L_a}i_a + \frac{1}{L_a}e_a \qquad (10)$$

Let the state variables

$$x_1 = \omega_L$$

$$x_2 = \dot{x}_1 = \frac{dx_1}{dt} = \frac{d\omega_L}{dt} = \theta_L$$

$$\dot{x}_2 = \frac{d\theta_L}{dt}$$

$$x_3 = i_a$$

$$\dot{x}_3 = \frac{di_a}{dt}$$

Substituting the state variables in eqns (5), (6) and (10), we get

$$\dot{x}_1 = -\frac{B}{J}x_1 + \frac{k_m}{J}\frac{N_1}{N_2}x_3 \qquad (11)$$

$$\dot{x}_2 = x_1 \qquad (12)$$

$$\dot{x}_3 = -\frac{k_b}{L_a}\frac{N_2}{N_1}x_1 - \frac{R}{L_a}x_3 + \frac{1}{L_a}e_a \qquad (13)$$

The state equation in vector matrix notation as obtained from eqns (11), (12) and (13) is given below:

$$\begin{bmatrix} \dot{x}_1 \\ \dot{x}_2 \\ \dot{x}_3 \end{bmatrix} = \begin{bmatrix} \dfrac{-B}{J} & 0 & \dfrac{k_m}{J}\dfrac{N_1}{N_2} \\ 1 & 0 & 0 \\ \dfrac{-k_b}{L_a}\dfrac{N_2}{N_1} & 0 & \dfrac{-R}{L_a} \end{bmatrix}\begin{bmatrix} x_1 \\ x_2 \\ x_3 \end{bmatrix} + \begin{bmatrix} 0 \\ 0 \\ \dfrac{1}{L_a} \end{bmatrix}e_a$$

Assuming output $y = \theta_L$, the output equation is $y = \theta_L = x_2$, or

$$y = \begin{bmatrix} 0 & 1 & 0 \end{bmatrix} \begin{bmatrix} x_1 \\ x_2 \\ x_3 \end{bmatrix}$$

If we assume $y = \theta_m$ as the output, then

$$y = \theta_m = \frac{N_1}{N_2}\theta_L = \frac{N_1}{N_2}x_2$$

and the output equation is

$$y = \begin{bmatrix} 0 & \dfrac{N_1}{N_2} & 0 \end{bmatrix} \begin{bmatrix} x_1 \\ x_2 \\ x_3 \end{bmatrix}$$

Problem 7.14

Draw a signal flow graph for the following state equations

(a) $\dot{x} = \begin{bmatrix} 0 & 1 & 0 \\ 0 & 0 & 1 \\ -4 & -6 & -8 \end{bmatrix} x + \begin{bmatrix} 0 \\ 0 \\ 1 \end{bmatrix} u$

$\quad y = \begin{bmatrix} 1 & 1 & 0 \end{bmatrix} \begin{bmatrix} x_1 \\ x_2 \\ x_3 \end{bmatrix}$

(b) $\dot{x} = \begin{bmatrix} 0 & 1 & 0 \\ 0 & -2 & 1 \\ -2 & -3 & -4 \end{bmatrix} x + \begin{bmatrix} 0 \\ 2 \\ 2 \end{bmatrix} u$

$\quad y = \begin{bmatrix} 2 & 2 & 0 \end{bmatrix} \begin{bmatrix} x_1 \\ x_2 \\ x_3 \end{bmatrix}$

Solution

(a) The state equations are

$$\dot{x}_1 = x_2 \tag{1}$$

$$\dot{x}_2 = x_3 \tag{2}$$

$$\dot{x}_3 = -4x_1 - 6x_2 - 8x_3 + u \tag{3}$$

$$y = x_1 + x_2 \tag{4}$$

Considering y as the output, u as the input and the nodes y, x_1, x_2, x_3, $\dot{x}_3$ and u from right to left as shown in Fig. 7.57, the state diagram is drawn.

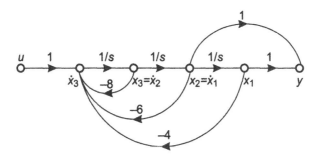

Fig. 7.57

(i) From eqn (3), it is seen that nodes x_1, x_2, x_3 and u having transmittance of -4, -6, -8 and 1 respectively are connected to node $\dot{x}_3$.

(ii) From eqn (4) it is seen that at node y, the inputs are from node x_1 and x_2, both having transmittance of 1.

(iii) Eqns (1) and (2) indicate that node $x_2 = \dot{x}_1$ and the node $x_3 = \dot{x}_2$.

(b) The state equations are

$$\dot{x}_1 = x_2 \tag{1}$$

$$\dot{x}_2 = -2x_2 + x_3 + 2u \tag{2}$$

$$\dot{x}_3 = -2x_1 - 3x_2 - 4x_3 + 2u \tag{3}$$

$$y = 2x_1 + 2x_2 \tag{4}$$

Let us consider u as the input and y as the output. The other nodes are $x_1, \dot{x}_1, x_2, \dot{x}_2, x_3$ and $\dot{x}_3$. These nodes have been shown in Fig. 7.58(a).

(i) From eqn (4), it is seen that input to the node y is from nodes x_1 and x_2 both having transmittance of 2. These are depicted in Fig. 7.58(b).

(ii) From eqn (3) it is seen that input to node $\dot{x}_3$ is from nodes x_1, x_2, x_3 and u, having transmittance of -2, -3, -4 and 2 respectively. These as depicted in Fig.7.58(c).

(iii) From eqn (2), it is seen that input to node $\dot{x}_2$ is from nodes x_2, x_3 and u, having transmittance of -2, 1 and 2 respectively. These are depicted in Fig. 7.58(d).

(iv) Similarly from eqn (1), it is seen that input to node $\dot{x}_1$ is from node x_2, having unit transmittance. This has been shown in Fig. 7.58(d)

The remaining paths i.e. $\dot{x}_3 - x_3$, $\dot{x}_2 - x_2$ and $\dot{x}_1 - x_1$ are marked with transmittance of $1/s$. The complete state-diagram is shown in Fig. 7.58(e)

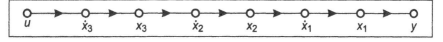

(a)

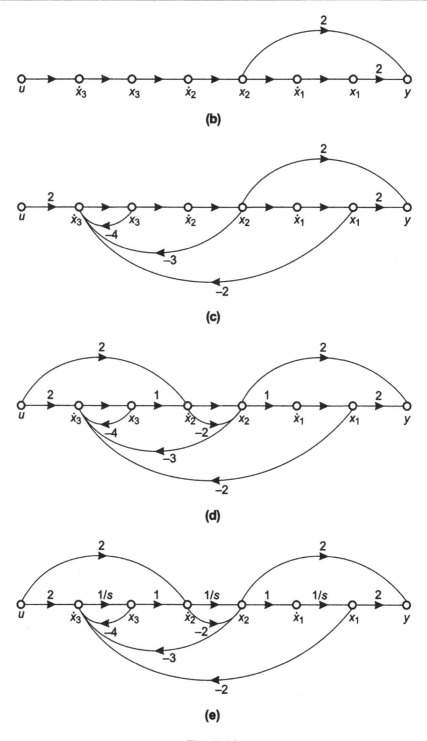

(b)

(c)

(d)

(e)

Fig. 7.58

Problem 7.15

If the open-loop transfer function of a control system

$$G(s) = \frac{-200(s^2 + 6s + 8)}{(s+10)(s-3)(s+5)},$$

Draw the signal flow diagram and determine the state-space representation in

(a) Phase-variable form   (b) Controller canonical form

(c) Observer canonical form   (d) Parallel form

Solution

(a) **Phase Variable Form**

$$G(s) = \frac{-200(s^2 + 6s + 8)}{(s+10)(s-3)(s+5)}$$

$$\frac{Y(s)}{U(s)} = G(s) = \frac{-200(s^2 + 6s + 8)}{s^3 + 12s^2 + 5s - 150}$$

$$\frac{Y(s)}{X(s)} \times \frac{X(s)}{U(s)} = \frac{-200(s^2 + 6s + 8)}{s^3 + 12s^2 + 5s - 150} = \frac{-200(s^{-1} + 6s^{-2} + 8s^{-3})}{1 + 12s^{-1} + 5s^{-2} - 150s^{-3}}$$

Let

$$\frac{Y(s)}{X(s)} = -200(s^{-1} + 6s^{-2} + 8s^{-3})$$

and

$$\frac{X(s)}{U(s)} = \frac{1}{1 + 12s^{-1} + 5s^{-2} - 150s^{-3}}$$

$$\frac{Y(s)}{U(s)} = G(s) = \frac{-200(s^{-1} + 6s^{-2} + 8s^{-3})}{1 - (-12s^{-1} - 5s^{-2} + 150s^{-3})}$$

According to Mason's gain formula, every term of the numerator enclosed in brackets indicate forward path gains, and the terms enclosed in the brackets also indicate summation of gain of all loops. This means there are three forward paths originating from input u having forward path gains of -200 times $\frac{1}{s}$, $\frac{6}{s^2}$ and $\frac{8}{s^3}$ respectively. The forward paths terminate at output y and are joined through intermediate nodes x_1, x_2, x_3, x_4. These are shown in Fig. 7.59.

(a)

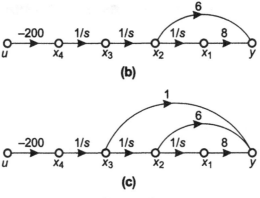

(b)

(c)

Fig. 7.59

Let us now add the loops to Fig. 7.60(c) defined by the terms enclosed in brackets in the denominator.

(a) the first term indicates a loop gain of $-12/s$. This is obtained by joining x_3 with x_4, with a branch having a transmittance of -12. The loop x_3–x_4–x_3 gives the desired loop gain of $-12/s$. The other two terms indicate the following loops:

(b) loop $x_2 - x_4 - x_3 - x_2$ formed by joining x_2, with x_4 having branch transmittance of -5 and loop gain of $-5/s^2$.

(c) loop $x_1 - x_4 - x_3 - x_2 - x_1$ formed by joining x_1 with x_4, having branch transmittance of 150 and loop gain of $150/s^3$.

The entire signal flow graph is shown in Fig. 7.60.

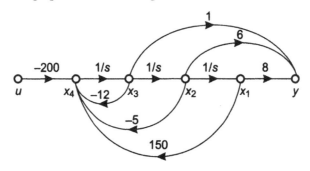

Fig. 7.60

$$x_1 = \frac{1}{s} x_2 \text{ or } \dot{x}_1 = x_2$$

$$x_2 = \frac{1}{s} x_3 \text{ or } \dot{x}_2 = x_3$$

$$x_3 = \frac{1}{s} x_4 \text{ or } \dot{x}_3 = x_4 = -12x_3 - 5x_2 + 150x_1 - 200u$$

and

$$y = 8x_1 + 6x_2 + x_3$$

The state equations in vector-matrix notation are

$$\begin{bmatrix} \dot{x}_1 \\ \dot{x}_2 \\ \dot{x}_3 \end{bmatrix} = \begin{bmatrix} 0 & 1 & 0 \\ 0 & 0 & 1 \\ 150 & -5 & -12 \end{bmatrix} \begin{bmatrix} x_1 \\ x_2 \\ x_3 \end{bmatrix} + \begin{bmatrix} 0 \\ 0 \\ -200 \end{bmatrix} u$$

$$y = \begin{bmatrix} 8 & 6 & 1 \end{bmatrix} \begin{bmatrix} x_1 \\ x_2 \\ x_3 \end{bmatrix}$$

(b) **Controller Canonical Form**

To represent the state-model in controller canonical form, the nodes y, x_3, x_2, x_1, x_0 and u are located from right to left, as shown in Fig. 7.61.

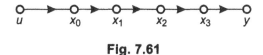

Fig. 7.61

The forward paths are

(a) $u - x_0 - x_1 - x_2 - x_3 - y$ with forward path gain $-200 \times \dfrac{1}{s} \times \dfrac{1}{s} \times \dfrac{1}{s} \times 8$

(b) $u - x_0 - x_1 - x_2 - y$ with forward path gain $-200 \times \dfrac{1}{s} \times \dfrac{1}{s} \times 6$

(c) $u - x_0 - x_1 - y$ with forward path gain of $-200 \times \dfrac{1}{s} \times 1$

These forward paths are depicted in Fig. 7.62.

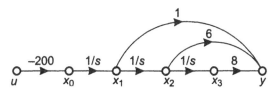

Fig. 7.62

The feedback loops are as follows:

(a) $x_1 - x_0 - x_1$ have loop gain $= -12 \times \dfrac{1}{s}$

(b) $x_2 - x_0 - x_1 - x_2$ have loop gain $= -5 \times \dfrac{1}{s} \times \dfrac{1}{s}$

(c) $x_3 - x_0 - x_1 - x_2 - x_3$ have loop gain $= 150 \times \dfrac{1}{s} \times \dfrac{1}{s} \times \dfrac{1}{s}$

The feedback loops are shown in Fig. 7.63

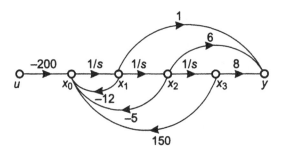

Fig. 7.63

$$x_1 = \frac{1}{s}x_0 \text{ or } \dot{x}_1 = x_0 = -12x_1 - 5x_2 + 150x_3 - 200u$$

$$x_2 = \frac{1}{s}x_1 \text{ or } \dot{x}_2 = x_1$$

$$x_3 = \frac{1}{s}x_2 \text{ or } \dot{x}_3 = x_2$$

and

$$y = x_1 + 6x_2 + 8x_3$$

The state equation are

$$\begin{bmatrix} \dot{x}_1 \\ \dot{x}_2 \\ \dot{x}_3 \end{bmatrix} = \begin{bmatrix} -12 & -5 & 150 \\ 1 & 0 & 0 \\ 0 & 1 & 0 \end{bmatrix} \begin{bmatrix} x_1 \\ x_2 \\ x_3 \end{bmatrix} + \begin{bmatrix} -200 \\ 0 \\ 0 \end{bmatrix} u$$

$$y = \begin{bmatrix} 1 & 6 & 8 \end{bmatrix} \begin{bmatrix} x_1 \\ x_2 \\ x_3 \end{bmatrix}$$

(c) Observer Canonical Form

$$\frac{Y(s)}{U(s)} = \frac{-200(s^2 + 6s + 8)}{s^3 + 12s^2 + 5s - 150}$$

or

$$\frac{Y(s)}{U(s)} = \frac{\dfrac{-200}{s} - \dfrac{1200}{s^2} - \dfrac{1600}{s^3}}{1 + \dfrac{12}{s} + \dfrac{5}{s^2} - \dfrac{150}{s^3}}$$

$$\left(\frac{-200}{s} - \frac{1200}{s^2} - \frac{1600}{s^3} \right) U(s) = \left(1 + \frac{12}{s} + \frac{5}{s^2} - \frac{150}{s^3} \right) Y(s)$$

Rearranging,

$$Y(s) = \frac{1}{s}\{-200U(s) - 12Y(s)\} + \frac{1}{s^2}\{-1200U(s) - 5Y(s)\} + \frac{1}{s^3}\{-1600U(s) + 150Y(s)\} \quad (1)$$

Locate the nodes y, x_1, x_2, x_3 and u from right to left. Insert additional nodes as given below:

(a) node a between nodes x_1 and x_2

(b) node b between nodes x_2 and u_3

(c) node c between nodes x_3 and u

The nodes are shown in Fig. 7.64(a)

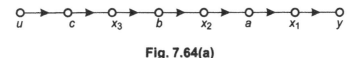

Fig. 7.64(a)

Consider eqn (1). There are three terms which will be considered one by one.

(a) **First term** $\dfrac{1}{s}\{-200U(s) - 12Y(s)\}$. It indicates input to the output node y from

(i) input node u having forward path gain of $-200 \times \dfrac{1}{s}$, which is depicted by joining input node u with node a, having transmittance of -200.

(ii) from output node y by joining output node y with node a, having transmittance of -12.

(iii) factor $\dfrac{1}{s}$, which is added on the branch connecting nodes a and x_1, and

(iv) nodes x_1 and y are connected with a branch transmittance of unity.

Addition of the first term is shown in Fig. 7.64(b)

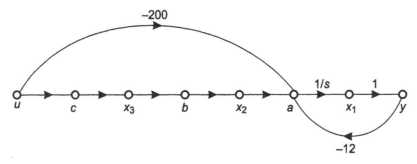

Fig. 7.64(b)

(b) **Second term** $\dfrac{1}{s^2}\{-1200U(s)-5Y(s)\}$. It indicates input to the output node y from

(i) input node u having forward path gain of $-1200 \times \dfrac{1}{s} \times \dfrac{1}{s}$, which is depicted by joining node u with node b, having transmittance of -1200.

(ii) output node y, by joining output node y with node b having transmittance of -5.

(iii) factor $1/s$, which is added on branch connecting nodes b and x_2, and

(iv) nodes x_2 and a are connected with a branch transmittance of unity.

Addition of the second term to Fig. 7.64(b) is shown in Fig. 7.64(c).

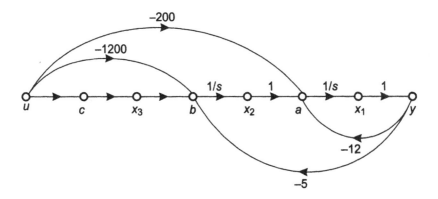

Fig. 7.64(c)

(c) **Third term** $\dfrac{1}{s^3}\{-1600U(s)+150Y(s)\}$. It indicates input to the output node y form

(i) input node u having forward path gain of $-1600 \times \dfrac{1}{s} \times \dfrac{1}{s} \times \dfrac{1}{s}$, which is depicted by joining nodes u and c with branch transmittance of -1600.

(ii) output node y by joining output node y with node c, having transmittance of 150.

(iii) factor $1/s$, which is added on the branch connecting nodes c and x_3, and

(iv) nodes x_3 and b are connected with a branch transmittance of unity.

Addition of the third term to Fig. 7.64(c) gives the complete signal flow graph shown in Fig. 7.64(d).

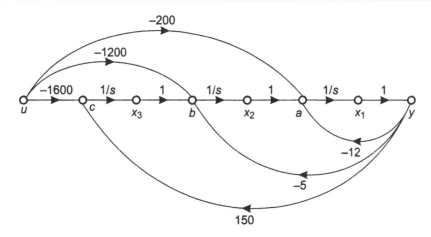

Fig. 7.64(d)

Writing the state equations from the signal flow graph shown in Fig. 7.64(d) gives

$$y = x_1 \tag{1}$$

$$x_1 = \frac{a}{s} = \frac{1}{s}(-12y - 200u + x_2) \tag{2}$$

or $\qquad \dot{x}_1 = -12y - 200u + x_2$

or $\qquad \dot{x}_1 = -12x_1 + x_2 - 200u \tag{3}$

$$x_2 = \frac{b}{s} = \frac{1}{s}(-5y - 1200u + x_3)$$

or $\qquad \dot{x}_2 = -5y - 1200y + x_3$

or $\qquad \dot{x}_2 = -5x_1 + x_3 - 1200u \tag{4}$

$$x_3 = \frac{c}{s} = \frac{1}{s}(150y - 1600u)$$

or $\qquad \dot{x}_3 = 150y - 1600u$

or $\qquad \dot{x}_3 = 150x_1 - 1600u \tag{5}$

The state equations in matrix notation are given by eqns (2), (3), (4) and (5)

$$\begin{bmatrix} \dot{x}_1 \\ \dot{x}_2 \\ \dot{x}_3 \end{bmatrix} = \begin{bmatrix} -12 & 1 & 0 \\ -5 & 0 & 1 \\ 150 & 0 & 0 \end{bmatrix} \begin{bmatrix} x_1 \\ x_2 \\ x_3 \end{bmatrix} + \begin{bmatrix} -200 \\ -1200 \\ -1600 \end{bmatrix} u$$

$$y = \begin{bmatrix} 1 & 0 & 0 \end{bmatrix} \begin{bmatrix} x_1 \\ x_2 \\ x_3 \end{bmatrix}$$

(d) Parallel form

$$\frac{Y(s)}{U(s)} = \frac{-200(s^2 + 6s + 8)}{(s+10)(s-3)(s+5)}$$

$$= \frac{-147.7}{s+10} + \frac{-67.31}{s-3} + \frac{15}{s+5}$$

or

$$Y(s) = \frac{-147.7}{s+10} U(s) - \frac{67.31}{s-3} U(s) + \frac{15}{s+5} U(s)$$

Input to the output node y is from three different paths, each originating from input node u. These are shown in Fig. 7.65.

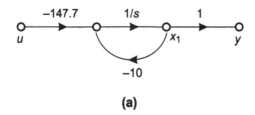

(a)

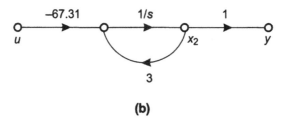

(b)

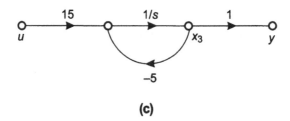

(c)

Fig. 7.65

The three paths are joined together to form the complete signal flow graph shown in Fig. 7.66.

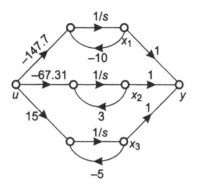

Fig. 7.66

The state equations are obtained from the Fig. 7.67 are

$$y = x_1 + x_2 + x_3$$

$$x_1 = \frac{1}{s}(-10x_1 - 147.7u)$$

or $\qquad \dot{x}_1 = -10x_1 - 147.7u$

$$x_2 = \frac{1}{s}(3x_2 - 67.31u)$$

or $\qquad \dot{x}_2 = 3x_2 - 67.31u$

$$x_3 = \frac{1}{s}(-5x_3 + 15u)$$

or $\qquad \dot{x}_3 = -5x_3 + 15u\ .$

State equations in vector matrix notation are

$$\begin{bmatrix} \dot{x}_1 \\ \dot{x}_2 \\ \dot{x}_3 \end{bmatrix} = \begin{bmatrix} -10 & 0 & 0 \\ 0 & 3 & 0 \\ 0 & 0 & -5 \end{bmatrix} \begin{bmatrix} x_1 \\ x_2 \\ x_3 \end{bmatrix} + \begin{bmatrix} -147.7 \\ -67.31 \\ 15 \end{bmatrix} u$$

$$y = \begin{bmatrix} 1 & 1 & 1 \end{bmatrix} \begin{bmatrix} x_1 \\ x_2 \\ x_3 \end{bmatrix}$$

Problem 7.16

State-space representation of a control system is described by the state equations

$$\dot{x} = Ax + Bu$$
$$y = Cx + Du$$

where matrices A, B, C and D are given as

$$A = \begin{bmatrix} -6 & -11 & -6 \\ 1 & 0 & 0 \\ 0 & 1 & 0 \end{bmatrix} \qquad B = \begin{bmatrix} 1 & 0 & 0 \end{bmatrix}^T \qquad C = \begin{bmatrix} 0 & 0 & 1 \end{bmatrix} \qquad D = [0]$$

Obtain the conventional phase variable representation.

Solution   Let the transformation matrix

$$P = \begin{bmatrix} 0 & 0 & 1 \\ 0 & 1 & 0 \\ 1 & 0 & 0 \end{bmatrix}$$

$$P^{-1} = \frac{adj[P]}{|P|}$$

$$adj[P] = \begin{bmatrix} 0 & 0 & 1 \\ 0 & 1 & 0 \\ 1 & 0 & 0 \end{bmatrix} \quad \text{and} \quad |P| = 1$$

Therefore

$$P^{-1} = \begin{bmatrix} 0 & 0 & 1 \\ 0 & 1 & 0 \\ 1 & 0 & 0 \end{bmatrix}$$

Matrix transformation into phase variable form is achieved as given below:

$$\bar{A} = P^{-1}AP = \begin{bmatrix} 0 & 0 & 1 \\ 0 & 1 & 0 \\ 1 & 0 & 0 \end{bmatrix} \begin{bmatrix} -6 & -11 & -6 \\ 1 & 0 & 0 \\ 0 & 1 & 0 \end{bmatrix} \begin{bmatrix} 0 & 0 & 1 \\ 0 & 1 & 0 \\ 1 & 0 & 0 \end{bmatrix}$$

$$= \begin{bmatrix} 0 & 0 & 1 \\ 0 & 1 & 0 \\ 1 & 0 & 0 \end{bmatrix} \begin{bmatrix} -6 & -11 & -6 \\ 0 & 0 & 1 \\ 0 & 1 & 0 \end{bmatrix}$$

$$= \begin{bmatrix} 0 & 1 & 0 \\ 0 & 0 & 1 \\ -6 & -11 & -6 \end{bmatrix}$$

$$\bar{B} = P^{-1}B = \begin{bmatrix} 0 & 0 & 1 \\ 0 & 1 & 0 \\ 1 & 0 & 0 \end{bmatrix} \begin{bmatrix} 1 \\ 0 \\ 0 \end{bmatrix} = \begin{bmatrix} 0 \\ 0 \\ 1 \end{bmatrix}$$

$$\bar{C} = CP = \begin{bmatrix} 0 & 0 & 1 \end{bmatrix} \begin{bmatrix} 0 & 0 & 1 \\ 0 & 1 & 0 \\ 1 & 0 & 0 \end{bmatrix} = \begin{bmatrix} 1 & 0 & 0 \end{bmatrix}$$

$$\bar{D} = 0$$

In vector matrix form

$$\dot{x} = \begin{bmatrix} 0 & 1 & 0 \\ 0 & 0 & 1 \\ -6 & -11 & -6 \end{bmatrix} x + \begin{bmatrix} 0 \\ 0 \\ 1 \end{bmatrix} u$$

$$y = \begin{bmatrix} 1 & 0 & 0 \end{bmatrix} x$$

Problem 7.17

A cascaded open-loop system having transfer functions G_1 and G_2 is shown in Fig. 7.67. (If state-space representation of) G_1 is

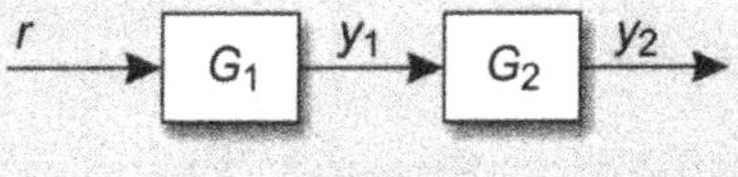

Fig. 7.67

$$\dot{x}_1 = A_1 x_1 + B_1 r$$

$$y_1 = C_1 x_1,$$

and that of G_2 is

$$\dot{x}_2 = A_2 x_2 + B_2 y_1$$

$$y_2 = C_2 x_2,$$

find the state-space representation of the entire system

Solution

$$\dot{x}_1 = A_1 x_1 + B_1 r \tag{1}$$

$$y_1 = C_1 x_1 \tag{2}$$

$$\dot{x}_2 = A_2 x_2 + B_2 y_1 \tag{3}$$

$$y_2 = C_2 x_2 \tag{4}$$

Substituting eqn (2) in eqn (3), we get

$$\dot{x}_1 = A_1 x_1 + B_1 r$$

$$\dot{x}_2 = B_2 C_1 x_1 + A_2 x_2$$

In matrix form

$$\begin{bmatrix} \dot{x}_1 \\ \cdots \\ x_1 \end{bmatrix} = \begin{bmatrix} A_1 & \vdots & 0 \\ ---- & + & --- \\ B_2 C_1 & \vdots & A_2 \end{bmatrix} \begin{bmatrix} x_1 \\ \cdots \\ x_2 \end{bmatrix} + \begin{bmatrix} B_1 \\ \cdots \\ 0 \end{bmatrix} r$$

and

$$y_2 = \begin{bmatrix} 0 & \vdots & C_2 \end{bmatrix} \begin{bmatrix} x_1 \\ \cdots \\ x_2 \end{bmatrix}$$

Problem 7.18

The transfer function of control system is given as

$$\frac{Y(s)}{U(s)} = \frac{s+2}{s^2 + 7s + 12}$$

Represent the system in the following forms (a) Phase variable (b) Controller canonical (c) Observer canonical (d) Parallel and (e) Cascade

Solution Representation of the system in the required forms is tabulated below:

Ser. No	Form	Transfer function	Signal flew model	State equation
1	Phase variable	$(s+2) \times \dfrac{1}{(s^2+7s+12)}$		$\dot{x} = \begin{bmatrix} 0 & 1 \\ -12 & -7 \end{bmatrix} x + \begin{bmatrix} 0 \\ 1 \end{bmatrix} u$ $y = \begin{bmatrix} 2 & 1 \end{bmatrix} x$
2	Controller canonical	$(s+2) \times \dfrac{1}{(s^2+7s+12)}$		$\dot{x} = \begin{bmatrix} -7 & -12 \\ 1 & 0 \end{bmatrix} x + \begin{bmatrix} 1 \\ 0 \end{bmatrix} u$ $y = \begin{bmatrix} 1 & 2 \end{bmatrix} x$
3	Observer canonical	$\dfrac{\dfrac{1}{s} + \dfrac{2}{s^2}}{1 + \dfrac{7}{s} + \dfrac{12}{s^2}}$		$\dot{x} = \begin{bmatrix} -7 & 1 \\ -12 & 0 \end{bmatrix} x + \begin{bmatrix} 1 \\ 2 \end{bmatrix} u$ $y = \begin{bmatrix} 1 & 0 \end{bmatrix} x$
4	Parallel	$\dfrac{-1}{(s+3)} + \dfrac{2}{(s+4)}$		$\dot{x} = \begin{bmatrix} -3 & 0 \\ 0 & -4 \end{bmatrix} x + \begin{bmatrix} -1 \\ 2 \end{bmatrix} u$ $y = \begin{bmatrix} 1 & 1 \end{bmatrix} x$
5	Cascade	$\dfrac{1}{(s+3)} \times \dfrac{(s+2)}{(s+4)}$		$\dot{x} = \begin{bmatrix} -4 & 1 \\ 0 & -3 \end{bmatrix} x + \begin{bmatrix} 0 \\ 1 \end{bmatrix} u$ $y = \begin{bmatrix} -2 & 1 \end{bmatrix} x$

Problem 7.19

A Controller and a plant having transfer functions of $\dfrac{1}{s+3}$ and $\dfrac{1}{s^2+2s+4}$ respectively constitutes a closed-loop control system with unity feedback.

    (a) Determine state variable representation of the controller.

    (b) Repeat para (a) for the plant

    (c) Having obtained the state-variable representation of the control system, use *series* and *cloop* functions to compute state-variable representation of the closed-loop system and plot its impulse response.

```
'Problem 7.19'                               %display enclosed text.
'State Variable Representation controller'    %display enclosed text.
numc=1;                                       %input numerator of controller
                                              %& suppress output.
denc=[1,3];                                   %input denominator of
[A,B,C,D]=tf2ss(numc,denc);                   %convert to state variable
                                              %representation & suppress
                                              %output.
printsys(A,B,C,D)                             %print readable form.
'State Variable Representation Plant'         %display enclosed text.
nump=1;                                       %input numerator of plant
                                              %& suppress output.
denp=[1 2 4];                                 %input denominator of plant
                                              %& suppress output.
[a,b,c,d]=tf2ss(nump,denp);                   %convert to state variable
                                              %representation & suppress
                                              %output.
printsys(a,b,c,d)                             %print readable form.
'Combine State Variable Representation'       %display enclosed text.
[A1,B1,C1,D1]=series(A,B,C,D,a,b,c.d);        %join transfer function of
                                              %controller & plant in series.
[As,Bs,Cs,Ds]=cloop(A1,B1,C1,D1,-1);          %eliminate feedback loop.
printsys(As,Bs,Cs,Ds)                         %print readable form.
'Convert to Transfer Function'                %display enclosed text.
[num,den]=ss2tf(As,Bs,Cs,Ds),                 %convert to transfer function
                                              %representation.
t=[0:0.001:10.0];                             %specify time interval.
y=impulse(As,Bs,Cs,Ds,1,t);                   %impulse response &
                                              %suppress output.
'Plot Impulse Resonse'                        %display enclosed text.
plot(t,y)grid                                 %plot response.
ylabel('y(t)'),xlabel('Time(sec)')            %label y and x axis.
'Alternatively the response can also be plotted'
sys=tf(num,den);                              %form transfer function
                                              %& assign to sys & suppress
                                              %output.
impulse(sys);                                 %plot impulse response.
```

```
Output

Problem 7.19
State Variable Representation Controller
a =
                      x1
           x1    -3.00000
b =
                      u1
           x1     1.0000

c =                   x1
           y1     1.00000
d =
                      u1
           y1         0

State Variable Representation Plant
a =
                      x1          x2
           x1    -2.00000    -4.00000
           x2     1.00000           0
b =
                      u1
           x1     1.00000
           x2          0
c =
                      x1          x2
           y1          0     1.00000
d =
                      u1
           y1          0

Combine State Variable Representations

a =
                      x1          x2          x3
           x1    -2.00000    -4.00000     1.00000
           x2     1.00000           0           0
           x3          0    -1.00000    -3.00000
b =
                      u1
           x1          0
           x2          0
           x3     1.00000
c =
                      x1          x2          x3
           y1          0     1.00000           0

d =
                      u1
           y1          0

Convert to Transfer Function
num =
        0     0.0000         0     1.0000

den =
    1.0000     5.0000    10.0000    13.0000

Plot Impulse Response
```

Note: The impulse response due to command $y=impulse(As, Bs, Cs, Ds, 1, t)$ is shown in Fig. 7.68 and impulse response due to command *impulse(sys)* is shown in Fig. 7.69.

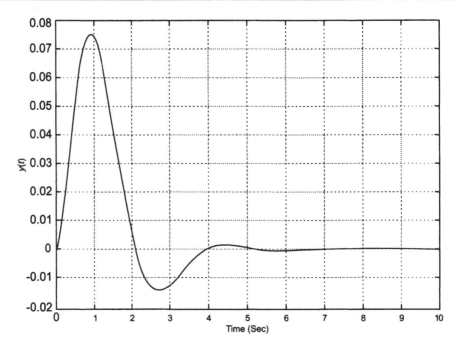

Fig. 7.68 *Impulse response (Problem 7.19). due to command*

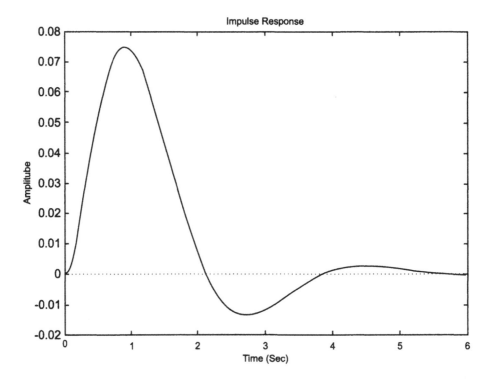

Fig. 7.69 Impulse response due to command *impulse (sys).*

Problem 7.20

Consider two systems represented in state-space as follows:

(a) First System (b) Second System

$$\dot{x}_1(t) = \begin{bmatrix} 0 & 1 & 0 \\ 0 & 0 & 1 \\ -4 & -5 & -8 \end{bmatrix} x_1(t) + \begin{bmatrix} 0 \\ 0 \\ 0 \end{bmatrix} u(t) \quad \dot{x}_2(t) = \begin{bmatrix} 0.5000 & 0.5000 & 0.7071 \\ -0.5000 & -0.5000 & 0.7071 \\ -6.3640 & -0.7071 & -8.0000 \end{bmatrix} x_2(t) + \begin{bmatrix} 0 \\ 0 \\ 4 \end{bmatrix} u(t)$$

$$y = \begin{bmatrix} 1 & 0 & 0 \end{bmatrix} x_1(t) \qquad\qquad y = \begin{bmatrix} 0.7071 & -0.7071 & 0 \end{bmatrix} x_2(t)$$

Determine transfer functions for the two systems. Compare the results and comment

```
'Problem 7.20'                                    %display enclosed text.
'First System'                                    %display enclosed text.
A=[0 1 0;0 0 1;-4 -5 -8];                          %input matrix A & suppress output.
B=[0;0;4];                                         %input matrix B & suppress output.
C=[1 0 0];                                         %input matrix C & suppress output.
D=0;                                               %input matrix D & suppress output.
[num,den]=ss2tf(A,B,C,D)                           %convert to transfer function form.
'Second System'                                    %display enclosed text.
a=[0.5000 0.5000 0.7071;-0.5000 -0.5000 0.7071;-6.3640 -0.7071 -8.0000];
                                                   %Input matrix a & suppress output.
b=[0;0;4];                                         %Input matrix b & suppress output.
c=[0.7071 -0.7071 0];                              %Input matrix c & suppress output.
d=0;                                               %Input matrix d & suppress output.
[n,d]=ss2tf(a,b,c,d)                               %convert to transfer function form.


Output
Problem 7.20
First System
num =
        0    -0.0000    -0.0000    4.0000
den =
    1.0000    8.0000    5.0000    4.0000
Second System
n =
        0    -0.0000    -0.0000    3.9999

d =
    1.0000    8.0000    5.0000    4.0000
```

Comments.
First system represented by num and den in the output has transfer function
which can be represented by $\dfrac{4}{s^3+8s^2+5s+4}$

Second system represented by n and d in the output has transfer function
which can be represented by $\dfrac{4}{s^3+8s^2+5s+4}$

Students should note that transfer function of both the systems is same where
as the state space representation in matrix notation is different which
implies that for a control system having a transfer function, the state-space
representation is not unique.

Problem 7.21

Determine the state variable representation of an open-loop control system whose transfer function is given by

(a) $G(s) = \dfrac{1}{s+9}$

(b) $G(s) = \dfrac{5s^2 + 12s + 1}{s^2 + 7s + 5}$

Solution We will use MATLAB function *tf2ss* to obtain state-space representation. MATLAB program is given below:

```
'Problem 7.21(a)'                      %display enclosed text.
'numa=1;                               %input mumerator
                                       %& suppress output.
dena=[1 9];                            %input denominator
                                       %& suppress output.
[A,B,C,D]=tf2ss(numa,dena);            %convert to state space
                                       %representation.
printsys(A,B,C,D)                      %print in readable form.
'problem 7.21(b)'                      %display enclosed text.
numb=[5 12 1];                         %input numerator
                                       %& suppress output.
denb=[1 7 5];                          %input denominator
                                       %& suppress output.
[A,B,C,D]=tf2ss(numb,denb);            %convert to state space
                                       %representation.
printsys(A,B,C,D)                      %print in readable form.

Output
Problem 7.21(a)
a =
                    x1
          x1    -9.00000
b =
                    u1
          x1     1.00000
c =
                    x1
          y1     1.00000
d =
                    u1
          y1        0

Problem 7.21(b)

a =
                    x1          x2
          x1    -7.00000    -5.00000
          x2     1.00000        0
b =
                    u1
          x2     1.00000
          x2        0
c =
                    x1          x2
          y1   -23.00000   -24.00000
                    u1
d =      y1      5.00000
```

Problem 7.22

The state-model of a control system given by

$$A = \begin{bmatrix} 1 & 1 & 0 \\ -2 & 0 & 4 \\ 6 & 2 & 10 \end{bmatrix}, \qquad B = \begin{bmatrix} 0 \\ 0 \\ 0 \end{bmatrix}, \qquad C = \begin{bmatrix} 0 & 1 & 0 \end{bmatrix}; \qquad D = 0$$

Determine the transfer function representation.

Solution We will use MATLAB function *ss2tf* to obtain transfer function representation. MATLAB program is given below:

```
'Problem 7.22'                          %display enclosed text.
'A=[1 1 0;-2 0 4;6 2 10];              %input matrix A
                                        %& suppress output.
B=[0 0 1]';                            %input matrix B
                                        %& suppress output.
C=[0 1 0];                             %input matrix C
                                        %& suppress output.
D=0;                                   %input matrix D
                                        %& suppress output.
[num,den]=ss2tf(A,B,C,D)               %convert to transfer
                                        %function form.
printsys (num,den)                     %print in readable form.

Output
Problem 7.22

num =
         0      -0.0000      4.0000      -4.0000
den =
    1.0000    -11.0000      4.0000     -36.0000

Note: This represents transfer function

          4s-4/s^3 - 11 s^2 + 4 s - 36

num/den =

   -1.954e-014 s^2 + 4 s - 4
   -------------------------------------
    s^3 - 11 s^2 + 4 s - 36

Note: This is the output due to printsys command.s^2 term in
the numerator is realised as zero.
```

Problem 7.23

A control system is represented as

$$A = \begin{bmatrix} -3 & 1 \\ 1 & -3 \end{bmatrix}, \qquad B = \begin{bmatrix} 1 \\ 2 \end{bmatrix}, \qquad C = \begin{bmatrix} 2 & 3 \end{bmatrix}; \quad \text{and} \quad D = 0$$

Create the diagonal form by two methods.

Solution   We will utilise transformation matrix method and then use *canon* command to obtain the result. MATLAB program is given below:

```
'Problem 7.23'                  %display enclosed text.
A=[-3 1;1 -3]                   %input matrix A.
B=[1;2]                         %input matrix B.
C=[2 3]                         %input matrix C.
D=0                             %input matrix D.
'Transformation Method'         %display enclosed text.
[V,d]=eig(A)                    %create transformation matrix
                                %V of eigenvectors and d of
                                %eigenvalues.
Ad=inv(V)*A*V                   %transform matrix A.
Bd=inv(V)*B                     %transform matrix B.
Cd=C*V                          %transform matrix C.
'Canon Command'                 %display enclosed text.
sys=ss(A,B,C,D);                %create state space object.
syscanon=canon(sys,'modal')     %create diagonal form.


Output
Problem 7.23
A =
    -3    1
     1   -3
B =
     1
     2
C =
     2    3
D =
     0


Transformation Method
V =
     0.7071    0.7071
    -0.7071    0.7071
d =
    -4     0
     0    -2
Ad =
    -4.0000         0
          0   -2.0000
```

```
Bd =
   -0.70701
    2.1213

Cd =
   -0.7071      3.5355

Canon Command
a =
        x1   x2
  x1    -4    0
  x2     0   -2
b =
             u1
  x1    -0.7071
  x2     2.121
c =
             x1        x2
  y1    -0.7071     3.536
d =
        u1
  y1     0

Continuous-time model.
>>
```

Problem 7.24

Given the system represented in state-space as follows:

$$\dot{x}(t) = \begin{bmatrix} 0 & 1 & 0 \\ 0 & 0 & 1 \\ -6 & -11 & -6 \end{bmatrix} x(t) + \begin{bmatrix} 0 \\ 0 \\ 6 \end{bmatrix} u(t)$$

$$y(t) = \begin{bmatrix} 1 & 0 & 0 \end{bmatrix} x(t) + \begin{bmatrix} \ \end{bmatrix} u(t)$$

Convert the system matrices into diagonal form by use of Vandermonde matrix formed with eigenvalues.

```
'Problem 7.24'                          %display enclosed text.
'A=[0 1 0;0 0 1;-6 -11 -6]              %input matrix A.
B=[0 0 6]'                              %input matrix B.
C=[1 0 0]                               %input matrix C.
D=0                                     %input matrix D.
'Eigenvalues'                           %display enclosed text.
M=eig(A)                                %find eigenvalues.
'Vandermonde Matrix                     %display enclosed text.
T=[1 1 1; M(1,1) M(2,1) M(3,1);         %create vandermonde
M(1,1)^2 M(2,1)^2 M(3,1)^2]             %matrix from eigenvalues.
Ac=inv(T)*A*T                           %convert matrix A to.
                                        %diagonal form.
Bc=inv(T)*B                             %convert matrix B.
Cc=C*T                                  %convert matrix C.


Solution
Problem
A =
      0      1      0
      0      0      1
     -6    -11     -6
B =
      0
      0
      6
C =
      1      0      0
D =
      0


Eigenvalues
M =
    -1.0000
    -2.0000
    -3.0000
Vandermonde Matrix
T =
     1.0000   1.0000   1.0000
    -1.0000  -2.0000  -3.0000
     1.0000   4.0000   9.0000
Ac =
    -1.0000   0.0000  -0.0000
    -0.0000  -2.0000   0.0000
     0.0000   0.0000  -3.0000
Bc =
     3.0000
    -6.0000
     3.0000
Cc =
      1      1      1
```

Problem 7.25

A control system with unity feedback is shown in Fig. 7.71

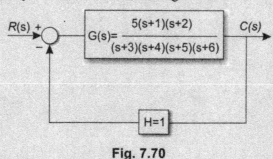

Fig. 7.70

Obtain phase variable and diagonal forms of representations.

```
'Problem 7.25'                          %display enclosed text.
'Transfer Function'                     %display enclosed text.
G=zpk([-1 -2],[-3 -4 -5],5 -1)          %input G(s) & output.
TF=feedback(G,1,-1)                     %returns LTI model TF
                                        %for G,feedback H=1
                                        %& feedback polarity
                                        %of negative in nature.
[num,den]=tfdata(TF,'V')                %returns the numerator
                                        %& denominator as row
                                        %vectors.
'Conversion to Phase Variable Form'     %display enclosed text.
Al=flipud(A)                            %flip rows of matrix A
                                        %up-down.
Ap=fliplr(Al)                           %flip columns of matrix Al
                                        %in left-right direction.
Bp=fliplr(B)                            %flip columns of matrix B
                                        %in left-right direction.
Cp=fliplr(C)                            %flip columns of matrix B
                                        %in left-right direction.
Dp=D                                    %equate Dp=D.
'Conversion to modal/diagonal form'     %display enclosed text.
[A,B,C,D,]=canon(A,B,C,D,'modal')

Output
Problem 7.25
Transfer Function

Zero/pole/gain:
   5 (s+1) (s+2)
--------------------- This is due to command 'zpk'
(s+3) (s+4) (s+5)

Zero/pole/gain:
        5 (s+1) (s+2)
-------------------------------------- This is due to command 'feedback'
(s+12.48)(s^2 + 4.518s + 5.608)
```

```
num =
         0    5.0000   15.0000   10.0000
den =
    1.0000  17.0000   62.0000   70.0000

This is due to command 'tfdata'

State Space Representation
A =
  -17.0000  -62.0000  -70.0000
    1.0000        0         0
        0    1.0000         0

B =
    1
    0
    0

C =
    5.0000   15.0000   10.0000

D =
    0

Conversion to Phase Variable Form

Al =
        0    1.0000        0
    1.0000        0         0
  -17.0000  -62.0000  -70.0000

Note: If we see the outputs 'Al' and 'A' we find that
rows are flipped up-down.

Ap =
        0    1.0000        0
        0        0    1.0000
  -70.0000  -62.0000  -17.0000

Bp =
    1
    0
    0
Cp =
   10.0000   15.0000    5.0000
Dp =
    0

Note: The output shows columns of matrices Al,B,C and D are
flipped left-right
```

```
Conversion to modal/diagonal form
A =
  -12.4822          0          0
         0    -2.2589     0.7109
         0    -0.7109    -2.2589
B =
    1.4883
    0.5319
   -0.6589
C =
    3.8501    -0.3816     0.8002
D =
    0
```

Problem 7.26

Obtain the state-space representation for the control system having transfer function

$$G(s) = \frac{s+1}{s^3 + 5s^2 + 5s + 1}$$

Solution   The state-space representation is obtained by expressing the transfer function $G(s)$ by using MATLAB command **tf(num, den)** where *num* and *den* are the numerator and denominator polynomials respectively. The state-space representation is obtained by creating state-space object by using MATLAB command $ss(G(s)$. MATLAB program is given below:

```
'Problem 7.26'              %display enclosed text.
num=[1 1];                  %represent numerator
                            %& suppress output.
den=[1 5 5 1];             %represent denominator.
                            %& suppress output.
G=tf(num,den)              %form the transfer &
                            %display output.
ss(G)                       %create state-space
                            %object & display output.


Output
Problem 7.26
Transfer function:
      s + 1
---------------------------
s^3 + 5 s^2 + 5 s + 1

a =
         x1     x2      x3
  x1     -5   -1.25  -0.125
  x2      4      0       0
  x3      0      2       0
```

```
b =
            u1
    x1   0.5
    x2    0
    x3    0


c =
        x1     x2     x3
    y1   0    0.5   0.25


d =
            u1
    y1    0


Continuous-time model.
```

Problem 7.27

Obtain the state-space representation by using MATLAB for a linear time-invariant control system described by the state-space equations.

$$\dot{x} = \begin{bmatrix} 0 & 1 & 0 \\ 0 & 0 & 1 \\ -6 & -11 & -6 \end{bmatrix} x + \begin{bmatrix} 2 \\ 1 \\ 3 \end{bmatrix} u$$

$$y = \begin{bmatrix} 4 & 2 & 3 \end{bmatrix} x + Du$$

Matrix D may be assumed as null matrix. Also, find the number of input, output and state variables associated with the given state-space model

Solution The state-space representation is obtained by specifying the matrix A, B, C and D and then creating the state-space object by using the MATLAB command $ss(A, B, C, D)$

MATLAB program is given below:

```
'Problem 7.27'                    %display enclosed text.
A=[0 1 0;0 0 1;-6 -11 -6];        %input matrix A.
                                  %& suppress output.
B=[2;1;3];                        %input matrix B
                                  %& suppress output.
C=[4 2 3];                        %input matrix C
                                  %& suppress output.
D=0;                              %input matrix D
                                  %& suppress output.
SSR=ss(A,B,C,D)                   %create state-space
                                  %object & display
                                  %output.

size(SSR)                         %display number of
                                  %inputs, outputs
                                  %& states.
```

```
Solution
Problem 7.27
a =
                x1          x2          x3
    x1           0           1           0
    x2           0           0           1
    x3          -6         -11          -6
b =
             u1
    x1        2
    x2        1
    x3        3
c =
                x1          x2          x3
    y1           4           2           3
d =
          u1
    y1     0
```

Problem 7.28

Write a MATLAB program to convert the matrices

$$A = \begin{bmatrix} 1 & 2 & 1 \\ 0 & 1 & 3 \\ 1 & 1 & 1 \end{bmatrix} \qquad B = [1 \quad 0 \quad 1]^T \qquad C = [1 \quad 1 \quad 0] \qquad D = [0]$$

into controllability and observability canonical forms.

Solution MATLAB program is given below:

```
'Problem 7.29'                     %display enclosed text.
A=[1 2 1;0 1 3;1 1 1];             %input matrix A
                                   %& suppress output.
B=[1;0;1];                         %input matrix B
                                   %& suppress output.
C=[1 1 0];                         %input matrix C
                                   %& suppress output.
D=0;                               %input matrix D
                                   %& suppress output.
'CCF Form'                         %display enclosed text.
E=poly(A);                         %form polynomial
                                   %vector & store in E.
a1=E(1,3);                         %define al from row
                                   %vector E & store.
a2=E(1,2);                         %define a2 from row
                                   %vector E & store.
M=[a1 a2 1,a2 1 0;1 0 0]           %form modal matrix M.
Qc=[B A*B A^2*B];                  %form controllability
                                   %matrix.
```

```
detQc=det(Qc)                    %find determinant of Qc
                                 %so as to ascertain Qc
                                 %is non-singular.Qc is
                                 %non-singular if
                                 %determinant is non-zero.
P=Qc*M;                          %form matrix P.
Abar=inv(P)*A*P                  %transform matrix A
                                 %into CCf form.
Bbar=inv(P)*B                    %transform matrix B
                                 %into CCF form.
Cbar=C*P                         %transform matrix C
                                 %into CCF form.
Dbar=D                           %transform matrix D
                                 %into CCF form.
'OCF Form'                       %display enclosed text.
Qo=[C;C*A;C*A^2];                %form observability
                                 %matrix.
detQo=det(Qo)                    %find determinant of Qo
                                 %so as to ascertain Qo
                                 %is non-singular Qo is
                                 %non-singular if
                                 %determinant is non-zero
P1=inv(M*Qo)                     %form matrix P1.
A1bar=inv(P1)*A*P1               %transform matrix A
                                 %into OCF form.
B1bar=inv(P1)*B                  %transform matrix B
                                 %into OCF form.
C1bar=C*P1                       %transform matrix C
                                 %into OCF form.
Dbar=D                           %transform matrix D
                                 %into OCF form.


Output
CCF Form
detQc=
    -9
Note: Since determinant of Qc is non-zero,
    CCF transformation is possible.
Abar =
    0.0000    1.0000    0.0000
    0.0000    0.0000    1.0000
    3.0000    1.0000    3.0000
Bbar =
    0.0000
    0.0000
    1.0000
Cbar =
    3.0000    2.0000    1.0000
Dbar =
    0
```

```
OCF Form
detQo=
    12
Not: Since determinant of Qc is non-zero,
    OCF transformation is possible.
Albar =
    0.0000    0.0000    3.0000
    1.0000    0.0000    1.0000
    0.0000    1.0000    3.0000
Note: Matrix Albar is transpose of Abar.
Blbar =
    3.0000
    2.0000
    1.0000
Clbar =
        0.0000    0.0000    1.0000
Note: Matrix Clbar  is transpose of Cbar.
Dbar =
    0
```

REVIEW EXERCISE

1. List the guidelines which help in assigning state variables.
2. What are the various forms in which state model can be formed. What are their advantages and disadvantages.
3. What is the general procedure to evaluate a state model.
4. Compare phase variable forms of state model.
5. What is similarity transformation. Why it is carried out.
6. Prove that characteristic equation and transfer function matrix do not alter during similarity transformation.
7. What are the conditions linked to CCF and OCF transformations.
8. List properties associated with Jordan form.
9. Compare CCF and OCF form of transformations.
10. Define decomposition. List three methods of decomposition.
11. If the dynamics of a control system is described by

$$\dot{x}(t) = Ax(t) + Bu(t)$$
$$y(t) = Cx(t) + Du(t)$$

where

(a) $A = \begin{bmatrix} -2 & 1 & 0 \\ 0 & -2 & 0 \\ 0 & 0 & -1 \end{bmatrix}$; $B = \begin{bmatrix} 0 \\ 1 \\ 1 \end{bmatrix}$; $C = \begin{bmatrix} -1 & -1 & 1 \end{bmatrix}$ and $D = 0$

(b) $A = \begin{bmatrix} -1 & 0 \\ 0 & -2 \end{bmatrix}$; $B = \begin{bmatrix} 1 \\ 1 \end{bmatrix}$; $C = \begin{bmatrix} 4 & -2 \end{bmatrix}$ and $D = 0$

(c) $A = \begin{bmatrix} 0 & 1 & 0 \\ 0 & 0 & 1 \\ -2 & -5 & -4 \end{bmatrix}$; $B = \begin{bmatrix} 0 \\ 0 \\ 1 \end{bmatrix}$; $C = \begin{bmatrix} 5 & 6 & 2 \end{bmatrix}$ and $D = 0$

(d) $A = \begin{bmatrix} 0 & 1 & 0 \\ -6 & -11 & -6 \\ -2 & -11 & 5 \end{bmatrix}$; $B = \begin{bmatrix} 0 \\ 0 \\ 1 \end{bmatrix}$; $C = \begin{bmatrix} 1 & 1 & 1 \end{bmatrix}$ and $D = 0$

Transform the above into CCF, OCF, DCF or JCF where feasible

12. *Derive* state model diagram of the above systems.

13. *Derive* phase variable and Jordan canonical form for the system described by the following differential equation

(a) $\dddot{y} + 4\ddot{y} + 5\dot{y} + 2y = 2\ddot{u} + 6\dot{u} + 5u$

(b) $\dddot{y} + 6\ddot{y} + 11\dot{y} + 10y = 3u$

(c) $4\ddot{y} + 3\ddot{y} + \dot{y} + 2y = 5u$

14. Draw the state diagram for a system described by $\dot{x}(t) = Ax(t) + Bu(t)$, where

$$A = \begin{bmatrix} -3 & 2 & 0 \\ -1 & 0 & 1 \\ -2 & -3 & -4 \end{bmatrix}; \quad B = \begin{bmatrix} 0 \\ 0 \\ 1 \end{bmatrix}$$

15. Draw the state diagrams for the systems described by the following transfer functions

(a) $G(s) = \dfrac{10(s+1)}{s^2(s+2)(s+3)}$

(b) $G(s) = \dfrac{5(s+2)}{s(s+1)(s+6)}$

(c) $G(s) = \dfrac{1}{s(s+2)(s^2+2s+2)}$

8

CONTROLLABILITY AND OBSERVABILITY

8.1 INTRODUCTION

In the preceding chapter, we had evolved state-space description of linear time-invariant control systems. The state equations evolved were

$$\dot{x}(t) = Ax(t) + Bu(t) \tag{8.1}$$

$$y(t) = Cx(t) + Du(t)$$

The physical form of the matrices A, B, C and D depends upon the type and nature of state-variables selected. The matrices may have time dependence or otherwise. Transformation from one form to any other desirable form of representation through various methods is feasible to ensure easy analysis and description of system dynamics. Equation (8.1) which describes the system dynamics also shows dependence of state vector $x(t)$ on the input $u(t)$. The output equation describes dependence of output $y(t)$ on the state vector $x(t)$ and input $u(t)$. If the input $u(t)$ is zero, the system dynamics and output of a control system depends on the state vector $x(t)$ specified by matrices A and C. The matrix B and D specifies any direct connection between the input and the output. The matrix B specifies the dependence of system dynamics on input $u(t)$ and matrix D represents transition from input $u(t)$ to output $y(t)$.

LEARNING OBJECTIVES

- ◆ To understand concept of controllability and output controllability.
- ◆ Define controllability and ascertain controllability of control systems.
- ◆ To understand the concept of observability.
- ◆ Define observability and ascertain observability of control systems.
- ◆ Design through state-space approach
  - • State feedback
  - • Pole placement
  - • Design of controllers and observers.

8.2 CONTROLLABILITY AND OBSERVABILITY

Kalman was first to introduce the concept of controllability in 1960. These concepts are used in optimal control theory to arrive at an optimal control solution. The concept of controllability is linked with the equation

$$\dot{x}(t) = Ax(t) + Bu(t) \tag{8.2}$$

and the concept of observability is linked with the equation

$$y(t) = Cx(t) + Du(t) \tag{8.3}$$

The concept basically depends on the input $u(t)$ and its affect on state vector $x(t)$ and output $y(t)$.

8.2.1 Controllability

Ability of input $u(t)$ to exercise control on the state-variables forming state vector $x(t)$ describes the controllability of control systems. Controllability means total or complete control on the system. Let us consider a control system whose signal flow diagram is shown in Fig. 8.1

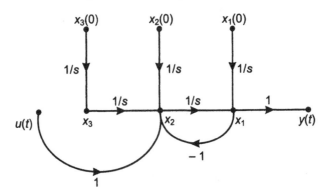

Fig. 8.1

In the signal flow diagram under consideration $u(t)$ and $y(t)$ are the input and output respectively and x_1, x_2 and x_3 are the state-variables. The state equations are

$$x_1 = \frac{1}{s}x_2 + \frac{1}{s}x_1(o) \tag{8.4}$$

$$x_2 = \frac{1}{s}x_3 - x_1 + u(t) + \frac{1}{s}x_2(o) \tag{8.5}$$

$$x_3 = \frac{1}{s}x_3(o) \tag{8.6}$$

It can be seen that the states x_1 and x_2 only are affected by input $u(t)$ and the system state x_3 is not affected by input $u(t)$. This implies that system state x_3 is not controllable. The dynamic system is thus constrained by input $u(t)$. In comparison, the signal flow diagram of the control system illustrated in Fig. 8.2, shows that the three states x_1, x_2 and x_3 are affected by input $u(t)$. The system is thus not constrained by input $u(t)$.

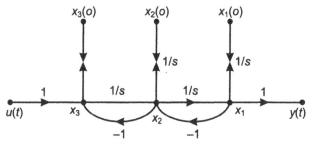

Fig. 8.2

8.2.2 Observability

Observability is the concept coupled with the states and the output, and is specified by the matrices C and D. The states x_1, x_2 and x_3 in Fig. 8.1 and Fig. 8.2 are connected with the input; directly or indirectly. However, the control system described by the signal flow diagram in Fig. 8.3 is not completely observable as the state x_3 is not coupled with the output $y(t)$. The state x_3 is thus not completely observable.

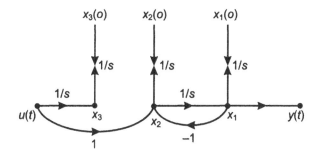

Fig. 8.3

8.3 DEFINITION

Having understood the basic concepts of controllability and observability, we will now define them.

8.3.1 Controllability

A system is said to be *controllable* if all the states are completely controllable. A system can also be considered as controllable, if every state of the system can be exercised control in such a manner that the states are transferred from an initial state to a desired final state in some finite time. It can also be said that if all the system states are controllable, then the system is controllable. However, the precise definition of controllability widely accepted in the control system engineering for a linear time-invariant control system described by the state equations

$$\dot{x}(t) = Ax(t) + Bu(t) \tag{8.7}$$

$$y(t) = Cx(t) + Du(t) \tag{8.8}$$

where, $x(t)$ = state vector $(n \times 1)$

$u(t)$ = input vector $(p \times 1)$

$y(t)$ = output vector $(q \times 1)$

    A = system matrix $(n \times n)$

    B = driving matrix or control matrix $(n \times p)$

    C = output or observation matrix $(q \times n)$

    D = transmission matrix $(q \times p)$

is *"The state $x(t)$ at $t = t_o$ is said to be controllable, if the state can be driven to a desired state $x(t_f)$ in some finite time $t = t_f$ by application of continuous control input $u(t)$".*

The controllability defined above is associated with the states and should not be confused with the output controllability, which is the property of input-output relationship. *State controllability* involves dependence of the state-variables on the inputs. In Fig. 8.1, since, the state x_3 is independent of the control $u(t)$, the state x_3 cannot be driven to a desired state by the control $u(t)$ and hence, the state x_3 is uncontrollable which makes the system uncontrollable.

8.3.2 Observability

A system is considered to be observable if the system states are observable. This implies that every state variable of the system affects some of the outputs. Measurements of outputs need not indicate any information of the system states in some systems. Such systems are considered to be unobservable. In Fig. 8.2, the system output $y(t)$ is not affected by the system state $x_3(t)$ and hence the system is unobservable. The precise definition of observability widely accepted in control system engineering for the linear time *invariant* system described by the dynamic eqns (8.7) and (8.8) is

"The state $x(t_0)$ at $t = t_0$ for a system subjected to control input $u(t)$ is said to be observable, if for a desired finite time $t = t_f \geq t_0$, knowledge of $u(t)$ and output $y(t)$ over the interval $t_0 \leq t \leq t_f$ determines the state (x_{t0})".

8.4 CONTROLLABILITY TESTS

Let us consider a linear time-invariant continuous system described by eqn (8.7). The solution of the equation is

$$x(t) = \phi(t - t_0)x(t_0) + \int_{t_0}^{t} \phi(t - \tau)Bu(\tau)\,d\tau \quad \text{where } t \geq t_0. \tag{8.9}$$

$$x(t) = e^{At}x(t_0) + \int_{t_0}^{t} e^{A(t - \tau)}Bu(\tau)d\tau \tag{8.10}$$

Let us assume that the system is controllable and desired final state is $t = t_f = 0$.

then $$x(t_f) = 0 \tag{8.11}$$

which gives $$0 = x(t_0) + \int_{t_0}^{t_f} e^{-A\tau} Bu(\tau) d\tau \tag{8.12}$$

or $$x(t_0) = -\int_{t_0}^{t_f} e^{-A\tau} Bu(\tau) d\tau \tag{8.13}$$

or $$x(t_0) = -\int_{t_0}^{t_f} \phi(t_0 - \tau) Bu(\tau) d\tau \tag{8.14}$$

The state transition matrix as obtained from Cayley-Hamilton theorem

$$e^{At} = \phi(t) = \sum_{k=o}^{n-1} \alpha_k(t) A^k \tag{8.15}$$

Here
$$\begin{aligned} x &= (n \times 1) \quad \text{matrix} \\ A &= (n \times n) \quad \text{matrix} \\ u &= (p \times 1) \quad \text{input vector} \\ B &= (n \times p) \quad \text{matrix} \end{aligned}$$

where $\alpha_\kappa(t)$ is a scaler function of t.

Substitution yields

$$x(t_0) = -\int_{t_0}^{t_f} \sum_{k=0}^{n-1} \alpha_k(t_0 - \tau) A^K Bu(\tau) d\tau$$

since the matrices A and B are not functions of τ

$$x(t_0) = -\sum_{k=0}^{n-1} A^k B \int_{t_0}^{t_f} \alpha_k(t_0 - \tau) u(\tau) d\tau \tag{8.16}$$

which can be written as

$$= -[B \quad AB \quad A^2 B \cdots A^{n-1} B] \begin{bmatrix} P_0 \\ P_1 \\ \vdots \\ P_{n-1} \end{bmatrix} \tag{8.17}$$

where $$P_k = \int_{t_0}^{t_f} \alpha_k(t_0 - \tau) u(\tau) d\tau \tag{8.18}$$

$$Q_c = \begin{bmatrix} B & AB & A^2 B & \dots & A^{n-1} B \end{bmatrix} \tag{8.19}$$

$$P = [P_0 \quad P_1 \quad P_2 \quad \dots \quad P_{n-1}]' \tag{8.20}$$

If we define Q_c and P as given in eqns (8.19) and (8.20) respectively; where

$$Q_c = (n \times np) \text{ Controllability matrix}$$

$$P = (np \times 1) \text{ vector}$$

then $\qquad\qquad\qquad\qquad x(t_0) = -Q_c P \qquad\qquad\qquad\qquad\qquad\qquad (8.21)$

Input $u(t)$ can be ascertained for given initial state $x(t_0)$, which drives the state to $x(t_f) = 0$ for the finite time interval $t_f - t_0 \geq 0$, if eqn (8.21) has a solution. A unique solution exists if matrix Q_c has a set of n linearly independent column vectors. If u is scaler, then Q_c is a $(n \times n)$ square matrix and relation $x(t_0) = -Q_c P$, represents a set of n linear independent equations. Such equations will have a solution, if Q_c is nonsingular i.e., determinant is not zero. The controllability criterion thus states that a linear time invariant system described by eqns (8.7) and (8.8) is controllable if

- Matrix Q_c has n linearly independent column vectors, or
- If u is scaler, Q_c is nonsingular i.e., determinant is not zero.

8.4.1 Theorems

8.4.1.1 Basic Theorem The theorem is based on the concept explained in Section 8.4. It states that *"a linear time invariant system described by dynamic equation $\dot{x}(t) = Ax(t) + Bu(t)$ is controllable, if and only the controllability matrix Q_c $(n \times np)$ is of rank n, the order of the system"*. The Q_c is

$$Q_c = [B \quad AB \quad A^2B \ \cdots \ A^{n-1}B]$$

If u is scaler, then Q_c is a $(n \times n)$ square matrix and relation $x(t_0) = -Q_c P$, represents a set of n linear independent equations. Such equations will have a solution, if Q_c is nonsingular i.e., determinant is not zero.

Alternatively *"necessary and sufficient controllability condition for a linear time-invariant system described by dynamic equations $\dot{x}(t) = Ax(t) + Bu(t)$ and $y(t) = Cx(t) + Du(t)$ is that the controllability matrix $Q_c(n \times np)$*

$$Q_c = [B \quad AB \quad A^2B \ \cdots \ A^{n-1}B]$$

is nonsingular". For a single-input system Q_c is a $(n \times n)$ square matrix.

The controllability depends upon matrices A and B and is also referred to as *pair* $[A, B]$. When it is stated that the matrix that *pair* $[A, B]$ is controllable, it implies that matrix Q_c is of rank n. The rank of matrix Q_c relates to the dimension of largest nonzero determinant formed out of the matrix Q_c. For a square matrix, the largest determinant is $|Q_c|$. If matrix Q_c is not a square matrix then the following procedure may be adopted.

- Consider all the rows and equal number of columns and compute the determinant.
- Consider all the columns and equal number of rows and compute the determinant.
- Compute $Q_c Q_c'$ and find the determinant. The product $Q_c Q_c'$ yields a square matrix.

The Theorem lays down an easy criterion for ascertaining controllability. However, for higher order systems and systems with number of outputs, manual test because tedious and time consuming.

8.4.1.2 Distinct Eigenvalues The Jordan canonical form of representation of linear time-invariant system given by eqns (7.59) and (7.60) is

$$\dot{\bar{x}}(t) = \bar{A}\bar{x}(t) + \bar{B}u(t) \tag{8.22}$$

$$y(t) = \bar{C}\bar{x}(t) + \bar{D}u(t) \tag{8.23}$$

where $\qquad\qquad x = P\bar{x}$

$\bar{A} = P^{-1}AP$ (diagonal form consisting of eigenvalues)

$\bar{B} = P^{-1}B$

$\bar{C} = CP$

$\bar{D} = D$

$P =$ Modal matrix of eigenvectors

The theorem states that *"if a linear time-invariant control system represented by eqns (8.22) and (8.23) is such that A has distinct eigenvalues then the pair [A,B] is considered to be controllable if and only if there are no zero rows of $\bar{B} = P^{-1}B$, where P is modal matrix of eigenvectors."*

Example 8.1 Ascertain the controllability condition of

$$\dot{x}(t) = \begin{bmatrix} 0 & 1 \\ -2 & 3 \end{bmatrix} x(t) + \begin{bmatrix} b_1 \\ b_2 \end{bmatrix}$$

Solution Eigenvalues of A are 1 and 2 and the eigenvectors are $[1 \quad 1]^T$ and $[1 \quad 2]^T$. Therefore

$$P = \begin{bmatrix} 1 & 1 \\ 1 & 2 \end{bmatrix} \text{ and } \bar{A} = \begin{bmatrix} 1 & 0 \\ 0 & 2 \end{bmatrix}$$

It can be verified that

$$P\bar{A} = AP = \begin{bmatrix} 1 \\ 4 \end{bmatrix}$$

Thus, by application of theorem (8.4.1.2)

$$\bar{B} = P^{-1}B$$

or $\qquad\qquad \bar{B} = \begin{bmatrix} 2 & -1 \\ -1 & 1 \end{bmatrix}\begin{bmatrix} b_1 \\ b_2 \end{bmatrix} = \begin{bmatrix} 2b_1 & -b_2 \\ -b_1 & b_2 \end{bmatrix}$

The condition of controllability is

♦ $2b_1 - b_2 \neq 0$; and

♦ $-b + b_2 \neq 0$

The system will become uncontrollable, if $2b_1 = b_2$ i.e., $b_1 = b_2/2$; and $-b_1 + b_2 = 0$ i.e., $b_1 = b_2$.

8.4.1.3 Controllable Canonical Form　The theorem states that *"for a single-input single-output system described by state equation $\dot{x}(t) = Ax(t) + Bu(t)$ the system is said to be controllable if matrices A and B are in controllable canonical forms or if by similarity transformation it can be transformed into controllable canonical forms.* Similarity transformation considers the matrix P be nonsingular (ref. Sec 7.5).

8.4.1.4 Diagonal/Jordan Form　The controllability of the system can also be ascertained from the form of matrix A of the system described by equation $\dot{x}(t) = Ax(t) + Bu(t)$. The theorem states that *"if matrix A of the system described by state equation $\dot{x}(t) = Ax(t) + Bu(t)$ is in diagonal or Jordan canonical forms, the system is controllable if all the elements in the rows of matrix B that relates to the last row of each Jordan block of matrix A are non zero".*

Example 8.2　Let us consider a linear time-invariant system.

$$\dot{x}(t) = \begin{bmatrix} 2 & 1 & 0 & 0 \\ 0 & 2 & 1 & 0 \\ 0 & 0 & 2 & 0 \\ 0 & 0 & 0 & 3 \end{bmatrix} x(t) + \begin{bmatrix} b_{11} & b_{12} \\ b_{21} & b_{22} \\ b_{31} & b_{32} \\ b_{41} & b_{42} \end{bmatrix}$$

Ascertain the condition of controllability.

Solution　In Section 7.6.4, properties associated with Jordan form are enumerated. The Jordan blocks are marked as shown below

$$A = \begin{bmatrix} 2 & 1 & 0 & 0 \\ 0 & 2 & 1 & 0 \\ 0 & 0 & 2 & 0 \\ \hline 0 & 0 & 0 & 3 \end{bmatrix} \text{ and } B = \begin{bmatrix} b_{11} & b_{12} \\ b_{21} & b_{22} \\ b_{31} & b_{32} \\ b_{41} & b_{42} \end{bmatrix} \begin{matrix} \text{" row 1'} \\ \text{" row 2'} \\ \text{" row 3'} \\ \text{" row 4'} \end{matrix}$$

The controllability depends upon on the last row of matrix that corresponds to the last row of the Jordan block. There are two Jordan blocks

- Elements of third row of matrix B i.e., b_{31} and b_{32} correspond to the last row of first Jordan block. Hence, condition for controllability is $b_{31} \neq 0$ and $b_{32} \neq 0$.
- Elements of fourth row of matrix B i.e., b_{41} and b_{42} correspond to last row of second Jordan block. Hence, controllability condition is $b_{41} \neq 0$ and $b_{42} \neq 0$.

8.5　RANK OF A MATRIX

"A square matrix of order (n × n) is said to possess rank r, if and only if it has atleast one nonsingular sub-matrix of order r but has no nonsingular sub-matrix of order more than r". The aim is to find out the largest square matrix in the given matrix whose determinant is not zero. The process of calculating determinants for higher order matrices becomes tedious and time consuming as all

possible sub-matrices have to be identified for calculation of determinants. Rank of higher order matrices can be ascertained by carrying out the following elementary transformations:

- Check for a nonzero element in the first row and the first column. The nonzero element preferably be a 1 (one). If required, perform elementary operations.
- If the nonzero element as given above is not a 1 (one), divide the first row by the nonzero element existing in the matrix.
- Subtract appropriate multiples of the first row from other rows to obtain zeros in the remainder of the first row.
- The steps need to be repeated for the second element in the second row.
- Continue till calculation of determinant becomes easier.

Elementary transformations so performed do not alter the rank of a matrix. Elementary transformation involves:

- Interchanging of two rows or column.
- Multiplication of column by a non-zero constant.
- Addition or subtraction of elements of a row or column with another row or column.
- Performing operations like multiplications of rows or columns by a nonzero constant and then performing addition or subtraction operation as given above.

The following emerges from the definition of rank of a matrix:

- The rank of a nonsingular matrix of order n is n.
- If the matrix is singular, its rank is less than n.
- If the rank of a $m \times n$ matrix can at most be equal to the smaller of m or n. It can also be less.
- The rank of a nonzero matrix ≥ 1
- The rank of a null matrix is zero.
- Rank of product of two matrices cannot be more than the rank of either matrix.

Example 8.3   Find the rank of the following matrices

$$A = \begin{bmatrix} 2 & 4 & 5 & 3 \\ 0 & 3 & 0 & 0 \\ 0 & 3 & 2 & 2 \\ 0 & 2 & 0 & 1 \end{bmatrix}$$

Solution   Let us compute determinants of sub-matrices:

(a) $|2| \neq 0$

(b) $\begin{vmatrix} 2 & 4 \\ 0 & 3 \end{vmatrix} \neq 0$

(c) $\begin{vmatrix} 3 & 0 & 0 \\ 3 & 2 & 2 \\ 2 & 0 & 1 \end{vmatrix} = 3 \begin{vmatrix} 2 & 2 \\ 0 & 1 \end{vmatrix} \neq 0$

(d) $\begin{bmatrix} 2 & 4 & 5 & 3 \\ 0 & 3 & 0 & 0 \\ 0 & 3 & 2 & 2 \\ 0 & 2 & 0 & 1 \end{bmatrix} = 2 \begin{vmatrix} 3 & 0 & 0 \\ 3 & 2 & 2 \\ 2 & 0 & 1 \end{vmatrix} \neq 0$

Since the largest square matrix in A is of order (4×4); the rank is 4.

Example 8.4 Find the rank of following matrices

(a) $A = \begin{bmatrix} 2 & 2 & 0 \\ 2 & 1 & 1 \end{bmatrix}$

(b) $B = \begin{bmatrix} 1 & 3 & -1 \\ 2 & 4 & 2 \end{bmatrix}$

(c) $C = \begin{bmatrix} 1 & 2 & 3 \\ 1 & 4 & 2 \\ 2 & 6 & 5 \end{bmatrix}$

(d) $D = \begin{bmatrix} 1 & 2 & 3 \\ 2 & 4 & 7 \\ 3 & 6 & 10 \end{bmatrix}$

Solution The determinant of matrices A & B cannot be determined. Therefore, the highest rank possible is 2, if we are to find a nonsingular sub-matrix of order 2×2.

(a) Since

$$|2| \neq 0 \quad \text{or} \quad \begin{vmatrix} 2 & 2 \\ 2 & 1 \end{vmatrix} \neq 0 \quad \text{or} \quad \begin{vmatrix} 2 & 0 \\ 2 & 1 \end{vmatrix} \neq 0 \quad \text{or} \quad \begin{vmatrix} 2 & 0 \\ 1 & 1 \end{vmatrix} \neq 0$$

Hence the rank is 2.

(b) Since

$$|1| \neq 0 \quad \text{or} \quad \begin{vmatrix} 1 & 3 \\ 2 & 4 \end{vmatrix} \neq 0 \quad \text{or} \quad \begin{vmatrix} 1 & -1 \\ 2 & 2 \end{vmatrix} \neq 0 \quad \text{or} \quad \begin{vmatrix} 3 & -1 \\ 4 & 2 \end{vmatrix} \neq 0$$

Hence the rank is 2.

(c) $C = \begin{bmatrix} 1 & 2 & 3 \\ 1 & 4 & 2 \\ 2 & 6 & 5 \end{bmatrix}$

Performing $R_2 - R_1$, we get

$$C \sim \begin{bmatrix} 1 & 2 & 3 \\ 0 & 2 & -1 \\ 1 & 2 & -1 \end{bmatrix}$$

and its determinant is not equal to zero, therefore, the rank of the matrix C is 3.

(d) $D = \begin{bmatrix} 1 & 2 & 3 \\ 2 & 4 & 7 \\ 3 & 6 & 10 \end{bmatrix}$

Performing $R_2 - 2R_1$ gives

$$D \sim \begin{bmatrix} 1 & 2 & 3 \\ 0 & 0 & 1 \\ 3 & 6 & 10 \end{bmatrix}$$

Performing $R_3 - 3R_1$ gives

$$C \sim \begin{bmatrix} 1 & 2 & 3 \\ 0 & 0 & 1 \\ 0 & 0 & 1 \end{bmatrix}$$

Since $|C| = 0$, the rank of matrix C is less than the order of the matrix C which is 3. Choosing a sub-matrix of order 2×2 and its determinant gives

$\begin{vmatrix} 4 & 7 \\ 6 & 10 \end{vmatrix} \neq 0$, therefore the rank is 2

8.6 OBSERVABILITY TESTS

Observability of a control system can be ascertained from the *observability matrix* (Section 7.6.2) and from matrices A and C. The following theorems provide criterion for ascertaining the observability of control systems.

8.6.1 Basic Theorem

A MIMO linear time-invariant system described by dynamic equations

$$\dot{x}(t) = Ax(t) + Bu(t)$$
$$y(t) = Cx(t) + Du(t)$$

is observable, if and only if observability matrix Q_0

$$Q_0 = \begin{bmatrix} C \\ CA \\ CA^2 \\ \vdots \\ CA^{n-1} \end{bmatrix} = \begin{bmatrix} C^T & A^T C^T & A^{2^T} C^T & \ldots & A^{n-1^T} C^T \end{bmatrix} \text{ posses rank } n, \text{ the order of the system}$$

Alternatively, *"a necessary and sufficient observability condition for a linear time-invariant system described by dynamic equations $\dot{x}(t) = Ax(t) + Bu(t)$ and $y(t) = Cx(t) + Du(t)$ is that the observability matrix Q_0 ($nq \times n$) is nonsingular"*.

For a SISO system where $p = 1$ and $q = 1$, Q_0 is a ($n \times n$) square matrix.

8.6.2 Distinct Eigenvalues

The theorem states that *if a linear time-invariant system described eqns (8.23) is such that A has distinct eigenvalues, "then the pair [A, C] is send to be observable if and only if*

$$\overline{C} = CP$$

has no zero columns; where P is the modal matrix formed by the eigenvectors of A ".

8.6.3 Observable Canonical Form

The theorem states that *"for a single-input single-output system described by dynamic equations $\dot{x}(t) = Ax(t) + Bu(t)$ and $y(t) = Cx(t) + Du(t)$, is observable if matrices A and C are in observable canonical forms or if by similarity transformation they can be transformed into observable canonical form".* Similarity transformation considers the matrix P be nonsingular (ref. Sec 7.5).

8.6.4 Diagonal/Jordan Canonical Form

The theorem states that *"A system described by dynamic equations $\dot{x}(t) = Ax(t) + Bu(t)$ and $y(t) = Cx(t) + Du(t)$ are observable if matrix A is in diagonal/Jordan canonical form and if the elements of matrix C that relates to first row of each Jordan block are non zero".*

Example 8.5 A linear time-invariant system described by

$$\dot{x}(t) = \begin{bmatrix} 0 & 1 & 0 \\ 0 & 0 & 1 \\ -6 & -11 & -6 \end{bmatrix} x(t) + \begin{bmatrix} 1 & 0 \\ 0 & 1 \\ 1 & 1 \end{bmatrix} \begin{matrix} u_1(t) \\ u_2(t) \end{matrix}$$

and

$$y = \begin{bmatrix} 1 & 0 & 1 \end{bmatrix} x(t)$$

Find if it is observable.

Solution: Modal matrix consisting of eigenvectors of matrix A is

$$P = \begin{bmatrix} 1 & 1 & 1 \\ -1 & -1 & -3 \\ 1 & 4 & 9 \end{bmatrix}$$

Now

$$\overline{C} = CP$$

$$= \begin{bmatrix} 1 & 0 & 1 \end{bmatrix} \begin{bmatrix} 1 & 1 & 1 \\ -1 & -1 & -3 \\ 1 & 4 & 9 \end{bmatrix} = \begin{bmatrix} 2 & 5 & 10 \end{bmatrix}$$

Since $\overline{C}$ has nonzero columns, the system is observable.

Example 8.6 Ascertain the condition of observability for a linear time-invariant system described by the state equation

$$\dot{x} = \begin{bmatrix} 3 & 1 & 0 & 0 \\ 0 & 3 & 1 & 0 \\ 0 & 0 & 3 & 0 \\ 0 & 0 & 0 & 4 \end{bmatrix} x(t) + \begin{bmatrix} b_{11} & b_{12} \\ b_{21} & b_{22} \\ b_{31} & b_{32} \\ b_{41} & b_{42} \end{bmatrix} u(t)$$

Solution The Jordan blocks are marked as shown below

$$A = \begin{bmatrix} 3 & 1 & 0 & 0 \\ 0 & 3 & 1 & 0 \\ 0 & 0 & 3 & 0 \\ \hline 0 & 0 & 0 & 4 \end{bmatrix} \quad \text{and} \quad B = \begin{bmatrix} b_{11} & b_{12} \\ b_{21} & b_{22} \\ b_{31} & b_{32} \\ b_{41} & b_{42} \end{bmatrix} \begin{matrix} \text{``row 1''} \\ \text{``row 2''} \\ \text{``row 3''} \\ \text{``row 4''} \end{matrix}$$

The observability depends upon the first row of the Jordan block. There are two Jordan blocks:

- Elements of first row of matrix B i.e., b_{11} and b_{12} correspond to the first row of first Jordan block. Hence, condition of observability is $b_{11} \neq 0$ and $b_{12} \neq 0$.
- Elements of fourth row of matrix B i.e., b_{41} and b_{42} correspond to first row of second Jordan block. Hence, condition of observability is $b_{41} \neq 0$ and $b_{42} \neq 0$.

8.7 OUTPUT CONTROLLABILITY

There may be need to control the output instead of the states of the system. In such a case ascertaining the output controllability may become necessary. We had seen that the state controllability is a property of the state-space, where as the output controllability is a property of input-output relationship. We can define the output controllability as we had done for state controllability.

Let us consider the system described by

$$\dot{x} = Ax + Bu$$
$$y = Cx$$

where

$$x = \text{state vector } (n \times 1)$$
$$u = \text{input or control vector } (p \times 1)$$
$$y = \text{output vector } (q \times 1) \quad \text{where } q \leq n$$
$$A = (n \times n) \text{ matrix}$$
$$B = (n \times p) \text{ matrix}$$
$$C = (q \times n) \text{ matrix}$$

The system described by above mentioned equations is said to be completely output controllable, *if it is possible to from an unconstrained control vector u(t) that will transfer any given initial output at time $t = t_0$ to any final output at time $t = t_f$ in a finite time interval $t_0 \le t \le t_f$.*

Output Controllability

$$Q_{op} = [CB \quad CAB]$$

$$= \left[\begin{bmatrix} 1 & 0 \\ 0 & 0 \end{bmatrix} \begin{bmatrix} 1 & 0 \\ 0 & 1 \end{bmatrix} \vdots \begin{bmatrix} 1 & 0 \\ 0 & 0 \end{bmatrix} \begin{bmatrix} -1 & 1 \\ 0 & -1 \end{bmatrix} \begin{bmatrix} 1 & 0 \\ 0 & 1 \end{bmatrix} \right]$$

$$= \begin{bmatrix} 1 & 0 & -1 & 0 \\ 0 & 0 & 0 & 0 \end{bmatrix}$$

The determinant of any (2×2) sub-matrix is zero. Hence, the rank of the matrix is one where as for output controllabaility it should be 2. Therefore, the system is not output controllable.

It can be proved that a linear time-invariant system described by equations above is output controllable is and only if the rank of $(q \times np)$ matrix.

$$Q_{op} = [CB \quad CAB \quad CA^2B \quad ....CA^{n-1}B] \text{ is } q.$$

However, if $y = Cx + Du$; where D is $(q \times p)$ matrix, the system is completely output controllable, if and only if

$$\underset{(qx(n+1)p)}{Q_{op}} = [CB \quad CAB \quad CA^2B \quad ....CA^{n-1}B \quad D] \text{ possess rank } q.$$

Example 8.7 Consider a system described as

$$\dot{x} = \begin{bmatrix} -1 & 1 \\ 0 & -1 \end{bmatrix} \begin{bmatrix} x_1 \\ x_2 \end{bmatrix} + \begin{bmatrix} 1 & 0 \\ 0 & 1 \end{bmatrix} u \quad \text{and} \quad y = \begin{bmatrix} 1 & 0 \\ 0 & 0 \end{bmatrix} x$$

Find out the controllability

Solution **State Controllability**

$$Q_0 = [B \quad AB] = \left[\begin{bmatrix} 1 & 0 \\ 0 & 0 \end{bmatrix} \begin{bmatrix} -1 & 1 \\ 0 & -1 \end{bmatrix} \begin{bmatrix} 1 & 0 \\ 0 & 1 \end{bmatrix} \right]$$

$$= \begin{bmatrix} 1 & 0 & \vdots & -1 & 1 \\ 0 & 1 & \vdots & 0 & -1 \end{bmatrix}$$

The rank of $Q_0 = 2$ because determinant of sub-matrix, $\det \begin{vmatrix} -1 & 1 \\ 0 & -1 \end{vmatrix} = 1$ i.e., $\ne 0$. Hence, it is state controllable

Output Controllability

$$Q_{op} = [CB \quad CAB] = \left[\begin{bmatrix} 1 & 0 \\ 0 & 0 \end{bmatrix} \begin{bmatrix} 1 & 0 \\ 0 & 1 \end{bmatrix} \begin{bmatrix} 1 & 0 \\ 0 & 0 \end{bmatrix} \begin{bmatrix} -1 & 1 \\ 0 & -1 \end{bmatrix} \begin{bmatrix} 1 & 0 \\ 0 & 1 \end{bmatrix} \right]$$

$$= \begin{bmatrix} 1 & 0 & \vdots & -1 & 0 \\ 0 & 1 & \vdots & 0 & 0 \end{bmatrix}$$

Since determinant of (2×2) sub-matrix is zero, the rank of $Q_{op} = 1$ and not equal to 2. Hence, it is not output controllable.

8.8   POLE-ZERO CANCELLATION IN TRANSFER FUNCTION

The concepts of controllability and observability are closely related to the properties of the transfer function. It the input-output relationship of the nth-order system having distinct eigenvalues is represented as

$$\frac{Y(s)}{X(s)} = \frac{k(s-z_1)(s-z_2)\ldots\ldots(s-z_m)}{(s-p_1)(s-p_2)\ldots\ldots(s-p_n)}; \ m<n$$

$$= \sum_{k=1}^{n} \frac{C_k}{(s-p_k)}$$

where $\qquad C_k$ = residue of poles at $s = p_k$.

Let us assume that state-space representations of the control system are of the following forms:

First form

$$\dot{x} = \begin{bmatrix} p_1 & 0 & \cdots & 0 \\ 0 & p_2 & \cdots & 0 \\ \vdots & \vdots & & \vdots \\ 0 & 0 & \cdots & p_n \end{bmatrix} \begin{bmatrix} x_1 \\ x_2 \\ \vdots \\ x_n \end{bmatrix} + \begin{bmatrix} c_1 \\ c_2 \\ \vdots \\ c_n \end{bmatrix} u$$

$$y = \begin{bmatrix} 1 & 1 & \cdots & 1 \end{bmatrix} \begin{bmatrix} x_1 \\ x_2 \\ \vdots \\ x_n \end{bmatrix} + Du$$

Second Form

$$\dot{x} = \begin{bmatrix} p_1 & 0 & \cdots & 0 \\ 0 & p_2 & \cdots & 0 \\ \vdots & \vdots & & \vdots \\ 0 & 0 & \cdots & p_n \end{bmatrix} \begin{bmatrix} x_1 \\ x_2 \\ \vdots \\ x_n \end{bmatrix} + \begin{bmatrix} 1 \\ 1 \\ \vdots \\ 1 \end{bmatrix} u$$

$$y = \begin{bmatrix} c_1 & c_2 & \cdots & c_n \end{bmatrix} \begin{bmatrix} x_1 \\ x_2 \\ \vdots \\ x_n \end{bmatrix} + Du$$

Let us assume that the transfer function has identical pole and zero, i.e., $p_i = z_i$ which makes $c_i = 0$. In such a case, if the state-space representation of the control system is as per 'First form'; then the matrix B will have the element $c_i = 0$; making the state variable x_i uncontrollable. On the other hand, if the state-space representation is as per 'Second Form', then the element $c_i = 0$ will appear in matrix C making the state variable x_i unobservable.

Therefore, depending upon how the state-variables are defined and state-space represented, the pole-zero cancellation due to identical pole and zero, will make the system either not state-controllable or unobservable or both; otherwise the state-space model representation can be completely controllable and observable.

Example 8.8 The transfer function of a control system is given as

$$\frac{Y(s)}{X(s)} = \frac{s+a}{s^2 + 7s^2 + 14s + 8}$$

(a) Determine the values of 'a' so that system is either uncontrollable or unobservable.
(b) With the value (s) of 'a' found as above, obtain state-space representation so that one of the states is uncontrollable.
(c) Repeat part (b) so that one of states is unobservable.

Solution The transfer function in the factored form is

$$\frac{Y(s)}{X(s)} = \frac{s+a}{(s+1)(s+2)(s+4)}$$

(a) The system can become either uncontrollable unobservable, if there is a pole-zero cancellation and that is possible, if $a = 1, 2$ or 4
(b) The transfer function is expanded by partial fraction expansion.

$$\frac{Y(s)}{X(s)} = \frac{(a-1)}{3(s+1)} + \frac{-(a-2)}{2(s+2)} + \frac{(a-4)}{6(s+4)}$$

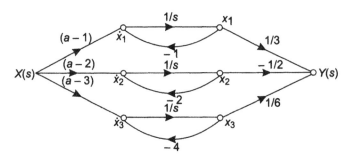

Fig. 8.4

$$\dot{x}_1 = -x_1 + (a_1 - 1)u$$

$$\dot{x}_2 = -2x_2 + (a - 2)u$$

$$\dot{x}_3 = -3x_3 + (a - 4)u$$

$$y = \frac{1}{3}x_1 - \frac{1}{2}x_2 + \frac{1}{6}x_3$$

In vector-matrix form

$$\dot{x} = \begin{bmatrix} -1 & 0 & 0 \\ 0 & -2 & 0 \\ 0 & 0 & -3 \end{bmatrix} \begin{bmatrix} x_1 \\ x_2 \\ x_3 \end{bmatrix} + \begin{bmatrix} a-1 \\ a-2 \\ a-4 \end{bmatrix} u$$

$$y = \begin{bmatrix} \dfrac{1}{3} & -\dfrac{1}{2} & \dfrac{1}{6} \end{bmatrix} \begin{bmatrix} x_1 \\ x_2 \\ x_3 \end{bmatrix}$$

The system will become uncontrollable for

	$a - 1 = 0$	i.e.	$a = 1$
or	$a - 2 = 0$	i.e.	$a = 2$
or	$a - 4 = 0$	i.e.	$a = 4$

(c) The state-space representation can be of the form

$$\dot{x} = \begin{bmatrix} -1 & 0 & 0 \\ 0 & -2 & 0 \\ 0 & 0 & -4 \end{bmatrix} \begin{bmatrix} x_1 \\ x_2 \\ x_3 \end{bmatrix} + \begin{bmatrix} 1/3 \\ -1/2 \\ 1/6 \end{bmatrix} u$$

$$y = \begin{bmatrix} a-1 & a-2 & a-4 \end{bmatrix}$$

The system will become uncontrollable for

	$a - 1 = 0$	i.e.	$a = 1$
or	$a - 2 = 0$	i.e.	$a = 2$
or	$a - 4 = 0$	i.e.	$a = 4$

8.9 DUALITY THEOREM/PRINCIPLE/PROPERTY

The principle of duality was first given by *kalman*. We have seen the conditions of controllability and observability. The pair $[A,B]$ is controllable if rank of

$$Q_c = [B \quad AB ...... A^{n-1}B] \text{ is } n$$

and the pair $[A,C]$ is considered to observable if rank of

$$Q_o = [C^T \quad A^T C^T ......(A^{n-1})^T C^T] \text{ is } n.$$

There seems to be a link between the two concepts. The duality theorem states that if the dynamics of a system is described in state-space with the help of matrices (A, B, C, D), then the system is completely state controllable if and only if the dual system (A^T, C^T, B^T, D^T) is completely observable.

The duality principle specifies that

- Pair $[A,B]$ is controllable also means that the pair $[A^T, B^T]$ is observable.
- Pair $[A,C]$ is observable also means that the pair $[A^T, C^T]$ is controllable.

The use of the theorem lays down a concept of performing check tests. State controllability of a given system can be checked by performing observability test of its dual. Similarly, the observability of a given system can be checked by performing state controllability test of its dual.

Example 8.9 System 1 (one) described by

$$\dot{x} = \begin{bmatrix} 0 & 1 \\ -2 & -3 \end{bmatrix} x + \begin{bmatrix} 1 \\ 1 \end{bmatrix} u, \quad y = \begin{bmatrix} 1 & 0 \end{bmatrix} x$$

has its dual system 2 (two) described by

$$\dot{x}_d = \begin{bmatrix} 0 & 2 \\ -1 & 3 \end{bmatrix} x_d + \begin{bmatrix} 1 \\ 0 \end{bmatrix} u_d, \quad y = \begin{bmatrix} 1 & 1 \end{bmatrix} x_d$$

Verify duality theorem.

Solution System 1

$$Q_c = \begin{bmatrix} B & AB \end{bmatrix} = \begin{bmatrix} \begin{bmatrix} 1 \\ 1 \end{bmatrix} & \begin{bmatrix} 0 & 1 \\ -2 & -3 \end{bmatrix} \begin{bmatrix} 1 \\ 1 \end{bmatrix} \end{bmatrix} = \begin{bmatrix} 1 & 1 \\ 1 & -5 \end{bmatrix}$$

$|Q_c| \neq 0$, hence, rank of Q_c is 2 (order of A). Therefore, The system is controllable

$$Q_o = \begin{bmatrix} C \\ CA \end{bmatrix} = \begin{bmatrix} \begin{bmatrix} 1 & 0 \end{bmatrix} \\ \begin{bmatrix} 1 & 0 \end{bmatrix} \begin{bmatrix} 0 & 1 \\ -2 & -3 \end{bmatrix} \end{bmatrix} = \begin{bmatrix} 1 & 0 \\ 0 & 1 \end{bmatrix}$$

$|Q_o| \neq 0$, hence, rank of $Q_o = 2$ (order of A). Therefore, the system is observable.

System 2
As per duality theorem system 1 is controllable (implying its dual should be observable) and observable (implying its dual should be controllable). Verification leads to

$$Q_c = \begin{bmatrix} B & AB \end{bmatrix} = \begin{bmatrix} \begin{bmatrix} 1 \\ 0 \end{bmatrix} & \begin{bmatrix} 0 & 2 \\ -1 & 3 \end{bmatrix} \begin{bmatrix} 1 \\ 0 \end{bmatrix} \end{bmatrix} = \begin{bmatrix} 1 & 0 \\ 0 & -1 \end{bmatrix}$$

$|Q_c| \neq 0$, hence, rank of Q_c is 2 (order of A). Therefore, the system is controllable

$$Q_o = \begin{bmatrix} C \\ CA \end{bmatrix} = \begin{bmatrix} \begin{bmatrix} 1 & 1 \end{bmatrix} \\ \begin{bmatrix} 1 & 1 \end{bmatrix} \begin{bmatrix} 0 & 2 \\ -1 & 3 \end{bmatrix} \end{bmatrix} = \begin{bmatrix} 1 & 1 \\ -1 & 5 \end{bmatrix}$$

$|Q_o| \neq 0$, hence rank of Q_o is 2 (order of A). therefore, the system is observable.

8.10   DESIGN

In Chapter 5, we discussed open-loop and closed-loop systems. The basic difference between the two was the presence of feedback in the closed-loop system which helps in restoration and does not let the system output, in the open-loop system, to get out of control and ensures its *maintenance* at desired value. The location of closed-loop poles govern the stability of a system. The system performance often needs to be modified after the stability has been analysed. The system parameters which generally define as to how good the system is; like open-loop gain, accuracy, transient response, sensitivity etc require fine tuning and modification by addition of certain types of devices to the basic feedback control system. The process of improving performance involves the following steps:

◆ *Stabilisation* to achieve the stability criterion.

◆ *Compensation* to meet increased accuracy requirements and accelerating of response.

   In short, design encompasses the process of stabilisation and accelerating. Design modification involves adjustment of known or unknown constants in the controllers transfer function or its state-space representation. The aim of modification is restricted to placing all the closed-loop poles at the desired and most suitable location, provided that the original system is completely state controllable, by introducing adjustable parameters through the state-space approach. Design of control system thus involves design of state variable feedback controller with specified characteristics root location. Design through state-space is supported by software designing and design validation, as it is based on matrix algebra.

8.10.1   State Feedback

Let us consider a linear time-invariant system described by the following set of equations

$$\dot{x}(t) = Ax(t) + Bu(t) \tag{8.24}$$

$$y(t) = Cx(t) \tag{8.25}$$

Where

r = reference input vector ($p \times 1$)

x = State vector ($n \times 1$)

u = Control vector ($p \times 1$)

y = Output vector ($q \times 1$)

A = Constant matrix ($n \times n$)

B = Constant matrix ($n \times p$)

C = Constant matrix ($q \times n$)

   The system represented in Fig. 8.5 depicts state variable feedback. The state equations describing the above system are

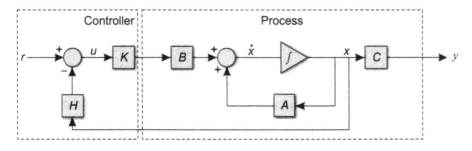

Fig. 8.5 *State variable feedback system (Matrix representation)*

$$\dot{x}(t) = Ax(t) + KBu(t) \tag{8.26}$$
$$y(t) = Cx(t) \tag{8.27}$$
$$u(t) = r(t) - Hx(t) \tag{8.28}$$

substitution of $u(t)$ gives

$$\dot{x}(t) = Ax(t) + KB(r(t) - Hx(t))$$

or
$$\dot{x}(t) = (A - KBH)x(t) + KBr(t) = A_K x(t) + KBr(t) \tag{8.29}$$

and
$$y(t) = Cx(t) \tag{8.30}$$

where $A_K = [A - KBH]$ is closed-loop system matrix

H = feedback constant-gain matrix $(p \times q)$ (8.31)

K = controller gain

If $K = 1$;

then $A_K = [A - BH]$

Comparing equations (8.29) and (8.30) with (8.24) and (8.25). We observe that they are identical. The important matrices which relate to the stability, controllability and observability are A_K, B, C

- Controllability depends upon on matrix pair $\{[A_K], [B]\}$
- Observability depends upon on matrix pair $\{[A_K], [C]\}$ or $\{[A_K], [C - DK]\}$ (if output equation depends upon D).
- Stability depends upon the eigenvalues of matrix $[A_K]$, also termed as regulator/controller poles.

Figure 8.5 shows matrix representation of the concept of linear state variable feedback. However, physical representation of a typical system which relates with the concept of state feedback shown in Fig. 8.5, is represented in Fig. 8.6, where it is assumed that all state-variables are measurable and available for feedback. Therefore u can be represented as

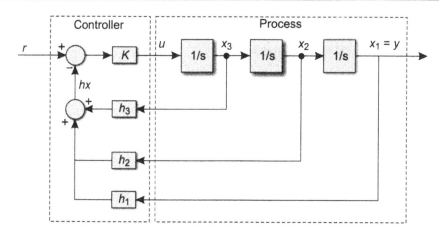

Fig. 8.6

$$u = K[r - (h_1x_1 + h_2x_2 + h_3x_3)] \tag{8.32}$$

or   in general

$$u = K[r - (h_1x_1 + h_2x_2 + \ldots + h_nx_n)] \tag{8.33}$$

$$= K[r - hx] \tag{8.34}$$

$$x = [x_1 \quad x_2 \quad \ldots \quad x_n]^T; \text{ and} \tag{8.35}$$

$$h = [h_1 \quad h_2 \quad \ldots \quad h_n]. \tag{8.36}$$

8.10.2   Relationship with Closed-loop Transfer Function

Equations (8.29) and (8.30) can be written as

$$\dot{x}(t) = A_k x(t) + KBr(t) \tag{8.37}$$

$$y(t) = Cx(t) \tag{8.38}$$

where $A_K = (A - KBh)$ is called closed-loop system matrix. $\tag{8.39}$

Taking Laplace transform of equations (8.37) and (8.38) assuming initial conditions to be zero, gives

$$sX(s) = A_K X(s) + KBR(s) \tag{8.40}$$

$$Y(s) = CX(s) \tag{8.41}$$

Solving for $X(s)$, gives

$$X(s) = K[sI - A_k]^{-1}BR(s) \tag{8.42}$$

$$= K\phi_k(s)BR(s) \tag{8.43}$$

where

$$\phi_k(s) = [sI - A_k]^{-1} \tag{8.44}$$

Substituting equation (8.43) in equation (8.41) gives

$$Y(s) = KC\phi_K(s)BR(s) \tag{8.45}$$

Therefore

$$\frac{Y(s)}{R(s)} = KC\phi_K(s)B \tag{8.46}$$

Characteristic Equation

$$\frac{Y(s)}{R(s)} = \frac{KC[adj(sI - A_K)]B}{\det(sI - A_K)} \tag{8.47}$$

The corresponding characteristic equation of the closed-loop system in terms of the closed-loop system matrix is obtained by equating denominator to zero

$$\det(sI - A_K) = 0 \tag{8.48}$$

An equivalent model is shown in Fig 8.7

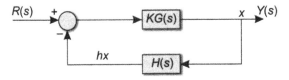

Fig. 8.7

From Fig 8.7, it is seen that

$$H(s) = \frac{hX(s)}{Y(s)} \tag{8.49}$$

Substituting equation (8.41) in equation (8.49), we get

$$H(s) = \frac{hX(s)}{CX(s)}$$

$$= \frac{h\phi_K(s)BR(s)}{C\phi_K(s)BR(s)} \tag{8.50}$$

or

$$H(s) = \frac{h\phi_K(s)B}{C\phi_K(s)B} \tag{8.51}$$

Also

$$\frac{Y(s)}{R(s)} = \frac{KG(s)}{1 + KG(s)H(s)} \tag{8.52}$$

Substituting eqns (8.46) and (8.51) in eqn (8.52), gives

or

$$G(s) = \frac{C\phi_K(s)B}{1 - K\phi_K(s)hB} \tag{8.53}$$

Combining equations (8.51) and (8.53) gives the open-loop transfer function

$$KG(s)H(s) = \frac{Kh\phi_K(s)B}{1 - K\phi_K(s)hB} \tag{8.54}$$

Following conclusions emerge on consideration of $G(s)$, $H(s)$, open-loop transfer functions $KG(s)H(s)$, and closed-loop transfer function $Y(s)/R(s)$

♦ Poles of open-loop transfer function $KG(s)H(s)$ are the poles of $G(s)$.

♦ Zeros of closed-loop transfer function $Y(s)/R(s)$ are the zeros of $G(s)$.

Example 8.10 The open-loop process required to be controlled has transfer function of $\dfrac{1}{s(s+40)}$. It is desired that the closed-loop characteristics of unity feedback control system be governed by $\omega_n = 40$ rad/sec, $K_v = 30$/sec, and $\xi = 0.707$. Design a suitable state feedback controller. Assume that transient response of the system is governed by dominant complex-conjugate poles.

Solution Closed-loop control system is given by

$$\frac{C(s)}{R(s)} = \frac{\omega_n^2}{s^2 + 2\xi\omega_n s + \omega_n^2} \tag{1}$$

Substitution of required parameters give

$$\frac{C(s)}{R(s)} = \frac{1600}{s^2 + 2 \times 0.707 \times 40 + 1600} \tag{2}$$

$$= \frac{1600}{s^2 + 56.56s + 1600} \tag{3}$$

The state-space representation of the process is shown in Fig. 8.8

Fig. 8.8

The corresponding state variable representation is

$$\dot{x}(t) = \begin{bmatrix} 0 & 1 \\ 0 & -40 \end{bmatrix} x(t) + \begin{bmatrix} 0 \\ 1 \end{bmatrix} u(t) \tag{4}$$

$$C(s) = \begin{bmatrix} 1 & 0 \end{bmatrix} x(t) \tag{5}$$

The state variable feedback representation is shown in Fig. 8.9

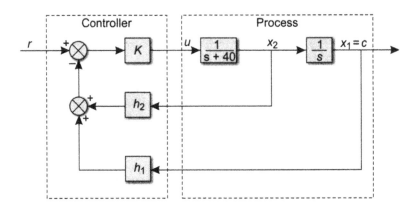

Fig. 8.9

Equivalent state-model is shown in Fig. 8.10

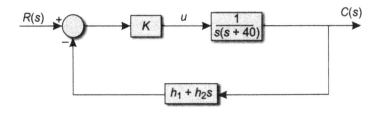

Fig. 8.10

Transfer function is given as

$$\frac{C(s)}{R(s)} = \frac{\dfrac{K}{s(s+40)}}{1 + \dfrac{K}{s(s+40)} \times (h_1 + h_2 s)} \tag{6}$$

$$= \frac{K}{s^2 + (40 + h_2 K)s + K h_1} \tag{7}$$

Comparing equations (7) and (3), gives

$$K = 1600 \tag{8}$$

$$40 + h_2 K = 56.56 \tag{9}$$

or $$h_2 = 0.01035; \text{ and} \tag{10}$$

$$K h_1 = 1600 \tag{11}$$

or $$h_1 = 1600/1600 = 1 \tag{12}$$

The state-diagram is represented in Fig. 8.11

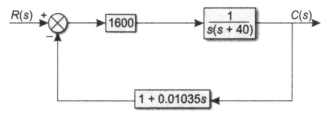

Fig. 8.11

Example 8.11 The state variable representation of a SISO system is

$$\dot{x}(t) = \begin{bmatrix} 0 & 1 \\ -1 & -2 \end{bmatrix} x(t) + \begin{bmatrix} 0 \\ 1 \end{bmatrix} u(t)$$

$$y(t) = \begin{bmatrix} 1 & 2 \end{bmatrix} x(t)$$

(a) Ascertain the controllability, observability and stability.

(b) If a feedback signal of $u(t) = y(t) - \begin{bmatrix} -2 & 1 \end{bmatrix} x(t)$ is introduced, ascertain the qualitative effects of feedback and comment.

(c) Introduce a feedback signal $u(t) = y(t) - \begin{bmatrix} 1 & 2.5 \end{bmatrix} x(t)$ and check for qualitative effects of feedback and comment.

Solution

(a) **Controllability** the controllability matrix

$$Q_c = \begin{bmatrix} B & AB \end{bmatrix} = \begin{bmatrix} 0 \\ 1 \end{bmatrix} \left\{ \begin{bmatrix} 0 & 1 \\ -1 & -2 \end{bmatrix} \begin{bmatrix} 0 \\ 1 \end{bmatrix} \right\}$$

$$= \begin{bmatrix} 0 & 1 \\ 1 & -2 \end{bmatrix}$$

Since det $Q_c \neq 0$, the rank of Q_c is 2 and hence the system is controllable

Observability

$$Q_0 = \begin{bmatrix} C \\ CA \end{bmatrix} = \begin{bmatrix} 1 & 2 \end{bmatrix}^1 \begin{bmatrix} 0 & 1 \\ -1 & -2 \end{bmatrix}^2 = \begin{bmatrix} 1 & 2 \\ -2 & -3 \end{bmatrix}$$

Since det $[Q_0] \neq 0$ the rank of Q_0 is 2 and hence the system is observable.

Stability

The characteristic equation of matrix A is $\lambda^2 + 2\lambda + 1 = 0$; which gives $\lambda_1 = -1$ and $\lambda_2 = -1$. Since, these are located in left half of s-plane, the system is stable.

(b) Here $H = \begin{bmatrix} -2 & 1 \end{bmatrix}$. The equations change to

$$\dot{x}(t) = \begin{bmatrix} A - BH \end{bmatrix} x(t) + Br(t)$$

$$y(t) = Cx(t)$$

or

$$\dot{x}(t) = \left\{ \begin{bmatrix} 0 & 1 \\ -1 & -2 \end{bmatrix} - \begin{bmatrix} 0 \\ 1 \end{bmatrix} [-2 \quad 1] \right\} x(t) + \begin{bmatrix} 0 \\ 1 \end{bmatrix} r(t)$$

$$= \left\{ \begin{bmatrix} 0 & 1 \\ -1 & -2 \end{bmatrix} - \begin{bmatrix} 0 & 0 \\ -2 & 1 \end{bmatrix} \right\} x(t) + \begin{bmatrix} 0 \\ 1 \end{bmatrix} r(t)$$

$$= \begin{bmatrix} 0 & 1 \\ 1 & -3 \end{bmatrix} x(t) + \begin{bmatrix} 0 \\ 1 \end{bmatrix} r(t)$$

and

$$y(t) = Cx(t) = [1 \quad 2]x(t).$$

Controllability

$$Q_c = \begin{bmatrix} 0 & \begin{bmatrix} 0 & 1 \\ 1 & -3 \end{bmatrix} \begin{bmatrix} 0 \\ 1 \end{bmatrix} \end{bmatrix}$$

$$= \begin{bmatrix} 0 & 1 \\ 1 & -3 \end{bmatrix}$$

Since det $Q_c \neq 0$, the rank of Q_c is 2 and hence the system is controllable.

Observability

$$Q_0 = \begin{bmatrix} C \\ CA \end{bmatrix} = \begin{bmatrix} 1 & 2 \\ [1 \quad 2] \begin{bmatrix} 0 & 1 \\ 1 & -3 \end{bmatrix} \end{bmatrix} = \begin{bmatrix} 1 & 2 \\ 2 & -5 \end{bmatrix}$$

Since det $Q_0 \neq 0$, the rank of Q_0 is 2 and hence the system is observable.

Stability

The characteristic equation of matrix $\begin{bmatrix} 0 & 1 \\ 1 & -3 \end{bmatrix}$ is $\lambda^2 + 3\lambda - 1 = 0$. The roots are -3.3 and 0.3. There is one root 0.3 which is positive and hence is located on the right half of s-plane. This is confirmed on application of Routh's Criteria.

$$\begin{array}{c|cc} s^2 & 1 & -1 \\ s^1 & 3 & 0 \\ s & -1 & \end{array}$$

There is one sign change which appears in the first column and confirms to one positive root. Hence, the feedback system is unstable.

Comments Introduction of feedback signal $y(t) - [-2 \quad 1]x(t)$ has caused no effect on controllability and observability but has driven the system to instability.

(c) Introduction of feedback signal $u(t) = y - [1 \quad 2.5]$ changes the system to

$$\dot{x}(t) = [A - BH]x(t) + Br(t)$$

$$y(t) = Cx(t)$$

Therefore

$$\dot{x}(t) = \left\{ \begin{bmatrix} 0 & 1 \\ -1 & -2 \end{bmatrix} - \begin{bmatrix} 0 \\ 1 \end{bmatrix} [1 \quad 2.5] \right\} x(t) + \begin{bmatrix} 0 \\ 1 \end{bmatrix} r(t)$$

$$= \left\{ \begin{bmatrix} 0 & 1 \\ -1 & -2 \end{bmatrix} - \begin{bmatrix} 0 & 0 \\ 1 & 2.5 \end{bmatrix} \right\} x(t) + \begin{bmatrix} 0 \\ 1 \end{bmatrix} u(t)$$

$$= \begin{bmatrix} 0 & 1 \\ -2 & -4.5 \end{bmatrix} x(t) + \begin{bmatrix} 0 \\ 1 \end{bmatrix} u(t)$$

Controllability

$$Q_c = \begin{bmatrix} \begin{bmatrix} 0 \\ 1 \end{bmatrix} & \left\{ \begin{bmatrix} 0 & 1 \\ -2 & -4.5 \end{bmatrix} \begin{bmatrix} 0 \\ 1 \end{bmatrix} \right\} \end{bmatrix}$$

$$= \begin{bmatrix} 0 & 1 \\ 1 & -4.5 \end{bmatrix}$$

Since det $Q_c \neq 0$ rank is 2 and hence feedback system is controllable

Observability

$$Q_c = \begin{bmatrix} C \\ CA \end{bmatrix} = \begin{bmatrix} 1 & 2 \\ [1 \quad 2] \begin{bmatrix} 0 & 1 \\ -2 & -4.5 \end{bmatrix} \end{bmatrix} = \begin{bmatrix} 1 & 2 \\ -4 & -8 \end{bmatrix}$$

Since det $Q_0 = 0$, the system is unobservable.

Stability

The characteristic equation $\lambda^2 + 4.5\lambda + 2$ yield roots -4 and -0.5. This is confirmed by Routh's Criteria

$$\begin{array}{c|cc} s^2 & 1 & 2 \\ s^1 & 4.5 & 0 \\ s^0 & 2 & \end{array}$$

Since the first row is all positive; means no sign change and hence all roots are negative and located on left side of s-plane; rendering the system stable.

Comments   The feedback signal $u(t) = y(t) - [1 \quad 2.5]x(t)$ has preserved the controllability and stability but made the system unobservable.

KEY POINTS LEARNT

- ◆ Design encompasses the process of stabilisation and compensation.
- ◆ The aim of design modification involves placing all the closed-loop poles at the desired locations.
- ◆ Design through state-feedback has many favourable reasons
  - ◆ Design through state-variables presents more options as state-variables hold important information.
  - ◆ State variable are generally available for measurement and hence design can be analytically analysed and altered suitably.
  - ◆ Design is supported by software designing and validation as state analysis is based on matrix algebra.
- ◆ If a system described by $\dot{x}(t) = Ax(t) + Bu(t)$ and $y(t) = Cx(t)$ is controllable then the state feedback system described by equation $\dot{x}(t) = (A_K)x(t) + KBr(t)$ and $y(t) = Cx(t)$ is also controllable. However, the observability and stability may not remain preserved.

8.10.3 Pole Placement/Assignment

State feedback results in altering the eigenvalues and its location on the s-plane and hence can be utilised to control the eigenvalues of the closed-loop control system. *The exercise of control on eigenvalues of a control system is termed as pole placement/assignment.* The necessary and sufficient condition for arbitrary pole placement is that *the system is completely state controllable.* The *necessary and sufficient condition for arbitrary pole placement is that the system is completely state controllable.* A number of algorithms proposed for closed-loop pole assignment assumes that open-loop system has been transformed to phase-variable also termed as controllable-canonical form (CCF).

If the control system is represented in the CCF, the following steps are followed for the pole-placements:

- ◆ Check for the controllability. If the system is completely state controllability, then proceed to next step.
- ◆ Phase variable form of or CCF a control system having characteristic equation $s^n + a_1 s^{n-1} + a_2 s^{n-2} + \ldots + a_{n-1} s + a_n = 0$ is given by

$$
A = \begin{bmatrix} 0 & 1 & 0 & \cdots & 0 \\ 0 & 0 & 1 & \cdots & 0 \\ \vdots & \vdots & \vdots & \vdots & \vdots \\ 0 & 0 & 0 & \cdots & 1 \\ -a_n & -a_{n-1} & -a_{n-2} & \cdots & -a_1 \end{bmatrix}; \quad B = \begin{bmatrix} 0 \\ 0 \\ \vdots \\ 0 \\ 1 \end{bmatrix} \tag{8.55}
$$

$$
C = [c_1 \quad c_2 \quad c_3 \quad \cdots \quad c_n]
$$

♦ Feedback each state variable to the input $u(t)$ through gain $H_i = [h_1 \ h_2 \ h_3 \ \dots \ h_n]$
(8.56)

♦ Determine matrix $[A - BH]$ for the closed-loop system.

$$= \begin{bmatrix} 0 & 1 & 0 & \dots & 0 \\ 0 & 0 & 1 & \dots & 0 \\ \vdots & \vdots & \vdots & \vdots & \vdots \\ 0 & 0 & 0 & \dots & 1 \\ -a_n & -a_{n-1} & -a_{n-2} & \dots & -a_1 \end{bmatrix} - \begin{bmatrix} 0 \\ 0 \\ \vdots \\ 0 \\ 1 \end{bmatrix} [h_1 \ h_2 \ \dots \ h_n]$$

$$= \begin{bmatrix} 0 & 1 & 0 & \dots & 0 \\ 0 & 0 & 1 & \dots & 0 \\ \vdots & \vdots & \vdots & \vdots & \vdots \\ 0 & 0 & 0 & \dots & 1 \\ -(a_n + h_1) & -(a_{n-1} + h_2) & -(a_{n-2} + h_3) & \dots & -(a_1 + h_n) \end{bmatrix}$$
(8.57)

♦ Ascertain characteristic equation of the closed-loop system

$$s^n + (a_1 + h_n)s^{n-1} + \dots + (a_{n-2} + h_3)s^{n-2} + (a_{n-1} + h_2)s + (a_n + h_1) = 0 \qquad (8.58)$$

♦ Determine an equivalent characteristic equation from the desired pole locations. Assuming the desired characteristic equation is

$$s^n + p_{n-1}s^{n-1} + \dots + p_{n-2}s^{n-2} + \dots + p_2 s^2 + p_1 s + p_0 = 0 \qquad (8.59)$$

♦ Equate the characteristic equations of equations (8.58) and (8.59) and solve for H_i.
♦ Substitute the value of H_i in equation (8.58) and obtain the desired characteristic equation.

Example 8.12 Phase variable representation of a control system is given as

$$\dot{x}(t) = \begin{bmatrix} 0 & 1 & 0 \\ 0 & 0 & 1 \\ -6 & -11 & -6 \end{bmatrix} \begin{bmatrix} x_1(t) \\ x_2(t) \\ x_3(t) \end{bmatrix} + \begin{bmatrix} 0 \\ 0 \\ 1 \end{bmatrix} u(t)$$

$$y(t) = \begin{bmatrix} 1 & 0 & 0 \end{bmatrix} \begin{bmatrix} x_1(t) \\ x_2(t) \\ x_3(t) \end{bmatrix}$$

Find the feedback matrix $h = [h_1 \ h_2 \ h_3]$ such that the closed-loop poles are placed at $-2, -5$ and -6.

Solution The desired characteristic equation is

$$|sI - A| = (s + 2)(s + 5)(s + 6) = 0$$

or
$$s^3 + 13s^2 + 52s + 60 = 0 \qquad (1)$$

Matrix $[A - BH]$ is

$$\begin{bmatrix} 0 & 1 & 0 \\ 0 & 0 & 1 \\ -6 & -11 & -6 \end{bmatrix} - \begin{bmatrix} 0 \\ 0 \\ 1 \end{bmatrix} [h_1 \quad h_2 \quad h_3]$$

$$= \begin{bmatrix} 0 & 1 & 0 \\ 0 & 0 & 1 \\ -(6+h_1) & -(11+h_2) & -(6+h_3) \end{bmatrix}$$

The characteristic equation is

$$s^3 + (6 + h_3)s^2 + (11 + h_2)s + (6 + h_1) \tag{2}$$

Equating the coefficients of eqns (1) and (2) give

$$6 + h_3 = 13 \qquad \therefore \qquad h_3 = 13 - 6 = 7$$
$$11 + h_2 = 52 \qquad \therefore \qquad h_3 = 52 - 11 = 41$$
$$6 + h_1 = 60 \qquad \therefore \qquad h_3 = 60 - 6 = 54$$

Therefore matrix

$$H = [h_1 \quad h_2 \quad h_3]$$
$$= [54 \quad 41 \quad 7]$$

Alternatively

Equation (1) in phase variable form is represented as

$$\begin{bmatrix} 0 & 1 & 0 \\ 0 & 0 & 1 \\ -60 & -52 & -13 \end{bmatrix} \tag{3}$$

$$-BH = A_d - A_g$$

Where

$$A_d = \text{A desired}$$
$$A_g = \text{A given}$$

Therefore

$$-\begin{bmatrix} 0 & 0 & 0 \\ 0 & 0 & 0 \\ h_1 & h_2 & h_3 \end{bmatrix} = \begin{bmatrix} 0 & 1 & 0 \\ 0 & 0 & 1 \\ -60 & -52 & -13 \end{bmatrix} - \begin{bmatrix} 0 & 1 & 0 \\ 0 & 0 & 1 \\ -6 & -11 & -6 \end{bmatrix}$$

$$= \begin{bmatrix} 0 & 1 & 0 \\ 0 & 0 & 1 \\ -54 & -41 & -7 \end{bmatrix}$$

$$\therefore \qquad H = [h_1 \quad h_2 \quad h_3] = [54 \quad 41 \quad 7]$$

8.10.4 Design of Controller by Transformation

At times the state-variable representation of the control systems is not in phase-variable form. In such cases, design steps adopted are:

- Check for controllability. If the system is completely state controllable then proceed further
- Transforming the given state-variable representation into phase variable form with the aid of transformation matrix
- Designing the feedback gains by matching the coefficients of the characteristic equations
- Converting the designed system representation into original representation.

Let us assume a control system whose state-variable representation is not in phase variable form

$$\dot{x}(t) = Ax(t) + Bu(t) \quad \text{or} \quad \dot{x} = Ax + Bu \tag{8.60}$$

$$y(t) = Cx(t). \quad \text{or} \quad y = Cx \tag{8.61}$$

The controllability matrix is given as

$$Q_c = [B \quad AB \quad A^2B \quad \dots \quad A^{n-1}B] \tag{8.62}$$

Let us assume that the given system can be transformed into phase variable form by

$$x = Pz \tag{8.63}$$

Where P is the transformation matrix. Substitute equation (8.63) in equations (8.60) and (8.61)

$$P\dot{z} = APz + Bu \tag{8.64}$$

or

$$\dot{z} = (P^{-1}AP)z + (P^{-1}B)u \tag{8.65}$$

and

$$y = (CP)z \tag{8.66}$$

The controllability matrix is

$$\begin{aligned}
Q'_c &= [P^{-1}B \quad (P^{-1}AP)(P^{-1}B) \quad (P^{-1}AP)^2(P^{-1}B) \dots (P^{-1}AP)^{n-1}(P^{-1}B)] \\
&= [P^{-1}B \quad (P^{-1}AP)(P^{-1}B) \quad (P^{-1}AP)(P^{-1}AP)(P^{-1}B) \dots] \\
&= [P^{-1}B \quad P^{-1}AB \quad P^{-1}AAB \dots] \\
&= [P^{-1}B \quad P^{-1}AB \quad P^{-1}A^2B \dots] \\
&= P^{-1}[B \quad AB \quad A^2B \dots A^{n-1}B]
\end{aligned} \tag{8.67}$$

Substituting equation (8.62) in equation (8.67) gives.

$$Q'_c = P^{-1}Q_c \tag{8.68}$$

or

$$P = Q_c Q'^{-1}_c \tag{8.69}$$

Considering state feedback $u = -K_z z + r$, equations (8.65) and (8.66) become

$$\dot{z} = P^{-1}APz - P^{-1}BK_z z + P^{-1}Br \tag{8.70}$$

or

$$\dot{z} = (P^{-1}APz - P^{-1}BK_z)z + P^{-1}Br \tag{8.71}$$

and

$$y = CPz \tag{8.72}$$

Equations (8.71) and (8.72) are then transformed from phase variable form back to original state variable representation by using the transformation

$$z = P^{-1}x$$

which gives

$$\dot{x} = Ax - BK_zP^{-1}x + Br \tag{8.73}$$

or

$$\dot{x} = (A - BK_zP^{-1})x + Br \tag{8.74}$$

or

$$\dot{x}(t) = (A - BK_zP^{-1})x(t) + Br(t) \tag{8.75}$$

and

$$y = Cz \tag{8.76}$$

or

$$y = CPP^{-1}x. \tag{8.77}$$

or

$$y = Cx. \tag{8.78}$$

or

$$y(t) = Cx(t) \tag{8.79}$$

Comparing coefficients of $x(t)$ in eqns (8.29) and (8.75), we see that state feedback gain H for the original system is

$$(A - BH) = A - BK_zP^{-1} \tag{8.80}$$

or

$$H = K_zP^{-1} \tag{8.81}$$

Example 8.13   The transfer function of the plant is given as $G(s) = \dfrac{s+4}{(s+1)\,(s+2)\,(s+3)}$, and its state variable diagram is as shown in Fig. 8.12.

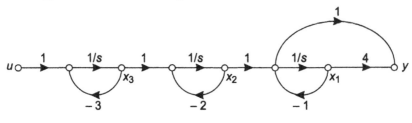

Fig. 8.12

Design a state variable feedback so that closed-loop poles are placed at –4, –5 and –6. Also, Obtain the transfer function.

Solution   The state equations are

$$\dot{x}_1 = -x_1 + x_2$$
$$\dot{x}_2 = -2x_2 + x_3$$
$$\dot{x}_3 = -3x_3 + u$$

$$y = 4x_1 + \dot{x}_1$$
$$= 4x_1 - x_1 + x_2$$
$$= 3x_1 + x_2$$

In matrix from

$$\begin{bmatrix} \dot{x}_1 \\ \dot{x}_2 \\ \dot{x}_3 \end{bmatrix} = \begin{bmatrix} -1 & 1 & 0 \\ 0 & -2 & 1 \\ 0 & 0 & -3 \end{bmatrix} x + \begin{bmatrix} 0 \\ 0 \\ 1 \end{bmatrix} u \qquad (1)$$

$$y = [3 \quad 1 \quad 0] \qquad (2)$$

The controllability matrix is

$$Q_c = [B \quad AB \quad A^2 B] = \begin{bmatrix} 0 & 0 & 1 \\ 0 & 1 & -5 \\ 1 & -3 & 9 \end{bmatrix} \qquad (3)$$

Characteristic equation is

$$\det(sI - A) = s^3 + 6s^2 + 11s + 6 = 0 \qquad (4)$$

The phase variable form of representation is

$$\dot{x} = \begin{bmatrix} 0 & 1 & 0 \\ 0 & 0 & 1 \\ -6 & -11 & -6 \end{bmatrix} x + \begin{bmatrix} 0 \\ 0 \\ 1 \end{bmatrix} u \qquad (5)$$

$$y = [3 \quad 1 \quad 0]x \qquad (6)$$

Controllability matrix is

$$Q'_c = [B \quad AB \quad A^2 B] = \begin{bmatrix} 0 & 0 & 1 \\ 0 & 1 & -6 \\ 1 & -6 & 25 \end{bmatrix} \qquad (7)$$

Transformation matrix P is

$$P = Q_c Q'^{-1}_c$$

$$= \begin{bmatrix} 0 & 0 & 1 \\ 0 & 1 & -5 \\ 1 & -3 & 9 \end{bmatrix} \begin{bmatrix} 0 & 0 & 1 \\ 0 & 1 & -6 \\ 1 & -6 & 25 \end{bmatrix}^{-1} = \begin{bmatrix} 0 & 0 & 1 \\ 0 & 1 & -5 \\ 1 & -3 & 9 \end{bmatrix} \begin{bmatrix} 11 & 6 & 1 \\ -1 & 1 & 0 \\ 1 & 0 & 0 \end{bmatrix} \qquad (8)$$

$$= \begin{bmatrix} 1 & 0 & 0 \\ 1 & 1 & 0 \\ 2 & 3 & 1 \end{bmatrix} \qquad (9)$$

The desired characteristic equation with closed-loop poles at $-4, -5$ and -6 is
$$(s + 4)(s + 5)(s + 6) = 0$$
or
$$s^3 + 15s^2 + 74s + 120 = 0 \qquad (10)$$
The state equations with state variable feedback for the phase variable form of equations (5) and (6) are

$$\dot{x} = \begin{bmatrix} 0 & 1 & 0 \\ 0 & 0 & 1 \\ -(6+h_1) & -(11+h_2) & -(6+h_3) \end{bmatrix} x \qquad (11)$$

$$y = \begin{bmatrix} 3 & 1 & 0 \end{bmatrix} x \qquad (12)$$

The characteristic equation is
$$s^3 + (6 + h_3)s^2 + (11 + h_2)s + (6 + h_1) = 0 \qquad (13)$$
Comparing eqns (13) and (10), we get
$$6 + h_3 = 15 \quad \text{or} \quad h_3 = 15 - 6 = 9$$
$$11 + h_2 = 74 \quad \text{or} \quad h_2 = 74 - 11 = 63$$
$$6 + h_1 = 120 \quad \text{or} \quad h_1 = 120 - 6 = 114$$
or
$$K_z = \begin{bmatrix} 114 & 63 & 9 \end{bmatrix} \qquad (14)$$
Also
$$H = K_z P^{-1}$$

$$= \begin{bmatrix} 114 & 63 & 9 \end{bmatrix} \begin{bmatrix} 1 & 0 & 0 \\ 1 & 1 & 0 \\ 2 & 3 & 1 \end{bmatrix}^{-1}$$

$$= \begin{bmatrix} 114 & 63 & 9 \end{bmatrix} \begin{bmatrix} 1 & 0 & 0 \\ -1 & 1 & 0 \\ 1 & -3 & 1 \end{bmatrix} \qquad (15)$$

$$= \begin{bmatrix} 60 & 36 & 9 \end{bmatrix}$$

The desired closed-loop system with state-variable feedback is shown in Fig. 8.13.

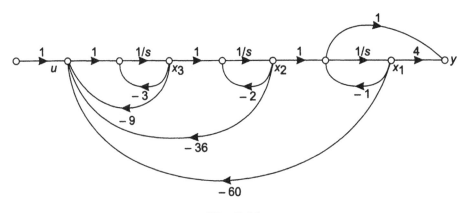

Fig. 8.13

The state equations are

$$\dot{x}_1 = -x_1 + x_2$$
$$\dot{x}_2 = -2x_2 + x_3$$
$$\dot{x}_3 = -3x_3 + u$$
$$= -3x_3 - 60x_1 - 36x_2 - 9x_3 + r$$
$$= -60x_1 - 36x_2 - 12x_3 + r$$
$$y = 4x_1 + \dot{x}_1$$
$$= 4x_1 - x_1 + x_2$$
$$= 3x_1 + x_2$$

or

$$\dot{x} = \begin{bmatrix} -1 & 1 & 0 \\ 0 & -2 & 1 \\ -60 & -36 & -12 \end{bmatrix} x + \begin{bmatrix} 0 \\ 0 \\ 1 \end{bmatrix} r$$

$$y = \begin{bmatrix} 3 & 1 & 0 \end{bmatrix}$$

Let us now the derive the transfer function

$$TF = C(sI - A)^{-1} B$$

$$= \begin{bmatrix} 3 & 1 & 0 \end{bmatrix} \begin{bmatrix} s+1 & -1 & 0 \\ 0 & s+2 & -1 \\ 60 & 36 & s+12 \end{bmatrix}^{-1} \begin{bmatrix} 0 \\ 0 \\ 1 \end{bmatrix}$$

$$= \frac{s+4}{s^3 + 15s^2 + 74s + 120}$$

$$= \frac{1}{(s+5)(s+6)}$$

Note: The effect of placing the pole at -4 is to cancel the zero at -4 existing in the transfer function of the plant.

8.10.5 Observer Design

The controller design through state variable feedback relied on the assumption that all the state-variables are available to be fed back for implementation of the control law

$$u = r - hx \tag{8.82}$$

The assumption of availability of the state-variables for feedback may not be realisable in practice due to non-accessibility of state-variables to measurement. The measurement of state-variables is performed with the help of devices whose availability may be limited from cost considerations point. In such cases of observable systems, states are estimated and these reasonable estimated states derived form the knowledge of the input $u(t)$, output $y(t)$ and the matrices A, B, C, and D.

An *observer*, also called an *estimator*, is utilised to calculate the state-variables that are not accessible. Estimation of states which are not measurable is called *observation*.

The function of the observer is to estimate the state-variables. The observation depends upon on the control variables and output. The necessary and sufficient condition for the state observation is that the system under consideration be completely observable.

Let us assume a plant whose dynamics are described by state equations:

$$\dot{x} = Ax + Bu \tag{8.83}$$

$$y = Cx \tag{8.84}$$

Since the knowledge of input, output and matrices A, B and C is known, a state model that has an accessible state vector and dynamics of original system; can be simulated. Let

$$\hat{x} = \text{state vector of the model}$$

$$\hat{y} = \text{model output}$$

$$u = \text{input}$$

then

$$\dot{\hat{x}} = A\hat{x} + Bu \tag{8.85}$$

$$\hat{y} = C\hat{x} \tag{8.86}$$

Let us consider the estimator model shown in Fig 8.14

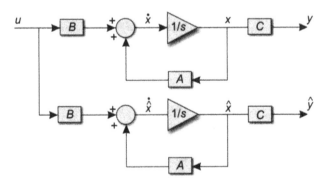

Fig. 8.14 *Open-loop state Observer/Estimator*

The state equations for the observer are

$$\dot{\hat{x}} = A\hat{x} + Bu \tag{8.87}$$

$$\hat{y} = C\hat{x} \tag{8.88}$$

The circumflex (^)denotes estimate of the variable. Subtraction of state equations for plant and observer yields

$$\dot{x}_e = \dot{x} - \dot{\hat{x}} = A(x - \hat{x}) = Ax_e \tag{8.89}$$

$$y - \hat{y} = C(x - \hat{x}) = Cx_e \tag{8.90}$$

where

x_e = error in estimate i.e., the error between the actual state vector and the estimated vector. (8.91)

$y - \hat{y}$ = Error between the actual output and the estimated output.

The error dynamics depends upon matrix A; over which no control can be exercised and is unforced. The solution of state eqn (8.89) is

$$x_e(t) = e^{A(t-t_0)}x_e(t_0) \tag{8.92}$$

If $x_e(t_0) = 0$; then $x_e(t) = 0$ i.e., the estimate of the state will exhibit no error and if $x_e(t_0) \neq 0$ then error will vary as $e^{A(t-t_0)}$ and will depend upon the eigenvalues of matrix A. If the eigenvalues are all negative or with negative real parts located on the left side of the s-plane, then the estimated error vector x_e will decay to zero exponentially. The estimate will become better with increase in time. However, if matrix A has eigenvalues positive or with positive real part, then any small error in the estimate of $\hat{x}(t_0)$ done initially will grow up to a large value, quickly no matter how shall the initial error is generate an unacceptable unstable model. Thus, the open-loop observer depending upon the matrix A needs improvement.

The improvement to the open-loop observer is affected by creation of closed-loop by comparing the ouput y and estimated $\hat{y}$ and using the difference $(y \sim \hat{y})$ to correct the anomaly described earlier. Fig 8.15 shows a closed-loop observer. The closed-loop observer has the following state equations:

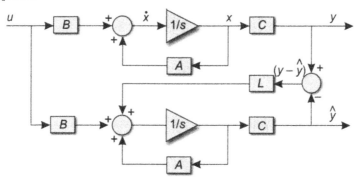

Fig. 8.15 *Closed-loop state observer/estimator*

$$\dot{\hat{x}} = A\hat{x} + Bu + L(y - \hat{y}) \tag{8.93}$$

where L is called feedback gain vector and has to be properly selected.

$$\hat{y} = C\hat{x} \tag{8.94}$$

or

$$\dot{\hat{x}} = [A - LC]\hat{x} + Bu + Ly \tag{8.95}$$

Subtracting equations (8.85) and (8.95) gives

$$\dot{x}_e = \dot{x} - \dot{\hat{x}} = Ax + Bu - [A - LC]\hat{x} - Bu - Ly \tag{8.96}$$

$$= Ax - [A - LC]\hat{x} - Ly \tag{8.97}$$

$$= A[x - \hat{x}] + LC\hat{x} - Ly \tag{8.98}$$

Substituting the value of y from equation (8.84) in equation (8.98) gives

$$\dot{x}_e = A[x - \hat{x}] + LC\hat{x} - LCx$$
$$= A[x - \hat{x}] + LC[x - \hat{x}]$$
$$= [A - LC][x - \hat{x}]$$
$$= [A - LC]x_e \qquad (8.99)$$

Equation (8.99) is unforced. The *behaviour* of x_e (estimate error) depends upon the eigenvalues of $(A - LC)$. If the eigenvalues of $[A-LC]$ can be chosen arbitrarily, the behaviour of x_e can be controlled. x_e can be made to decay if eigenvalues are properly selected. Negative eigenvalues will ensure decaying of x_e to zero.

8.10.6 Observer Design Through Phase Variable Form

Let us consider the characteristic equation for the plant which is given by det $(sI - A)$ and is of the form

$$s^n + a_1 s^{n-1} + a_2 s^{n-2} + \cdots\cdots + a_{n-1}s + a_n = 0 \qquad (8.100)$$

and is represented in vector-matrix notation on

$$A = \begin{bmatrix} -a_1 & 1 & 0 & 0 & \ldots & 0 \\ -a_2 & 0 & 1 & 0 & \ldots & 0 \\ \vdots & \vdots & \vdots & \vdots & \vdots & \vdots \\ -a_{n-1} & 0 & 0 & 0 & \ldots & 1 \\ -a_n & 0 & 0 & 0 & \ldots & 0 \end{bmatrix} \qquad (8.101)$$

$$C = [1 \quad 0 \quad 0 \quad 0 \quad \ldots \quad 0] \qquad (8.102)$$

the characteristic equation $A - LC$ is

$$s^n + (a_1 + l_1) s^{n-1} + (a_2 + l_2) s^{n-2} + \ldots + (a_{n-1} + l_{n-1})s + (a_n + l_n) \qquad (8.103)$$

The phase variable representation in vector-matrix notation is

$$A - LC = \begin{bmatrix} -a_1 & 1 & 0 & 0 & \ldots & 0 \\ -a_2 & 0 & 1 & 0 & \ldots & 0 \\ \vdots & \vdots & \vdots & \vdots & \vdots & \vdots \\ -a_{n-1} & 0 & 0 & 0 & \ldots & 1 \\ -a_n & 0 & 0 & 0 & \ldots & 0 \end{bmatrix} - \begin{bmatrix} l_1 \\ l_2 \\ \vdots \\ l_{n-1} \\ l_n \end{bmatrix} + [1 \quad 0 \quad 0 \quad 0 \ldots\ldots 0]$$

or

$$A - LC = \begin{bmatrix} -(a_1 + l_1) & 1 & 0 & 0 & \ldots & 0 \\ -(a_2 + l_2) & 0 & 1 & 0 & \ldots & 0 \\ \vdots & \vdots & \vdots & \vdots & \vdots & \vdots \\ -(a_{n-1} + l_{n-1}) & 0 & 0 & 0 & \ldots & 1 \\ -(a_n + l_n) & 0 & 0 & 0 & \ldots & 0 \end{bmatrix} \qquad (8.104)$$

Assume the desired characteristic equation formed from the desired parameters is

$$s^n + p_1 s^{n-1} + p_2 s^{n-2} + \ldots\ldots\ldots p_{n-1} s + p_n = 0 \qquad (8.105)$$

Then by matching the coefficients of equations (8.105) and (8.103), we get

$$
\begin{array}{lll}
p_1 = a_1 + l_1 & \text{or} & l_1 = p_1 - a_1 \\
p_2 = a_2 + l_2 & \text{or} & l_2 = p_2 - a_2 \\
\cdots\cdots\cdots\cdots & & \cdots\cdots\cdots\cdots \\
p_{n-1} = a_{n-1} + l_n & \text{or} & l_{n-1} = p_{n-1} - a_{n-1} \\
p_n = a_n + l_n & \text{or} & l_n = p_n - l_n
\end{array}
\qquad (8.106)
$$

Example 8.14   Transfer function of a plant is given by

$$G(s) = \frac{s+3}{s^3 + 7s^2 + 11s + 8}$$

It is desired to place the observer poles at -5, -6 and -10. Design an observer for the plant.

Solution   Let

$$\frac{X(s)}{U(s)} = \frac{s+3}{s^3 + 7s^2 + 14s + 8} \qquad (1)$$

Rearranging

$$Y(s)(s^3 + 7s^2 + 14s + 8) = (s+3)\,U(s)$$

or $\qquad s^3 Y(s) + 7s^2 Y(s) + 14s\,Y(s) + 8Y(s) = sU(s) + 3U(s)$

or $\qquad Y(s) + \dfrac{7}{s} Y(s) + \dfrac{1}{s^2}(14Y(s) - U(s)) + \dfrac{1}{s^3}(8Y(s) - 3U(s)) = 0$

or $\qquad Y(s) = \dfrac{1}{s^3}(3U(s) - 8Y(s)) + \dfrac{1}{s^2}(U(s) - 14Y(s)) - \dfrac{7}{s} Y(s) \qquad (2)$

The state diagram is shown in Fig. 8.16

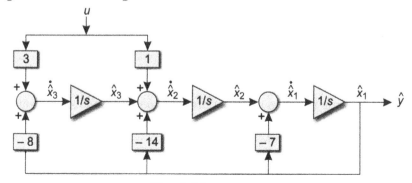

Fig. 8.16

The signal flow diagram is shown in Fig 8.17

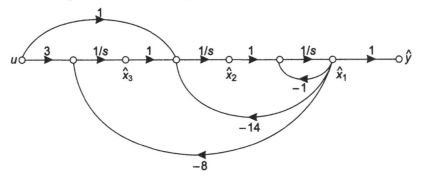

Fig. 8.17 *Signal flow graph with state-space model in OCF*

Output Equation

$$y = \hat{x}_1 \tag{3}$$

State Equation

$$\dot{\hat{x}}_1 = -7y + \hat{x}_2 = -7\hat{x}_1 + x_2 \tag{4}$$

$$\dot{\hat{x}}_2 = -14y + \hat{x}_3 + u = -14x_1 + \hat{x}_3 + u \tag{5}$$

$$\dot{\hat{x}}_3 = -8y + 3u = -8\hat{x}_1 + 3u \tag{6}$$

State model in vector matrix notation is

$$\dot{\hat{x}} = \begin{bmatrix} -7 & 1 & 0 \\ -14 & 0 & 1 \\ -8 & 0 & 0 \end{bmatrix} \hat{x} + \begin{bmatrix} 0 \\ 1 \\ 3 \end{bmatrix} u \tag{7}$$

$$\hat{y} = \begin{bmatrix} 1 & 0 & 0 \end{bmatrix} \hat{x} \tag{8}$$

The observer estimated error is

$$\dot{x}_e = (A - LC)\, x_e \tag{9}$$

$$= \begin{bmatrix} -(7+l_1) & 1 & 0 \\ -(14+l_2) & 0 & 1 \\ =(8+l_3) & 0 & 0 \end{bmatrix} x_e$$

The characteristic equation is given by $\det(sI - (A - LC)] = 0$ and is

$$s^3 + (7+l_1)s^2 + (14+l_2)s + 8 + l_3 = 0 \tag{10}$$

The desired characteristic equation is formed from the observer poles which are desired to be placed and is

$$(s+5)(s+6)(s+10) = 0$$

or
$$s^3 + 21s^2 + 140s + 300 = 0 \tag{11}$$

Equating equations (10) and (11), we get

$$7 + l_1 = 21 \qquad \text{or} \qquad l_1 = 21 - 7 = 14 \qquad (12)$$

$$14 + l_2 = 140 \qquad \text{or} \qquad l_2 = 140 - 14 = 126 \qquad (13)$$

$$8 + l_3 = 300 \qquad \text{or} \qquad l_3 = 300 - 8 = 292 \qquad (14)$$

Signal flow graph with feedback to create observer with desired poles is shown in Fig. 8.18.

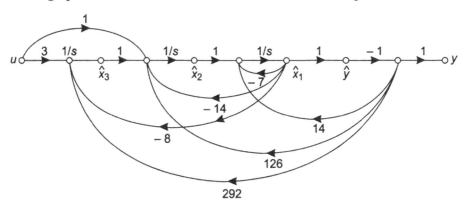

Fig 8.18   *Signal flow graph of observer with desired poles.*

8.10.7   Observer Design Through Transformation

There may be times when the state-space representation of the plant is not available in the observable canonical form. In that case, the following steps are performed for the observer design.

- ◆ Transform the state-space representation of the plant into observer canonical form.
- ◆ Design the observer.
- ◆ Transform the design back to the original state-space representation.

Let us assume a plant whose state space representation

$$\dot{x} = Ax + Bu \qquad (8.107)$$

$$y = Cx \qquad (8.108)$$

is not in observer canonical form.

The observability matrix equation (7.92) is given as

$$Q_0 = \begin{bmatrix} C \\ CA \\ CA^2 \\ \vdots \\ CA^{n-1} \end{bmatrix} \qquad (8.109)$$

Let us assume that the state-space representation of equations (8.110) and (8.108) can be transformed into observer canonical form, with the help of transformation

$$x = Pz \tag{8.110}$$

Substituting equation (8.110) into equations (8.107) and (8.108), we get the state equation in observer canonical form

$$\dot{z} = P^{-1}APz + P^{-1}Bu \tag{8.111}$$

$$y = CPz$$

which gives the observability matix

$$Q_0' = \begin{bmatrix} CP \\ CP(P^{-1}AP) \\ CP(P^{-1}AP)(P^{-1}AP) \\ \vdots \\ CP(P^{-1}AP)(P^{-1}AP)\dots(P^{-1}AP) \end{bmatrix} \tag{8.113}$$

$$= \begin{bmatrix} CP \\ C(PP^{-1}A)P \\ C(PP^{-1})A(PP^{-1})AP \\ \vdots \\ C(PP^{-1})A(PP^{-1})AP\dots(P^{-1}AP) \end{bmatrix} = \begin{bmatrix} CP \\ CAP \\ CAAP \\ \vdots \\ CA^{n-1}P \end{bmatrix} = \begin{bmatrix} C \\ CA \\ CA^2 \\ \vdots \\ CA^{n-1} \end{bmatrix} P \tag{8.114}$$

Substituting equation. (8.109) into equation (8.113), we get

$$Q_0' = Q_o P \tag{8.115}$$

or $$P = Q_0^{-1}Q_0' \tag{8.116}$$

Thus the state space representation of the plant not given in observer canonical form by equations (8.107) and (8.108), can be transformed into observer canonical form by the help of matrix P obtained from two observability matrices as given in equation (8.116).

Next step is to design the observer feedback gain L_z. This is achieved by using the matrices of equations (8.111) and (8.112) and substituting in equations (8.99) and (8.90), which gives

$$\dot{x}_c = (A - LC)x_c$$

$$= (P^{-1}AP - L_zCP)x_e \tag{8.117}$$

$$y - \hat{y} = Cx_e$$

$$= CPx_e \tag{8.118}$$

Since $\qquad z = P^{-1}x \qquad$ and $\qquad \hat{z} = P^{-1}\hat{x}$

Then $\qquad x_e = z - \hat{z}$

$$= P^{-1}x - P^{-1}\hat{x}$$

$$= P^{-1}(x - \hat{x})$$

$$= P^{-1}x_z \qquad (8.119)$$

Substituting x_e from equation (8.119) in equation (8.116) gives

$$\dot{x}_e = \left(P^{-1}AP - L_z(CP)\right)x_e$$

$$= \left(P^{-1}AP - L_z(CP)\right)P^{-1}x_e$$

$$= \left[(P^{-1}AP) - L_zCPP^{-1}\right]x_z$$

$$= \left[P^{-1}A - L_zC\right]x_z \qquad (8.120)$$

$$x_e = \left[P^{-1}A - L_zC\right]\int x_z \qquad (8.121)$$

Substituting the value of x_e from equation (8.119) in equation (8.121), we get

$$P^{-1}x_z = \left[P^{-1}A - L_zC\right]\int x_z$$

or $\qquad x_z = \left[A - PL_zC\right]\int x_z$

or $\qquad \dot{x}_z = \left[A - PL_zC\right]x_z$

Also $\qquad y - \hat{y} = CPx_e = CPP^{-1}x_z = Cx_z \qquad (8.122)$

8.10.8 State Variable Analysis of Output Feedback

Let us consider a control system whose system dynamics are defined by the state equations

$$\dot{x}(t) = Ax(t) + Bu(t) \qquad (8.123)$$

$$y(t) = Cx(t) \qquad (8.124)$$

The general state-variable diagram with the controller having the constant gain square matrices K_R and K_0 is shown in Fig. 8.19

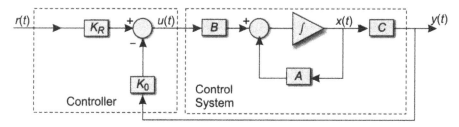

Fig. 8.19

The control law for the output feedback

$$u(t) = K_R r(t) - K_0 y(t)$$

Substituting the value of $u(t)$ in eqn (8.122) we get

$$\dot{x}(t) = Ax(t) + B[K_R r(t) - K_0 y(t)] \tag{8.125}$$

Substituting the value of $y(t)$ from eqn (8.12) in eqn (8.124), we get

$$\dot{x}(t) = Ax(t) + B[K_R r(t) - K_0 Cx(t)]$$

or

$$\dot{x}(t) = [A - BK_0 C]x(t) + BK_R r(t) \tag{8.126}$$

$$= A_{CL} x(t) + BK_R r(t)$$

where A_{CL} = closed-loop system matrix.

The design involves choosing matrics K_R and K_0.

Matrix Gain K_0

We had defined $A_{CL} = [A - BK_0 C]$ which depends upon K_0. The closed-loop system poles depend upon the eigenvalues of A_{CL}. Any variation in the gain K_0 will change the values of eigenvalues of A_{CL} and shift them to new locations. The aim is to ensure that change in gain K_0 meets the desired transient performance and stability criterion.

Reference Controller Matrix Gain K_R

Once the matrix gain K_0 is derived based or transient performance and stability criterion, the value of K_R is ascertained. Let us assume that the steady state value of $x(t)$ is x_{ss} for $r(t) = r_0$ and output $y(t) = y_{ss}$. Therefore equation (8.126) can be written on

$$\dot{x}_{ss} = A_{CL} x_{ss} + BK_R r_0 \tag{8.127}$$

Since x_{ss} is constant, $\dot{x}_{ss} = 0$. Therefore eqn (8.127) becomes

$$A_{CL} x_{ss} + BK_R r_0 = 0 \tag{8.128}$$

or

$$x_{ss} = [-A_{CL}]^{-1} BK_R r_0 \tag{8.129}$$

The output equation is written as

$$y_{ss} = Cx_{ss} \tag{8.130}$$

Substituting the value of x_{ss} from eqn (8.129) in eqn (8.130) , we get

$$y_{ss} = C[-A_{CL}]^{-1} BK_R r_0 \tag{8.131}$$

Since the y_{ss} is required to reach the input reference r_0, we have,

$$I = C[-A_{CL}]^{-1} BK_R \tag{8.132}$$

or

$$K_R = [C[-A_{CL}]^{-1} B]^{-1} \tag{8.133}$$

And the output control law is

$$u(t) = K_R r(t) - K_0 y(t) \tag{8.134}$$

The design steps are:

♦ Determine closed-loop system matrix $A_{CL} = A - BK_0C$ for control system described by a given state-space equation.

♦ Determine the polynomial of the given closed-loop control system in s-domain and K_0 by

$$|sI - A_{CL}| = 0$$

♦ Obtain the desired design polynomial from the desired transient performance parameters or from the desired closed-loop poles.

♦ Comparison of the two polynomials given above will yield the value of K_0 from which closed-loop matrix (A_{CL}) is determined.

♦ Determine the value of $K_R = [C(-A_{CL})^{-1}B]^{-1}$

8.11 ACKERMANN'S FORMULA

Ackermann's formula is an useful tool for determining the state variable feedback matrix for a single-input single output system

$$h = [h_1, h_2 \ldots\ldots h_n]$$

Where

$$u = -hx$$

If the desired characteristic equation

$$q(s) = s^n + a_1 s^{n-1} + a_2 s^{n-2} + \ldots + a_n$$

Then the state feedback gain matrix is given by

$$\boxed{h = [0 \quad 0 \ldots\ldots 1]Q_c^{-1}q(A)}$$ (8.135)

Where

$$q(A) = A^n + a_1 A^{n-1} + \ldots\ldots + a_{n-1}A + a_nI$$

and

$$Q_c = \text{the controllability matrix.}$$

Example 8.14 A control system is represented as

$$\begin{bmatrix} \dot{x}_1 \\ \dot{x}_2 \end{bmatrix} = \begin{bmatrix} 0 & 1 \\ 0 & 0 \end{bmatrix}\begin{bmatrix} x_1 \\ x_2 \end{bmatrix} + \begin{bmatrix} 0 \\ 1 \end{bmatrix}u$$

It is desired to place the closed-loop poles at $s = -1$ and $s = -2$. Determine the state feedback gain matrix by use of Ackermann's formula.

Solution The desired characteristic equation is

$$(s + 1)(s + 2) = s^2 + 3s + 2 = 0$$

which gives $\qquad a_1 = 3$ and $a_2 = 2$

the Controllability matrix is

$$Q_c = \begin{bmatrix} B & AB \end{bmatrix} = \begin{bmatrix} 0 \\ 1 \end{bmatrix} \begin{bmatrix} 0 & 1 \\ 0 & 0 \end{bmatrix} \begin{bmatrix} 0 \\ 1 \end{bmatrix} = \begin{bmatrix} 0 & 1 \\ 1 & 0 \end{bmatrix}$$

Now

$$h = \begin{bmatrix} 0 & 1 \end{bmatrix} Q_c^{-1} q(A)$$

$$= \begin{bmatrix} 0 & 1 \end{bmatrix} \begin{bmatrix} 0 & 1 \\ 1 & 0 \end{bmatrix}^{-1} q(A)$$

$$= \begin{bmatrix} 0 & 1 \end{bmatrix} \begin{bmatrix} 0 & 1 \\ 1 & 0 \end{bmatrix} q(A)$$

$$q(A) = \begin{bmatrix} 0 & 1 \\ 0 & 0 \end{bmatrix}^2 + 3 \begin{bmatrix} 0 & 1 \\ 0 & 0 \end{bmatrix} + 2 \begin{bmatrix} 1 & 0 \\ 0 & 1 \end{bmatrix}$$

$$= 0 + \begin{bmatrix} 0 & 3 \\ 0 & 0 \end{bmatrix} + \begin{bmatrix} 2 & 0 \\ 0 & 2 \end{bmatrix}$$

$$= \begin{bmatrix} 2 & 3 \\ 0 & 2 \end{bmatrix}$$

Therefore

$$h = \begin{bmatrix} 0 & 1 \end{bmatrix} \begin{bmatrix} 0 & 1 \\ 1 & 0 \end{bmatrix} \begin{bmatrix} 2 & 3 \\ 0 & 2 \end{bmatrix} = \begin{bmatrix} 2 & 3 \end{bmatrix}$$

8.12 MATLAB

MATLAB has a number of in-built commands with the help of which controllability, observability and state variable feedback design can be computed.

♦ $Q_c = ctrb$ **(AB).** *Ctrb* computes the controllability matrix for state-space systems. For matrices A and B of order $n \times n$ and $n \times m$ respectively, the function returns the controllability matrix

$$Q_c = \begin{bmatrix} B & AB & \ldots\ldots & A^{n-1} B \end{bmatrix} \quad \text{Where } Q_c \text{ has } n \text{ rows and } nm \text{ columns.}$$

♦ **[Abar, Bbar, Cbar, T, K]** = *ctrbf* **(A, B, C).** If the controllability matrix Q_c of pair $[A, B]$ has a rank $r \leq n$ where n is the size of A, then there exists a similarity transformation

$$\overline{A} = TAT^T, \ \overline{B} = TB, \ \overline{C} = CT^T$$

where T is unitary, and transformed system for a staircase form, in which if there are any uncontrollable modes, they are located in the upper left corner.

$$\overline{A} = \begin{bmatrix} A_{uc} & 0 \\ A_{21} & A_c \end{bmatrix}; \ \overline{B} = \begin{bmatrix} 0 \\ B_c \end{bmatrix}; \ \overline{C} = \begin{bmatrix} C_{nc} & C_c \end{bmatrix}$$

where

(A_C, B_C) is controllable; all eigenvalues of A_{uc} are uncontrollable; and

$$C_c(sI - A_c)^{-1}B_c = C(sI - A)^{-1}B$$

$T =$ similarity transformation matrix

$K =$ vector of length n, where n is order of the system represented by matrix A. Each entry of K represents the number of controllable states factored out during each step of the transformation matrix calculation. The number of nonzero elements of K indicates how many iterations were necessary to calculate T. $sum(K)$ is the number of states in A_c, the controllable portion of Abar.

- $Q_0 = obsv(A, C)$. It computes the observability matrix. If A is $(n \times n)$ and C is (p, n), then the command returns,

$$Q_0 = \begin{bmatrix} C \\ CA \\ CA^2 \\ \vdots \\ CA^{n-1} \end{bmatrix} \text{ with } n \text{ columns and } np \text{ rows}$$

- $[Abar, Bbar, Cbar, T, k] = obsvf(A, B, C)$. If the observability matrix $Q_0 =$ pair $[A, C]$ has a rank $r \leq n$ where n is the size of A, then the transformed system, as explained for the *ctrbf* command, has the unobservable mode (if any), in the upper left corner.

$$\overline{A} = \begin{bmatrix} A_{no} & A_{12} \\ 0 & A_0 \end{bmatrix}; \; \overline{B} = \begin{bmatrix} B_{no} \\ B_0 \end{bmatrix}; \; \overline{C} = \begin{bmatrix} O & C_0 \end{bmatrix}$$

where (C_0, A_0) is observable and the eigenvalues of A_{no}, are the unobservable modes

- $K = acker(A, B, p)$. It helps in pole placement design for a single-input single-output controllable system

$$\dot{x} = Ax + Bu$$

and a vector p of desired closed-loop pole locations. The command uses the Ackermann's formula to calculate a gain vector K such that the state feedback $u = -Kx$ places the closed-loop poles at the location p.

- $l = acker(A', C', p)$.' Used for estimator gain selection.

- $K = place(A, B, P)$. For a single or multi-input system $\dot{x} = Ax + Bu$ and vector p of desired self-conjugate closed-loop pole location, the command computes a feedback gain matrix K that achieves the desired closed-loop pole locations, assuming that all the inputs of the plant are control inputs.

- $l = place(A', C', p)$.' Used for estimator gain selection.

SUMMARY

- The concept of system controllability and observability is linked to matrices A, B, C, and D which describe system dynamics.

- A state controllability is linked to pair $[A,B]$ and observability is linked to pair $[A,C]$.

- State controllability can be checked by computing the rank of controllability matrix.

$$Q_c = [B \quad AB \ldots A^{n-1}B]$$

If Q_c is nonsingular and has rank n which is equal to the order of matrix A, then the system is state controllable.

- Observability can checked by computing the rank of observability matrix

$$Q_o = [C^T \quad A^T C^T \ldots (A^{n-1})^T C^T]$$

If Q_o is nonsingular and has rank n which is equal to the order of matrix A, then the system is observable.

- State controllability is a property of the state-space where as output controllability is linked to input-output relationship. A system is considered to be output controllable, if

$$Q_{op} = [CB \quad CAB \quad CA^2B \ldots CA^{n-1}B]$$
$$(q \times np)$$

possess rank q.

- The concepts of controllability and observability are closely related to the properties of the transfer function. Depending upon how the state variables are defined and represented in state-space, the pole-zero cancellation makes the system either not state controllable or unobservable or both.

- Design encompasses the process of stabilization and compensation to achieve stability criterion, increased accuracy requirements and accelerating of response.

- Design through state-space approach is supported by software designing and design validation as it is based on matrix algebra. It involves design of state variable feedback controller with specified characteristic roots location where the transient response meets the required designed response.

- State feedback can be used to control the eigenvalues of the closed-loop control system and this control exercised on eigenvalues is termed as pole placement/assignment.

- If the system controllable, closed-loop poles can be arbitrarily assigned by using state variable feedback

- Employment of state variable feedback requires all state variables are available for measurements. If not, then a state observer/estimated needs to be constructed.

- If the plant is observable, the poles of an observer can be arbitrarily placed.

PROBLEMS AND SOLUTIONS

Problem 8.1

Input-output relationship of linear system is described as

$$\ddot{y} + 2\dot{y} + y = \dot{u} + u$$

Choose two methods for state representation to show that controllability and observability depends upon the state model.

Solution Let

$$x_1 = y$$
$$x_2 = \dot{y} - u$$

or
$$\dot{x}_2 = \ddot{y} - \dot{u}$$

The given input-output relationship can be written as

$$\ddot{y} - \dot{u} = -2\dot{y} - y + u$$
$$\dot{x}_2 = -2(x_2 + u) - x_1 + u$$
$$= -2x_2 - 2u - x_1 + u$$
$$= -x_1 - 2x_2 - u$$

and
$$\dot{x}_1 = \dot{y} = x_2 + u.$$

In vector matrix form

$$\begin{bmatrix} \dot{x}_1 \\ \dot{x}_2 \end{bmatrix} = \begin{bmatrix} 0 & 1 \\ -1 & -2 \end{bmatrix} \begin{bmatrix} x_1 \\ x_2 \end{bmatrix} + \begin{bmatrix} 1 \\ -1 \end{bmatrix} u$$

and so
$$y = \begin{bmatrix} 1 & 0 \end{bmatrix} \begin{bmatrix} x_1 \\ x_2 \end{bmatrix}$$

$$Q_c = \begin{bmatrix} B & AB \end{bmatrix} = \begin{bmatrix} 1 & \begin{bmatrix} 0 & 1 \\ -1 & -1 \end{bmatrix} \begin{bmatrix} 1 \\ -1 \end{bmatrix} \end{bmatrix}$$

$$= \begin{bmatrix} 1 & -1 \\ -1 & 1 \end{bmatrix}$$

Since
$$\det |Q_c| = 1 - (-1)(-1) = 1 - 1 = 0$$

Hence, rank is one, therefore, the system is not controllable.

$$Q_0 = \begin{bmatrix} C \\ CA \end{bmatrix}$$

$$= \begin{bmatrix} 1 \\ \begin{bmatrix} 1 & 0 \end{bmatrix} \begin{bmatrix} 0 & 1 \\ -1 & -2 \end{bmatrix} \end{bmatrix} = \begin{bmatrix} 1 & 0 \\ 0 & 1 \end{bmatrix}$$

$\det |Q_0| = 1$ and hence, the rank of $Q_0 = 2$. Therefore, the system is observable.

Let us now represent the system in another state-space form. The differential equation can be written as

$$s^2 Y(s) + 2s Y(s) + Y(s) = sU(s) + U(s).$$

or

$$\frac{Y(s)}{U(s)} = \frac{s+1}{s^2+2s+1} = \frac{\left(\dfrac{1}{s}+\dfrac{1}{s^2}\right)X(s)}{\left(1+\dfrac{2}{s}+\dfrac{1}{s^2}\right)X(s)}$$

Therefore

$$Y(s) = \frac{1}{s} X(s) + \frac{1}{s^2} x(s) \tag{1}$$

and

$$U(s) = X(s) + \frac{2}{s} X(s) + \frac{1}{s^2} X(s) \tag{2}$$

or

$$X(s) = -\frac{2}{s} X(s) - \frac{1}{s^2} X(s) + U(s)$$

Let the variables by

$$x_2 = \frac{1}{s} x \Rightarrow \dot{x}_2 = x$$

$$x_1 = \frac{1}{s^2} x \Rightarrow \dot{x}_1 = \frac{1}{s} x = x_2$$

Therefore, the state equations are

$$\dot{x}_1 = x_2$$

$$\dot{x}_2 = x = -2x_2 - x_1 + u = -x_1 - 2x_2 + u.$$

Output equation is obtained from eqn(1).

$$y = x_1 + x_2$$

In vector matrix form

$$\begin{bmatrix} \dot{x}_1 \\ \dot{x}_2 \end{bmatrix} = \begin{bmatrix} 0 & 1 \\ -1 & -2 \end{bmatrix} \begin{bmatrix} x_1 \\ x_2 \end{bmatrix} + \begin{bmatrix} 0 \\ 1 \end{bmatrix}$$

$$y = \begin{bmatrix} 1 & 1 \end{bmatrix} \begin{bmatrix} x_1 \\ x_2 \end{bmatrix}$$

$$Q_c = \begin{bmatrix} B & AB \end{bmatrix}$$

$$= \begin{bmatrix} 0 & \begin{bmatrix} 0 & 1 \\ -1 & -2 \end{bmatrix} \begin{bmatrix} 0 \\ 1 \end{bmatrix} \\ 1 \end{bmatrix} = \begin{bmatrix} 0 & 1 \\ 1 & -2 \end{bmatrix}$$

Since det $|Q_c| \neq 0$, the rank of $Q_c = 2$ and hence, the system is controllable

$$Q_0 = \left[\begin{bmatrix} 1 & \overset{1}{1} \\ 0 & 0 \end{bmatrix} \quad \begin{bmatrix} 0 & \overset{1}{1} \\ -1 & -2 \end{bmatrix} \right] = \begin{bmatrix} 1 & \overset{1}{1} \\ -1 & -1 \end{bmatrix}$$

Since $|Q_0| = 0$, the rank of $Q_0 = 1$ and hence, the system is not observable

---

Problem 8.2

If state equation is given as

$$\begin{bmatrix} \dot{x}_1 \\ \dot{x}_2 \\ \dot{x}_3 \end{bmatrix} = \begin{bmatrix} -a_1 & 1 & 0 \\ -a_2 & 0 & 1 \\ -a_3 & 0 & 0 \end{bmatrix} \begin{bmatrix} x_1 \\ x_2 \\ x_3 \end{bmatrix}$$

and the eigenvalues are λ_1, λ_2 and λ_3; Find the transformation matrix P which will convert the state equation into diagonal form.

Solution Let us denote the $(n, m)th$ element of transformation matrix P as K_{n1m}. In the characteristic equation we substitute $\lambda = \lambda_i$.

$$|A - \lambda_i I| = \begin{vmatrix} -(a_1 + \lambda_i) & 1 & 0 \\ -a_2 & -\lambda_i & 1 \\ -a_3 & 0 & -\lambda_i \end{vmatrix}$$

The sum of the elements of any row and the cofactors of corresponding elements of a different row is zero. Considering the elements of the second row and cofactors of the first row and then the third row and cofactors of the first row, we get

$$A_{i12} = \left(\frac{-a_2}{\lambda_i} - \frac{a_3}{\lambda_i^2} \right) A_{i11} \tag{1}$$

$$A_{i13} = \frac{-a_3}{\lambda_i} A_{i11}$$

For A_{i11} we get

$$A_{i11} = \lambda_i^2$$

Substituting into eqn(1), we get

$$A_{i11} = \lambda_i^2$$

$$A_{i12} = -(a_2\lambda_i + a_3)$$

$$A_{i13} = -a_3\lambda_i$$

Then transformation matrix P can be written as

$$P = \begin{bmatrix} \lambda_1^2 & \lambda_2^2 & \lambda_3^2 \\ -(a_2\lambda_1 + a_3) & -(a_2\lambda_2 + a_3) & -(a_2\lambda_3 + a_3) \\ -a_3\lambda_1 & -a_3\lambda_2 & -a_3\lambda_3 \end{bmatrix}$$

Since

$$\lambda_i^3 + a_1\lambda_i^2 + a_2\lambda_i + a_3 = 0$$

We have

$$-(a_2\lambda_i + a_3) = \lambda_i^3 + a_1\lambda_i^2$$
$$-a_3\lambda_i = \lambda_i^4 + a_1\lambda_i^3 + a_2\lambda_i^2$$

Dividing each column of P by λ_i^2 we get the transformation matrix P'

$$P' = \begin{bmatrix} 1 & 1 & 1 \\ \lambda_1 + a_1 & \lambda_2 + a_1 & \lambda_3 + a_1 \\ \lambda_1^2 + a_1\lambda_1 + a_2 & \lambda_2^2 + a_1\lambda_1 + a_2 & \lambda_3^2 + a_1\lambda_1 + a_2 \end{bmatrix}$$

Problem 8.3

It is required to control a process in a unity feedback closed-loop system whose transfer function is given by

$$G(s) = \frac{1}{s(s+1)(s+10)}$$

The number of closed loop poles are four and it is assumed that response of the system is governed by dominant complex-conjugate poles which satisfies $K_v = 0.93$, $\xi = 0.707$ and $\omega_n = 1$ rad/sec. The other poles are located at 10 and 20 respectively. Design a suitable linear state variable feedback.

Solution Since

$$(N_{pc} - N_{zc}) = (N_{po} - N_{zo})$$

Where

$$N_{pc} = \text{Number of closed-loop poles} = 4$$
$$N_{zc} = \text{Number of closed-loop zeros} = ?$$
$$N_{po} = \text{Number of open-loop poles} = 3$$
$$N_{zo} = \text{Number of open-loop zeros} = 0$$

Therefore

$$4 - N_{zc} = 3 - 0$$

or

$$N_{zc} = 4 - 3 = 1$$

The value of zero is found from the following equation

$$\frac{1}{K_v} = \frac{2\xi}{\omega_n} + \frac{1}{P_3} + \frac{1}{P_4} - \frac{1}{Z}$$

$$\frac{1}{0.93} = \frac{2 \times 0.707}{1} + \frac{1}{20} + \frac{1}{10} - \frac{1}{Z}$$

$$\frac{1}{0.93} = \frac{1.414}{1} + \frac{1}{20} + \frac{1}{10} - \frac{1}{Z}$$

$$\frac{1}{Z} = \frac{1.414}{1} - \frac{1}{0.93} + \frac{1}{20} + \frac{1}{10}$$

or
$$Z \simeq 2$$

Since the resulting unity feedback, closed-loop transfer function has one zero and four poles, its general form is

$$\frac{Y(s)}{X(s)} = \frac{\omega_n^2 \, P_3 P_4}{z} \times \frac{s+z}{(s^2 + 2\xi\omega_n s + \omega_n^2)(s+P_3)(s+P_4)}$$

$$= \frac{1 \times 20 \times 10}{2} \frac{(s+2)}{(s^2 + 2 \times 0.707 \times 1s + 1)(s+20)(s+10)}$$

$$= \frac{100(s+2)}{(s^2 + 1.414s + 1)\,(s+20)(s+10)}$$

$$= \frac{100(s+2)}{s^4 + 31.414s^3 + 243.42s^2 + 312.8s + 200}$$

It can be seen that denominator of $G(s)$ is of third order and $Y(s)/X(s)$ has a denominator of fourth order. Therefore, we add $(s + a)$ to the denominator of $G(s)$. Also, we add $(s + 2)$ to $G(s)$ to ensure numerator of the $G(s)$ must be same as that of the $Y(s)/X(s)$. Hence, the compensating network to be added to $G(s)$ is $\dfrac{s+2}{s+a}$

The resulting linear state-variable feedback system is shown in Fig. 8.26

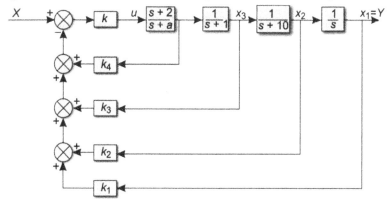

Fig. 8.26

An equivalent block diagram of Fig. Is shown in Fig. 8.27

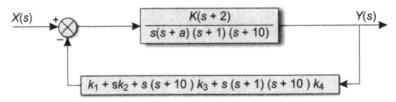

Fig. 8.27

The closed-loop transfer function is

$$\frac{K(s+2)}{[(1+Kk_4)s^4 + \{(11+a)+K(k_3+13k_4)\}s^2 + \{(10+11a)+K(k_2+12k_3+32k_4)\}s^2} \tag{2}$$
$$+ \{10a+K(k_1+2k_2+20k_3+20k_4)\}s+2Kk_1]}$$

Equating eqns(1) and (2), we get

$$K = 100$$

$$Kk_4 + 1 = 1 \Rightarrow k_4 = 0$$

$$(11 + a) + K(k_3 + 13k_4) = 31.414 \tag{3}$$

$$(10 + 11a) + K(k_2 + 12k_3 + 32k_4) = 243.42 \tag{4}$$

$$10a + K(k_1 + 2k_2 + 20k_3 + 20k_4) = 312.8$$

$$2Kk_1 = 200 \Rightarrow k_1 = 1 \tag{5}$$

Substituting the value of k_4 in eqns(3),(4) and (5), we get

$$100\, k_3 = 20.414 - a$$

$$100\, k_2 + 1200\, k_3 + 11a - 233.42 = 0$$

$$200\, k_2 + 2000\, k_3 + 10a = 212.8$$

which yields

$$k_2 = 0.1006;\ a = 21.548 \text{ and } k_3 = -0.01134 \approx 0$$

Problem 8.4

The transfer function of a plant is given as

$$G(s) = \frac{1}{s(s+3)(s+7)}$$

Design an observer to yield transient response specifications of $\xi = 0.4$ and $\omega_n = 75$. The third pole is required to be placed at $s = 200$.

Solution

$$G(s) = \frac{1}{s(s+3)(s+7)} = \frac{1}{s^3 + 10s^2 + 21s + 0}$$

The matrices A and C in OCF can be written as

$$A = \begin{bmatrix} -10 & 1 & 0 \\ -21 & 0 & 1 \\ 0 & 0 & 0 \end{bmatrix} \text{ and } C = \begin{bmatrix} 1 & 0 & 0 \end{bmatrix}$$

The observability matrix

$$Q_0 = \begin{bmatrix} C \\ CA \\ CA^2 \end{bmatrix} = \begin{bmatrix} 1 & 0 & 0 \\ -10 & 1 & 0 \\ 79 & -10 & 1 \end{bmatrix}$$

Since det $|Q_0| = 1$, and is not equal to zero, the rank of the matrix is equal to order of the matrix i.e. 3, and hence the system is observable.

The characteristic equation of the system is

$$s^3 + 10s^2 + 21s = 0 = s^2 + a_1 s^2 + a_2 x = 0 \tag{1}$$

The transient response specifications required to be met yield

$$s^2 + 2\xi\, \omega_n s + \omega_n^2 = 0$$

or $\qquad s^2 + 2 \times 0.4 \times 75s + 75^2 = 0$

or $\qquad (s^2 + 60s + 5625) = 0$

Also, the third pole is required to be placed at $s = 200$, therefore the desired characteristic equation is

$$(s + 200)(s^2 + 60s + 5625) = 0$$

or $\qquad s^3 + 260s^2 + 17625s + 1125000 = 0 = s^3 + p_1 s^2 + p_2 s + p_3 \tag{2}$

The observer gain can be calculated from eqn (8.106)

$$l_1 = p_1 - a_1 = 260 - 10 = 250$$

$$l_2 = p_2 - a_2 = 17625 - 21 = 17604$$

$$l_3 = p_3 - a_3 = 1125000 - 0 = 1125000$$

Problem 8.5

Obtain the state space representation of the system represented in Fig. 8.28. Determine whether the system is completely controllable and observable

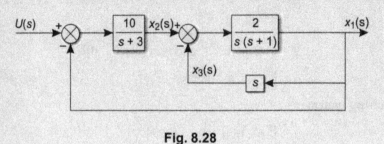

Fig. 8.28

Solution Writing the input-output relationship for each transfer function

$$\frac{X_3(s)}{X_1(s)} = s \quad \text{as} \quad sX_1(s) = X_3(s)$$

$$\frac{X_2(s)}{U(s) - X_1(s)} = \frac{10}{s+3} \quad \text{or} \quad (s+3)X_2(s) = 10[U(s) - X_1(s)]$$

$$\frac{X_1(s)}{X_2(s) - X_3(s)} = \frac{2}{s(s+1)} \quad \text{or} \quad s(s+1)X_1(s) = 2[X_2(s) - X_1(s)]$$

Inverse Laplace transform gives

$$\dot{x}_1 = x_3$$

$$\dot{x}_2 = -10x_1 - 3x_2 + 10u$$

$$\dot{x}_3 = 2x_2 - 3x_3$$

In the vector matrix form

$$\begin{bmatrix} \dot{x}_1 \\ \dot{x}_2 \\ \dot{x}_3 \end{bmatrix} = \begin{bmatrix} 0 & 0 & 1 \\ -10 & -3 & 0 \\ 0 & 2 & -3 \end{bmatrix} \begin{bmatrix} x_1 \\ x_2 \\ x_3 \end{bmatrix} + \begin{bmatrix} 0 \\ 10 \\ 0 \end{bmatrix} u$$

Output equation

$$Y(s) = X_1(s)$$

or
$$y = [1 \quad 0 \quad 0]x$$

Controllability

$$B = [0 \quad 10 \quad 0]^T$$

$$AB = \begin{bmatrix} 0 & 0 & 1 \\ -10 & -3 & 0 \\ 0 & 2 & 3 \end{bmatrix} \begin{bmatrix} 0 \\ 10 \\ 0 \end{bmatrix} = \begin{bmatrix} 0 \\ -30 \\ 20 \end{bmatrix}$$

$$A^2B = \begin{bmatrix} 0 & 0 & 1 \\ -10 & -3 & 0 \\ 0 & 2 & -3 \end{bmatrix} \begin{bmatrix} 0 & 0 & 1 \\ -10 & -3 & 0 \\ 0 & 2 & -3 \end{bmatrix} \begin{bmatrix} 0 \\ 10 \\ 0 \end{bmatrix} = \begin{bmatrix} 20 \\ 90 \\ -120 \end{bmatrix}$$

Therefore

$$Q_c = \begin{bmatrix} 0 & 0 & 20 \\ 11 & -30 & 90 \\ 0 & 20 & -120 \end{bmatrix}$$

$|Q_c| = 5800$ i.e., $\neq 0$; Hence, rank of Q_c is equal to 3, which proves that the system is controllable.

Observability

$$C = [1 \quad 0 \quad 0]$$

$$CA = [1 \quad 0 \quad 0] \begin{bmatrix} 0 & 0 & 1 \\ -10 & -3 & 0 \\ 0 & 2 & -3 \end{bmatrix} = [0 \quad 0 \quad 1]$$

$$CA^2 = [1 \quad 0 \quad 0] \begin{bmatrix} 0 & 0 & 1 \\ -10 & -3 & 0 \\ 0 & 2 & -3 \end{bmatrix} \begin{bmatrix} 0 & 0 & 1 \\ -10 & -3 & 0 \\ 0 & 2 & -3 \end{bmatrix} = [0 \quad 2 \quad -3]$$

Therefore

$$Q_0 = \begin{bmatrix} 1 & 0 & 0 \\ 0 & 0 & 1 \\ 0 & 2 & -3 \end{bmatrix}$$

$|Q_0| = -2$ i.e., $\neq 0$ hence rank of Q_0 is 3, which proves that the system observable.

Problem 8.6

A system is represented pictorially in Fig. 8.29. Obtain the state space representation and check whether the system is observable

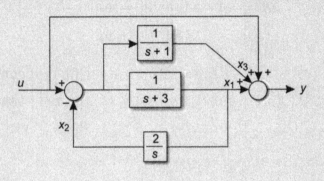

Fig. 8.29

Solution   Writing the input-output relationship for each transfer function, we get

$$\frac{2}{s} X_1(s) = X_2(s)$$

$$\frac{U - X_2(s)}{s+3} = X_1(s)$$

$$\frac{U - X_2(s)}{s+1} = X_3(s)$$

or

$$2X_1(s) = sX_2(s)$$

$$-3X_1(s) - X_2(s) + U(s) = sX_1(s)$$

$$- X_2(s) - X_3(s) + U(s) = sX_3(s)$$

or

$$\dot{x}_1 = -3x_1 - x_2 + u$$

$$\dot{x}_2 = 2x_1$$

$$\dot{x}_3 = -x_2 - x_3 + u$$

Also

$$Y(s) = X_1(s) + X_3(s) + U(s)$$

or

$$y = x_1 + x_3 + u$$

In vector matrix form

$$\begin{bmatrix} \dot{x}_1 \\ \dot{x}_2 \\ \dot{x}_3 \end{bmatrix} = \begin{bmatrix} -3 & -1 & 0 \\ 2 & 0 & 0 \\ 0 & -1 & -1 \end{bmatrix} \begin{bmatrix} x_1 \\ x_2 \\ x_3 \end{bmatrix} + \begin{bmatrix} 1 \\ 0 \\ 1 \end{bmatrix} u$$

and

$$y = \begin{bmatrix} 1 & 0 & 1 \end{bmatrix} \begin{bmatrix} x_1 \\ x_2 \\ x_3 \end{bmatrix} + u$$

Observability

$$C = \begin{bmatrix} 1 & 0 & 1 \end{bmatrix}$$

$$CA = \begin{bmatrix} 1 & 0 & 1 \end{bmatrix} \begin{bmatrix} -3 & -1 & 0 \\ 2 & 0 & 0 \\ 0 & -1 & -1 \end{bmatrix} = \begin{bmatrix} -3 & -2 & -1 \end{bmatrix}$$

$$CA^2 = \begin{bmatrix} 1 & 0 & 1 \end{bmatrix} \begin{bmatrix} -3 & -1 & 0 \\ 2 & 0 & 0 \\ 0 & -1 & -1 \end{bmatrix} \begin{bmatrix} -3 & 1 & 0 \\ 2 & 0 & 0 \\ 0 & -1 & -1 \end{bmatrix} = \begin{bmatrix} 5 & 4 & 1 \end{bmatrix}$$

Therefore

$$Q_0 = \begin{bmatrix} 1 & 0 & 1 \\ -3 & -2 & -1 \\ 5 & 4 & 1 \end{bmatrix}$$

$|Q_0| = 0$; Therefore, the rank of Q_0 is not equal to the order of Q_0 i.e., 3; Hence, the system is not observable.

Problem 8.7

Determine the controllability and observability of the system shown in Fig. 8.30 and determine the conditions on the:

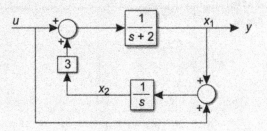

Fig. 8.30

(a) matrices A, B and C
(b) pole–zero cancellation of the transfer function.

Solution   Writing input-output relationship for each transfer function, we get

$$\frac{X_1(s)}{U(s) + 3X_2(s)} = \frac{1}{s+2}$$

and

$$\frac{X_2(s)}{U(s) + X_1(s)} = \frac{1}{s}$$

or

$$s\,X_1(s) = -2X_1(s) + 3X_2(s) + U(s)$$
$$sX_2(s) = X_1(s) + U(s)$$

and

$$Y(s) = X_1(s).$$

In vector-matrix form

$$\begin{bmatrix} \dot{x}_1 \\ \dot{x}_2 \end{bmatrix} = \begin{bmatrix} -2 & 3 \\ 1 & 0 \end{bmatrix} \begin{bmatrix} x_1 \\ x_2 \end{bmatrix} + \begin{bmatrix} x_1 \\ x_2 \end{bmatrix} u$$

$$y = \begin{bmatrix} 1 & 0 \end{bmatrix} \begin{bmatrix} x_1 \\ x_2 \end{bmatrix}$$

(a) **Controllability**

$$Q_c = [B \quad AB] = \begin{bmatrix} 1 & \begin{bmatrix} -2 & 3 \\ 1 & 0 \end{bmatrix}\begin{bmatrix} 1 \\ 1 \end{bmatrix} \\ 1 & \end{bmatrix} = \begin{bmatrix} 1 & 1 \\ 1 & 1 \end{bmatrix}$$

Since $|Q_c| = 0$; Q_c is singular and hence, the system is uncontrollable

Observability

$$Q_0 = \begin{bmatrix} C \\ CA \end{bmatrix} = \begin{bmatrix} 1 & 0 \\ \begin{bmatrix} 1 & 0 \end{bmatrix}\begin{bmatrix} -2 & 3 \\ 1 & 0 \end{bmatrix} \end{bmatrix} = \begin{bmatrix} 1 & 0 \\ -2 & 3 \end{bmatrix}$$

Since det $|Q_c| = 3$; Q_0 is nonsingular. The system is observable.

(b) **Transfer function**

$$\frac{Y(s)}{U(s)} = C(sI - A)^{-1} B$$

$$= \begin{bmatrix} 1 & 0 \end{bmatrix} \begin{bmatrix} s+2 & -3 \\ -1 & s \end{bmatrix}^{-1} \begin{bmatrix} 1 \\ 1 \end{bmatrix}$$

$$= \frac{s+3}{s^2 + 2s - 3} = \frac{s+3}{(s+3)(s-1)} = \frac{1}{s-1}$$

Since, there is a pole-zero cancellation due to identical pole and zero i.e., 3; the system is either uncontrollable or unobservable or both . Here, we have already found out that the system is uncontrollable.

Problem 8.8

If a control system is represented in state space as

$$\dot{x}(t) = \begin{bmatrix} 0 & 1 & 0 \\ 0 & 0 & 1 \\ -6 & -11 & -6 \end{bmatrix} x(t) + \begin{bmatrix} 0 \\ 0 \\ 2 \end{bmatrix} u(t)$$

$$y(t) = \begin{bmatrix} 1 & 0 & 0 \end{bmatrix} x(t)$$

Verify if the system is controllable and observable

```
'Problem 8.8'                                    %display enclosed text.
A=[0 1 0;0 0 1;-6 -11 -6];                       %input matrix A & suppress
                                                 %output.
B=[0;0;2];                                       %input matrix B & suppress
                                                 %output.
C=[1 0 0];                                        %input matrix C & suppress
                                                 %output.
Qc=ctrb(A,B)                                     %Compute controllability
                                                 %matrix & store in Qc.
DET=det(Qc)                                       %compute determinant of
                                                 %controllability matrix &
                                                 %store in DET.
'First If Loop'                                  %display enclosed text.
if abs(DET)==0                                    %perform if loop command.
    disp('System is UNCONTROLLABLE')
end
    disp('System is CONTROLLABLE')
end

Qo=obsv(A,C)                                     %compute observability
                                                 %matrix & store in Qo.
DETO=det(Qo)                                       %compute determinant of
                                                 %controllability matrix &
                                                 %store in DET.
 'Second If Loop'                                %display enclosed text.
 if abs(DETO)==0                                  %perform if loop command.
    disp('System is UNOBSERVABLE')
else
    disp('System is OBSERVABLE')
end

Output
Problem 8.8
Qc =
     0     0     2
     0     2   -12
     2   -12    50
```

```
DET =
    -8

First If Loop
System is CONTROLLABLE
Qo =
     1      0      0
     0      1      0
     0      0      1
DETO =
     1

Second If Loop
System is OBSERVABLE
```

Problem 8.9

Ascertain the controllability of the control system whose system dynamics is described as

$$\dot{x}(t) = \begin{bmatrix} -1 & 1 & 0 \\ 4 & 0 & -3 \\ -6 & 8 & 10 \end{bmatrix} x(t) + \begin{bmatrix} 1 \\ 0 \\ -1 \end{bmatrix} u(t)$$

$$y(t) = \begin{bmatrix} 1 & 2 & 1 \end{bmatrix} x(t)$$

```
'Problem 8.9'                          %display enclosed text.
A=[-1 1 0;4 0 -3;-6 8 10];             %input matrix A.
B=[1;0;-1];                            %input matrix B.
C=[1 2 1];                             %input matrix C.
Qc=ctrb(A,B)                           %compute controllability
                                       %matrix & store in Qc. If
                                       %Qc is square matrix ,then
                                       %we can ascertain
                                       %controllability from the
                                       %rank of the Qc i.e. the
                                       %rank has to be equal to
                                       %the order of Qc.
Rc=rank(Qc)                            %find rank of Qc.
Qo=obsv(A,C)                           %compute observability
                                       %matrix & store in Qo.
Ro=rank(Qo)                            %find rank of Qc.

Output 8.9
Problem
Qc =
     1     -1      8
     0      7     44
    -1    -16    -98
```

```
Rc =
     3
Qo =
     1      2      1
     1      9      4
    11     33     13
Ro =
     3
Note: Since the ranks of Qo and Qc are equal to the
order of Qc and Qo respectively, the system is
CONTROLLABLE AND OBSERVABLE.
```

Problem 8.10

Solve Example 8.12 Using MATLAB

```
'Problem 8.10'                          %display enclosed text.
A=[0 1 0;0 0 1;-6 -11 -6];              %input matrix A & suppress output.
B=[0;0;1];                              %input matrix B & suppress output.
C=[1 0 0];                              %input matrix C & suppress output.
Qc=ctrb(A,B)                            %compute controllability matrix.
Det=det(Qc)                             %determine determinant.If
                                        %determinant is not zero,then
                                        %the system is controllable &
                                        %arbitrary pole placement is
                                        %possible.
Pd=[-2 0 0;0 -5 0;0 0 -6];              %determine the desired matrix
                                        %formed with the desired poles
                                        %which are input at the diagonal
                                        %of the matrix.
P=poly(Pd)                              %determine the characteristic
                                        %equation in polynomial form
                                        %& store in P.
qA=polyvalm(P,A)                        %compute the matrix charateristic
                                        %polynomial & store in qA.
H=[0 0 1]*inv(Qc)*qA                    %determine the feedback matrix H.

Output
Problem 8.10
Qc =
     0      0      1
     0      1     -6
     1     -6     25
Det =
    -1
P =
     1     13     52     60
qA =
    54     41      7
   -42    -23     -1
     6    -31    -17
```

```
H =
    54      41      7
```

Note: The result is same as solved in Example 8.12.

Problem 8.11

A Control system is represented in state space as

$$\dot{x}(t) = \begin{bmatrix} 1 & 2 & 1 \\ 0 & 1 & 3 \\ 1 & 1 & 1 \end{bmatrix} x(t) + \begin{bmatrix} 1 \\ 0 \\ 1 \end{bmatrix} u(t)$$

Check if the system is controllable and also whether it is strongly controllable

```
'Problem 8.11'                      %display enclosed text.
A=[1 2 1;0 1 3;1 1 1];              %input matrix A
                                    %& suppress output.
B=[1;0;1];                          %input matrix B
                                    %& suppress output.
Qc=ctrb(A,B)                        %compute controllability matrix.
Det=det(Qc)                         %determine determinant.
CNo=cond(Qc)                        %compute condition number
                                    %of Qc.

Output
Problem 8.11
Qc =
    1       2       10
    0       3       9
    1       2       7
Det =
    -9
CNo =
    30.9630
Note: Since the determinant is not zero, the
system is controllable and since condition
number(CNo) is not very large, the system is
strongly controllable.
```

Problem 8.12

A Control system is described as

$$\dot{x}(t) = \begin{bmatrix} 0 & 1 \\ -1 & 2 \end{bmatrix} x(t) + \begin{bmatrix} 1 \\ -1 \end{bmatrix} u(t)$$

Find a gain matrix H so that the closed-loop poles of the system are −1 and −2. Use state feedback $u = -Hx$

```
'Problem 8.12'                      %display enclosed text.
A=[0 1;-1 2];                       %input matrix A & suppress
                                    %output.
B=[1;-1];                           %input matrix B & suppress
                                    %output.
V=[-1;-2];                          %form vector matrix with
Det=det(Qc)                         %desired poles & suppress
                                    %output.
H=acker(A,B,V)                      %compute gain matrix
                                    %which will place the desired poles.

Output
Problem 8.12
H =
      1      -4
```

Note: The pole placement for higher order plants may
report error as the transformation to CCF becomes
ill conditioned by use of Ackermann's formula and the
command 'acker' is based on the same. It is suitable
for SISO systems and the pair (A,B) must be controllable.
As said earlier, for problems of order more than five and
whose 'condition number' is large i.e. weakly
controllable systems, pole placement by this method
is not reliable. For MIMO systems
and for placing complex poles, command 'place(A,B,V)'
should be used.

Problem 8.13

Solve Example 8.13 using MATLAB

```
'Problem 8.13'                      %display enclosed text.
A=[-1 1 0;0 -2 1;0 0 -3];           %input matrix A
                                    %&suppress output.
B=[0;0;1];                          %input matrix B
                                    %&suppress output.
Qc=ctrb(A,B)                        %compute controllability
                                    %matrix & store in Qc.
CE=poly(A)                          %form characteristic equation
                                    %in polynomial form & store.
                                    %in CE.
an=-CE(4)                           %assign elements an,an1 &an2
an1=-CE(3)                          %for forming the matrix in
an2=-CE(2)                          %phase variable form.
Aph=[0 1 0;0 0 1;an an1 an2]        %form the matrix A in Phase
                                    %variable form and store
                                    %in Aph.
Qcdash=ctrb(Aph,B)                  %compute controllability
                                    %matrix & store in Qcdash.
```

```
P=Qc*inv(Qcdash)              %form transformation matrix
                              %and store in P.
Pd=[-4 0 0;0 -5 0;0 0 -6]     %form matrix Pd consisting of
                              %desired poles which are
                              %placed at the diagonal.
CEd=poly(Pd)                  %form characteristic equation
                              %of the desired poles
                              %in polynomial form & store
                              %in CEd.
and=CEd(4)                    %assign elements and, anld &
anld=CEd(3)                   %an2d for forming the
an2d=CEd(2)                   %matrix Kz.
Kz=[and+an anld+anl an2d+an2] %form matrix Kz.
H=Kz*inv(P)                   %compute feedback gain matrix.

Problem 8.13
Qc =
       0      0      1
       0      1     -5
       1     -3      9
CE =
       1      6     11      6
an =
      -6
anl =
     -11
an2 =
      -6
Aph =
       0      1      0
       0      0      1
      -6    -11     -6

Qcdash =
       0      0      1
       0      1     -6
       1     -6     25
P =
       1      0      0
       1      1      0
       2      3      1
Pd =
      -4      0      0
       0     -5      0
       0      0     -6
CEd =
       1     15     74    120
and =
     120
```

```
anld =
    74
an2d =
    15
Kz =
   114      63      9
H =
   60.0000   36.0000    9.0000
```

Problem 8.14

Solve Example 8.13 using transformation method

`'Problem 8.14'`	`%display enclosed text.`
`A=[-1 1 0;0 -2 1;0 0 -3];`	`%input matrix A`
	`%& suppress output.`
`B=[0;0;1];`	`%input matrix B`
	`%& suppress output.`
`Qc=ctrb(A,B)`	`%compute controllability`
	`%matrix & store in Qc.`
`CE=poly(A)`	`%form characteristic equation`
	`%in polynomial form & store.`
	`%in CE.`
`a1=-CE(2)`	`%assign elements a1,a2 & a3`
`a2=-CE(3)`	`%for forming the matrix in`
`a3=-CE(4)`	`%phase variable form.`
`Aph=[0 1 0;0 0 1;a1 a2 a3]`	`%form the matrix A in Phase`
	`%variable form and store`
	`%in Aph.`
`Qcdash=ctrb(Aph,B)`	`%compute controllability`
	`%matrix & store in Qcdash.`
`P=Qc*inv(Qcdash)`	`%form transformation matrix`
	`%and store in P.`
`Pd=[-4 0 0;0 -5 0;0 0 -6]`	`%form matrix Pd consisting of`
	`%desired poles which are`
	`%placed at the diagonal.`
`CEd=poly(Pd)`	`%form characteristic equation`
	`%of the desired poles`
	`%in polynomial form & store`
	`%in CEd.`
`a1d=CEd(2)`	`%assign elements`
`a2=CEd(3)`	`%a1d,a2d & a3d`
`a2d=CEd(3)`	`%for forming the`
`a3d=CEd(4)`	`%matrix Kz.`
`Kz=[CEd(4)-CE(4) CEd(3)-CE(3) CEd(2)-CE(2)]`	`%form matrix Kz.`
`H=Kz*inv(P)`	`%compute feedback gain matrix.`

```
Problem 8.14
Qc =
     0     0     1
     0     0    -5
     1    -3     9
```

```
Det =
    -1
CE =
    1      6     11      6
a1 =
    -6
a2 =
    -11
a3 =
    -6

Aph =
    0      1      0
    0      0      1
   -6    -11     -6
Qcdash =
    0      0      1
    0      1     -6
    1     -6     25
P =
    1      0      0
    1      1      0
    2      3      1
Pd =
   -4      0      0
    0     -5      0
    0      0     -6
CEd =
    1     15     74    120
a1d =
    15
a2d =
    74
a3d =
    120
Kz =
   114     63      9
H =
   60.0000   36.0000    9.0000
```

Problem 8.15

Solve Example 8.13 by another method using MATLAB

```
'Problem 8.15'                          %display enclosed text.
A=[-1 1 0;0 -2 1;0 0 -3];               %input matrix A
                                        %& suppress output.
B=[0;0;1];                              %input matrix B
                                        %& suppress output.
Qc=ctrb(A,B)                            %compute controllability
                                        %matrix & store in Qc.
CE=poly(A)                              %form characteristic equation
                                        %in polynomial form & store.
                                        %in CE.
an1=CE(3)                               %assign elements a1,a2 &a3
an2=CE(2)                               %& an for forming the
an=CE(4)                                %matrix M.
M=[an1 an2 1;an2 1 0; 1 0 0]            %form the matrix M
P=Qc*M                                  %form transformation martix
                                        %and store in P.
Pd=[-4 0 0;0 -5 0;0 0-6]                %form matrix Pd consisting of
                                        %desired poles which are
                                        %placed at the diagonal.
CEd=poly(Pd)                            %form characteristic equation.
                                        %form matrix Pd consisting of
                                        %of the desired poles
                                        %in polynomial form & store
                                        %in CEd.
and=CEd(4)                              %assign elements
an1d=CEd(3)                             %and, an1d & an2d
an2d=CEd(2)                             %for forming the matrix Kz.
Kz=[and-an an1d-an1 an2d-an2]           %form vector matrix Kz.
H= kz* inv(p)                           %compute feedback gain matrix
Output
Problem 8.15
Qc =
      0      0      1
      0      1     -5
      1     -3      9

CE =
      1      6     11      6
an1 =
     11
an2 =
      6
an =
      6
M =
     11      6      1
      6      1      0
      1      0      0
P =
      1      0      0
      1      1      0
      2      3      1
```

```
Pd =
    -4     0     0
     0    -5     0
     0     0    -6
CEd =
     1    15    74   120
and =
   120
anld =
    74
an2d =
    15
Kz =
   114    63     9
H =
   60.0000   36.0000    9.0000
```

Problem 8.16

A control system is represented as

$$G(s) = \frac{s+8}{(s+3)(s+5)(s+7)}$$

It is desired to place the poles to meet the following transient response specifications:

(a) Mp = 10% (b) Ts = 1 where Mp = overshoot in percentage and Ts is the settling time in seconds.

(b) Another pole is desired to be placed at a location to affect pole-zero cancellation.

Find the controller gain matrix. The following relationships hold good:

(a) $\xi = \dfrac{-\log (Mp/100)}{\sqrt{(\pi^2 + \log (Mp/100)^2)}}$ (b) $T_s = \dfrac{4}{\xi \omega_n}$

```
'Problem 8.16'                                    %display enclosed text.
A=[-3 1 0;0 -5 1;0 0 -7]                          %input matrix A
B=[0;0;1]                                         %input matrix B
C=[5 1 0]                                         %input matrix C
                                                  %&suppress output.
Mp=10                                             %input peak overshoot desired
Ts=1                                              %input settling time desired.
Geta=(-log(Mp/100))/(sqrt(pi^2+log(Mp/100)^2))    %input relation of damping
                                                  %ratio.
Wn=4/(Geta*Ts)                                    %compute natural frequency of.
                                                  %oscillations & store in Wn.
[n,d]=ord2(Wn,Geta)                               %form second order system with
                                                  %the desired Mp and Geta.
```

```
P=roots(d)                                  %store poles/roots of
                                            %denominator in P as obtained
                                            %from forming second order.
                                            %system based on transients
                                            %specifications.
P(3)=-8                                     %specify another pole sequence
                                            %ment for designing the
                                            %controller to cancel the
                                            %closed loop zero.
V=[P(1) P(2) P(3)]                          %store the desired pole
                                            %placements in V.
K=place(A,B,V)                              %place poles and compute
                                            %Controller gain matrix.
Output
Problem 8.16
A =
    -3     1     0
     0    -5     1
     0     0    -7

B =
     0
     0
     1
C =
     5     1     0
Mp =
    10
Ts =
     1
Geta =
    0.5912
Wn =
    6.7664
n =
     1
P =
  -4.0000 + 5.4575i
  -4.0000 - 5.4575i
P =
  -4.0000 + 5.4575i
  -4.0000 - 5.4575i
  -8.0000
V =
  -4.0000 + 5.4575i   -4.0000 - 5.4575i   -8.0000
K =
   153.9218    30.7844     1.0000
Mp =
    10
Ts =
     1
```

```
Wn =
    6.7664
n =
    1
d =
    1.0000    8.0000      45.7844
P =
   -4.0000 + 5.4575i
   -4.0000 + 5.4575i   -4.0000 - 5.4575i   -4.0000
K =
    30.7844    30.7844    -3.0000
```

Problem 8.17

The transfer function of a control system is

$$G(s) = \frac{s+3}{s^3 + 7s^2 + 14s + 8}$$

Ascertain feedback matrix. It is desired to place the observer poles at -5, -6, and -10.

```
'Problem 8.17'              %display enclosed text.
num=[1 3]                   %input numerator polynomial
                           %an store in num.
den=[1 7 14 8]             %input denominator polynomial
                           %and store in den.
G=tf(num,den)             %form transfer function from
                           %num and den and store in G.
[A,B,C,D]=tf2ss(num,den)   %obtain state space
                           %representation in CCF.
Aob=A'                     %convert matrices A, B,
Bob=C'                     %C and D in CCF to
Cob=B'                     %Aob, Bob, Cob and Dob in
Dob=D'                     %OCF.
Pobd=[-5 -6 -10]           %Specify desired observer
                           %poles and store in Pobd.
L=acker(Aob',Cob',Pobd)    %place poles and obtain
                           %feedback matrix L.

Output
Problem 8.17
num =
        1    3
den =
        1    7    14    8
Transfer function:
        S + 3
---------------------------
s^3 + 7 s^2 + 14 s + 8
```

```
A =
      -7    -14    -8
       1      0     0
       0      1     0
B =
       1
       0
       0
C =
       0      1     3
D =
       0
Aob =
      -7      1     0
     -14      0     1
      -8      0     0
Bob =
       0
       1
       3
Cob =
       1      0     0
Dob =
       0
Pobd =
      -5     -6   -10
L =
      14    126   292
```

Problem 8.18

Design observer gain matrix. The observer poles are required to be placed at 50, 200 and 800. The state space representation of the control system is given as

$$\dot{x}(t) = \begin{bmatrix} -15 & 1 & 0 \\ 0 & -10 & 1 \\ 0 & 0 & -7 \end{bmatrix} x(t) + \begin{bmatrix} 0 \\ 0 \\ 1 \end{bmatrix} u(t)$$

$$y(t) = \begin{bmatrix} 1 & 0 & 0 \end{bmatrix} x(t)$$

```
'Problem 8.18'                              %display enclosed text.
A=[-15 1 0;0 -10 1;0 0 -7]                  %input matrix A.
B=[0;0;1]                                   %input matrix B.
C=[1 0 0]                                   %input matrix C.
P1=50                                       %store desired pole in P1.
P2=200                                      %store desired pole in P2.
P3=800                                      %store desired pole in P3.
V=[P1 P2 P3]                                %store the desired Obsver.
                                            %pole placements in V.
K=acker(A',C',V)'                           %place poles and compute
                                            %observer gain matrix.


Output
Problem 8.18
A =
    -15     1      0
      0   -10      1
      0     0     -7
B =
      0
      0
      1
C =
      1     0      0
P1 =
     50
P2 =
    200
P3 =
    800
V =
     50    200    800
K =
          -1082
         228069
        -9521793
```

REVIEW EXERCISE

1. Define Controllability. State the basic theorem of controllability.
2. If a control system is represented in diagonal form with distinct eigenvalues, state how the controllability can be ascertained.
3. State the conditions of controllability linked to Jordan blocks.
4. Define controllability matrix. Derive the relation $x(t_0) = -Q_c P$. State Controllability Criteria associated with it.
5. Define observability. State the basic theorem of observability.

6. If a control system is represented in state-space in diagonal form with distinct eigenvalues, state how the observability can be ascertained.
7. State the conditions of observability linked to Jordan blocks.
8. Define rank of a matrix. If the rank of the controllability and observability matrix is less then the order of the matrices, is the system controllable and observable.
9. Explain output controllability. State output controllability criteria.
10. Are the concepts of controllability and observability related to the properties of transfer function. If there is an identical pole and zero in the transfer function, how does it affect controllability and observability.
11. State and explain the principle of duality. A control system and it dual is represented below

$$\dot{x}(t) = \begin{bmatrix} 1 & 1 & 0 \\ 0 & 1 & 0 \\ 0 & 0 & 2 \end{bmatrix} x(t) + \begin{bmatrix} 0 \\ 1 \\ 2 \end{bmatrix} u(t)$$

$$y(t) = [1 \quad 0 \quad 0] \, x(t)$$

and

$$\dot{x}_d(t) = \begin{bmatrix} -1 & 0 & 0 \\ -1 & 0 & 0 \\ 0 & 0 & -2 \end{bmatrix} x_d(t) + \begin{bmatrix} 0 \\ 1 \\ 2 \end{bmatrix} u_d(t)$$

$$y_d(t) = [0 \quad 1 \quad 2] \, x_d(t)$$

Apply duality principle and comment.
12. What do you understand by designing.
13. Explain the concept of linear state variable feedback. If a control system is represented as
$$\dot{x}(t) = Ax(t) + Bu(t)$$
$$y(t) = Cx(t)$$
Represent the matrix form of representation when employing state variable feedback. State the criteria for
(a) Controllability   (b) Observability   (c) Stability
14. Derive an expression for open-loop and closed-loop transfer functions for state variable feedback.
15. What do you understand by pole assignment
16. If a control system is represented in controllable canonical form, explain the steps involved for pole placement. If it is not represented in controllable canonical form, what is the method for pole placement.
17. Define an observer. What is the difference between full order state observer and reduced order state observer.
18. What are the steps involved in observer design when the control system is represented in
    (a) Phase variable form   (b) Any other form.

19. Explain the concept of output feedback. Explain the steps involved in the design through output feedback.

20. Determine whether the control system given below is controllable or observable

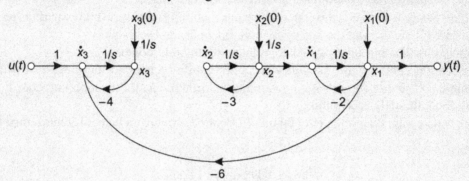

21. Show that the following system is completely state controllable and observable.

$$\dot{x}(t) = \begin{bmatrix} -1 & -2 & -2 \\ 0 & -1 & 1 \\ 1 & 0 & -1 \end{bmatrix} x(t) + \begin{bmatrix} 2 \\ 0 \\ 1 \end{bmatrix} u\,(t)$$

$$y(t) = \begin{bmatrix} 1 & 1 & 0 \end{bmatrix} x(t)$$

22. A third order control system is described by the following system matrices.

$$A = \begin{bmatrix} 1 & 2 & -1 \\ 0 & 1 & 0 \\ 1 & -4 & 3 \end{bmatrix} \text{ and } B = \begin{bmatrix} 0 & 0 & 1 \end{bmatrix}^{T}$$

Show that the system is not controllable
(a) by forming the matrix Q_c
(b) by forming the state equations into Jordan canonical form.

23. The open-loop process required to be controlled has transfer function of $1/s(s + 30)$. It is desired that the closed-loop characteristics of unity feedback system is governed by $\omega_n = 40$ rad/sec, $K_v = 25$/sec and $\xi = 0.707$. Assuming that the system transient response is governed by dominant complex-conjugate poles; design a suitable state feedback controller.

24. The state variable representation of a single-input single-output system is

$$\dot{x}(t) = \begin{bmatrix} 0 & 2 \\ -2 & -4 \end{bmatrix} x(t) + \begin{bmatrix} 0 \\ 1 \end{bmatrix} u(t)$$

$$y(t) = \begin{bmatrix} 1 & 2 \end{bmatrix} x(t)$$

(a) Find the controllability, observability and stability
(b) If a feedback signal $u(t) = y(t) - [-2 \quad 1]x(t)$ is introduced, ascertain the qualitative feedback and comment.

25. Phase-variable representation of a control system is given as

$$\dot{x}(t) = \begin{bmatrix} 0 & 1 & 0 \\ 0 & 0 & 1 \\ -5 & -11 & -5 \end{bmatrix} x(t) + \begin{bmatrix} 0 \\ 0 \\ 1 \end{bmatrix} u(t)$$

$$y(t) = \begin{bmatrix} 1 & 0 & 0 \end{bmatrix} x(t)$$

Find the feedback matrix $h = [\, h_1 \ h_2 \ h_3 \,]$ such that the closed-loop poles are placed at -3, $-7, -9$.

26. The transfer function of a plant is given as $G(s) = (s + 3)/(s + 1)(s + 2)(s + 5)$. Design a state variable feedback so that the closed-loop poles are placed at $-3, -4$, and -6. Also, obtain the transfer function.

27. Transfer function of a plant is given by $G(s) = (s + 3)/s_3 + 7s^2 + 14s + 8$. It is desired to place the observer poles at $-5, -6, -8$. Design an observer for the plant.

REFERENCES

Modern Control Engineering by K Ogata, Prentice Hall Linear Control System Analysis and Design by D'Azzo and CH Houpis, McGraw-Hill

Introduction to Linear Systems Analysis by G M Swisher, Matrix Publishers.

Modern Control Theory by W L Brogan Quantum Publishers

Elementary Linear Algebra by F E Hohn Macmillian Co.

Matrix Analysis by R Bellman McGraw-Hill Book Co.

Modern Control System by R C Dorf and R H Bishop, Addison – Wesley.

The Art of Control Engineering by K Dutton, S Thompson, B Barraclough Addison-Wesley

Matrix Algebra–A programmed Introduction by R C Dorf, John Wiley & Sons.

Modern Control System Theory and Application by S M Shinners, Addison-Wesley

Control Engineering—Introductory Course by J Wilkie, M Johnson, R Katebi, Palgrave

Control Systems Engineering by Nagraths and Gopal, Wiley Eastern Ltd.

Automatic Control System by B C Kuo, Prenctice Hall

Modern Control System Theory by M Gopal, Wiley Eastern Ltd.

Control Systems—Principles and Design by M Gopal, Tata McGraw Hill.

Control System Theory by S Dasgupta

Control Systems Engineering by N S Nise, John Nilcy and Sons.

Linear Systems and signal by B P Lathi, Oxford University Press

Using MATLAB—Simulink & Control Systems Toolbox by A Carallo, R Setola, F Vasca , Prentice Hall

Modern Control Design with MATLAB and Simulink by A Tewari, John Wiley and Sons.

Feedback Control of Dynamic Systems by G F Franklin, J D Powell, Abbas Emami–Maeini, Pearson Education

Control System Design by Goodwin, Graebe, Salgado, Prentice Hall

MATLAB Programming for Engineers by Chapman , Thomson Learning

Mastering MATLAB by Hanselman, Littlefield, Prentice Hall.

Network Analysis and Synthesis by Kuo, John Wiley & Sons.

Feedback and Control Systems by Schaum Series

INDEX